Instructor Solutions Manual

Jay R. Schaffer

University of Northern Colorado

Elementary Statistics
Picturing the World
Third Edition

Larson • Farber

Upper Saddle River, NJ 07458

Editor-in-Chief: Sally Yagan
Acquisitions Editor: Petra Recter
Supplements Editor: Joanne Wendelken
Vice President of Production and Manufacturing: David W. Riccardi
Executive Managing Editor: Kathleen Schiaparelli
Assistant Managing Editor: Becca Richter
Production Editor: Allyson Kloss
Assistant Manufacturing Manager: Michael Bell
Manufacturing Buyer: Ilene Kahn
Supplement Cover Designer: Robert Aleman

© 2006 Pearson Education, Inc.
Pearson Prentice Hall
Pearson Education, Inc.
Upper Saddle River, NJ 07458

All rights reserved. No part of this book may be reproduced in any form or by any means, without permission in writing from the publisher.

Pearson Prentice Hall™ is a trademark of Pearson Education, Inc.

The author and publisher of this book have used their best efforts in preparing this book. These efforts include the development, research, and testing of the theories and programs to determine their effectiveness. The author and publisher make no warranty of any kind, expressed or implied, with regard to these programs or the documentation contained in this book. The author and publisher shall not be liable in any event for incidental or consequential damages in connection with, or arising out of, the furnishing, performance, or use of these programs.

This work is protected by United States copyright laws and is provided solely for the use of instructors in teaching their courses and assessing student learning. Dissemination or sale of any part of this work (including on the World Wide Web) will destroy the integrity of the work and is not permitted. The work and materials from it should never be made available to students except by instructors using the accompanying text in their classes. All recipients of this work are expected to abide by these restrictions and to honor the intended pedagogical purposes and the needs of other instructors who rely on these materials.

Printed in the United States of America

10 9 8 7 6 5 4 3 2 1

ISBN 0-13-148318-8

Pearson Education Ltd., *London*
Pearson Education Australia Pty. Ltd., *Sydney*
Pearson Education Singapore, Pte. Ltd.
Pearson Education North Asia Ltd., *Hong Kong*
Pearson Education Canada, Inc., *Toronto*
Pearson Educación de Mexico, S.A. de C.V.
Pearson Education—Japan, *Tokyo*
Pearson Education Malaysia, Pte. Ltd.

CONTENTS

CONTENTS

Solutions to Odd-Numbered Problems

Chapter 1	Introduction to Statistics	1
Chapter 2	Descriptive Statistics	7
Chapter 3	Probability	39
Chapter 4	Discrete Probability Distributions	51
Chapter 5	Normal Probability Distributions	65
Chapter 6	Confidence Intervals	87
Chapter 7	Hypothesis Testing with One Sample	99
Chapter 8	Hypothesis Testing with Two Samples	119
Chapter 9	Correlation and Regression	137
Chapter 10	Chi-Square Tests and the F-Distribution	153
Chapter 11	Nonparametric Tests	173
Appendix		187

Solutions to Even-Numbered Problems

Chapter 1	Introduction to Statistics	191
Chapter 2	Descriptive Statistics	195
Chapter 3	Probability	219
Chapter 4	Discrete Probability Distributions	229
Chapter 5	Normal Probability Distributions	237
Chapter 6	Confidence Intervals	251
Chapter 7	Hypothesis Testing with One Sample	259
Chapter 8	Hypothesis Testing with Two Samples	275
Chapter 9	Correlation and Regression	291
Chapter 10	Chi-Square Tests and the F-Distribution	305
Chapter 11	Nonparametric Tests	321
Case Studies		331
Uses and Abuses		341
Real Statistics–Real Decisions		345
Technologies		353

Introduction to Statistics

1.1 AN OVERVIEW OF STATISTICS

1.1 Try It Yourself Solutions

1a. The population consists of the prices per gallon of regular gasoline at all gasoline stations in the United States.

   b. The sample consists of the prices per gallon of regular gasoline at the 900 surveyed stations.

   c. The data set consists of the 900 prices.

2a. Because the numerical measure of $2,130,863,461 is based on the entire collection of player's salaries, it is from a population.

   b. Because the numerical measure is a characteristic of a population, it is a parameter.

3a. Descriptive statistics involve the statement "76% of women and 60% of men had a physical examination within the previous year."

   b. An inference drawn from the study is that a higher percentage of women had a physical examination within the previous year.

1.1 EXERCISE SOLUTIONS

1. A sample is a subset of a population.

3. A parameter is a numerical description of a population characteristic. A statistic is a numerical description of a sample characteristic.

5. False. A statistic is a measure that describes a sample characteristic.

7. True

9. False. A population is the collection of all outcomes, responses, measurements, or counts that are of interest.

11. The data set is a population since it is a collection of the ages of all the state governors.

13. The data set is a sample since the collection of the 500 students is a subset within the population of the university's 10,000 students.

15. Sample, because the collection of the 20 patients is a subset within the population.

17. Population: Party of registered voters in Bucks County.

    Sample: Party of Bucks County voters responding to phone survey.

19. Population: Ages of adults in the United States who own computers.

    Sample: Ages of adults in the United States who own Dell computers.

21. Population: Collection of all infants in Italy.

    Sample: Collection of the 33,043 infants in the study.

23. Population: Collection of all women in the United States.

    Sample: Collection of the 546 U.S. women surveyed.

25. Statistic. The value $68,000 is a numerical description of a sample of annual salaries.

27. Statistic. 8% is a numerical description of a sample of computer users.

29. Statistic. 47% is a numerical description of a sample of people in the United States.

31. The statement "56% are the primary investor in their household" is an application of descriptive statistics.

    An inference drawn from the sample is that an association exists between U.S. women and being the primary investor in their household.

33. Answers will vary.

35. (a) An inference drawn from the sample is that senior citizens who own luxury cars have more of their natural teeth than senior citizens who do not own luxury cars.

    (b) It implies that if you purchase a luxury car, an individual will keep more of their natural teeth.

37. Answers will vary.

1.2 DATA CLASSIFICATION

1.2 Try It Yourself Solutions

1a. One data set contains names of cities and the other contains city populations.

b. City: Nonnumerical
Population: Numerical

c. City: Qualitative
Population: Quantitative

2. (1a) The final standings represent a ranking of hockey teams.

(1b) ordinal

(2a) The collection of phone numbers represent labels. No mathematical computations can be made.

(2b) Nominal, because you cannot make calculations on the data.

3. (1a) The data set is the collection of body temperatures.

(1b) Interval, because the data can be ordered and meaningful differences can be calculated, but it does not make sense writing a ratio using the temperatures.

(2a) The data set is the collection of heart rates.

(2b) Ratio, because the data can be ordered, can be written as a ratio, you can calculate meaningful differences, and the data set contains an inherent zero.

1.2 EXERCISE SOLUTIONS

1. Nominal and ordinal

3. False. A systematic sample is selected by ordering a population in some way and then selecting members of the population at regular intervals.

5. False. Data at the nominal level simply represent labels. Calculations cannot be performed.

7. Qualitative, because telephone numbers are merely labels.

9. Quantitative, because the scores of a class on an accounting exam are numerical measures.

11. Ordinal. Data can be arranged in order, but differences between data entries make no sense.

13. Nominal. No mathematical computations can be made and data are categorized using names.

15. Ordinal. The data can be arranged in order, but differences between data entries are not meaningful.

17. Ordinal

19. Nominal

21. An inherent zero is a zero that implies "none." Answers will vary.

1.3 EXPERIMENTAL DESIGN

1.3 Try It Yourself Solutions

1. (1a) Focus: Effect of exercise on senior citizens.

    (1b) Population: Collection of all senior citizens.

    (1c) Experiment

    (2a) Focus: Effect of radiation fallout on senior citizens.

    (2b) Population: Collection of all senior citizens.

    (2c) Sampling

2a. Example: start with the first digits 92630782 . . .

  b. 92|63|07|82|40|19|26

  c. 63, 7, 40, 19, 26

3. (1a) The sample was selected by only using available students.

    (1b) Because the students were readily available in your class, this is convenience sampling.

    (2a) The sample was selected by numbering each student in the school, randomly choosing a starting number, and selecting students at regular intervals from the starting number.

    (2b) Because the students were ordered in a manner such that every 25th student is selected, this is systematic sampling.

1.3 EXERCISE SOLUTIONS

1. False. Using stratified sampling guarantees that members of each group within a population will be sampled.

3. False. To select a systematic sample, a population is ordered in some way and then members of the population are selected at regular intervals.

5. In this study, you want to measure the effect of a treatment (using a fat substitute) on the human digestive system. So, you would want to perform an experiment.

4 CHAPTER 1 | INTRODUCTION TO STATISTICS

7. Because it is impractical to create this situation, you would want to use a simulation.

9. Each U.S. telephone number has an equal chance of being dialed and all samples of 1599 phone numbers have an equal chance of being selected, so this is a simple random sample. Telephone sampling only samples those individuals who have telephones, are available, and are willing to respond, so this is a possible source of bias.

11. Because the students were chosen due to their convenience of location (leaving the library), this is a convenience sample. Bias may enter into the sample because the students sampled may not be representative of the population of students. For example, there may be an association between time spent at the library and drinking habits.

13. Because a random sample of out-patients were selected and all samples of 1210 patients had an equal chance of being selected, this is a simple random sample.

15. Because a sample is taken from each one-acre subplot (stratum), this is a stratified sample.

17. Because every ninth name on a list is being selected, this is a systematic sample.

19. Census, since it is relatively easy to obtain the salaries of the 50 employees.

21. Question is biased since it already suggests that drinking fruit juice is good for you. The question might be rewritten as "How does drinking fruit juice affect your health?"

23. Question is unbiased since it does not imply how many hours of sleep are good or bad.

25. The households sampled represent various locations, ethnic groups, and income brackets. Each of these variables is considered a stratum.

27. Open Question
    Advantage: Allows respondent to express some depth and shades of meaning in the answer.
    Disadvantage: Not easily quantified and difficult to compare surveys.

    Closed Question
    Advantage: Easy to analyze results.
    Disadvantage: May not provide appropriate alternatives and may influence the opinion of the respondent.

CHAPTER 1 REVIEW EXERCISE SOLUTIONS

1. Population: Collection of all U.S. adult VCR owners.
   Sample: Collection of the 898 U.S. adult VCR owners that were sampled.

3. Population: Collection of all U.S. ATMs.
   Sample: Collection of 82,188 ATMs that were sampled.

5. The team payroll is a parameter since it is a numerical description of a population (entire baseball team) characteristic.

7. Since "89 students" is describing a characteristic of a population (University of Arizona students), it is a parameter.

9. The average surcharge of $1.48 for withdrawals from a competing bank is representative of the descriptive branch of statistics. An inference drawn from the sample is that all ATMs charge an average $1.48 for withdrawals from a competing bank.

11. Quantitative because monthly salaries are numerical measurements.

13. Quantitative because ages are numerical measurements.

15. Interval. It makes no sense saying that 100 degrees is twice as hot as 50 degrees.

17. Nominal. The data is categorical and cannot be arranged in a meaningful order.

19. Because judges keep accurate records of charitable donations, you could take a census.

21. In this study, you want to measure the effect of a treatment (plant hormone) on chrysanthemums. You would want to perform an experiment.

23. Because random telephone numbers were generated and called, this is a simple random sample.

25. Since each community is considered a cluster and every pregnant woman in a selected community is surveyed, this is a cluster sample.

27. Since grades are considered strata and 25 students are sampled from each stratum, this is a stratified sample.

29. Telephone sampling only samples individuals who have telephones, are available, and are willing to respond.

31. The selected communities may not be representative of the entire area.

CHAPTER 1 QUIZ SOLUTIONS

1. Population: Collection of all individuals with sleep disorders.
   Sample: Collection of 254 patients in study.

2. (a) Statistic. 53% is a characteristic of a sample of parents.

   (b) Parameter. 67% is a characteristic of the entire union (population).

3. (a) Qualitative, since student identification numbers are merely labels.

   (b) Quantitative, since a test score is a numerical measure.

4. (a) Nominal. Players may be ordered numerically, but there is no meaning in this order and no mathematical computations can be made.

   (b) Ratio. It makes sense to say that the number of products sold during the 1st quarter was twice as many as sold in the 2nd quarter.

   (c) Interval because meaningful differences between entries can be calculated, but a zero entry is not an inherent zero.

5. (a) In this study, you want to measure the effect of a treatment (low dietary intake of vitamin C and iron) on lead levels in adults. You want to perform an experiment.

   (b) Since it would be difficult to survey every individual within 500 miles of your home, sampling should be used.

6. (a) Because people were chosen due to their convenience of location (on the beach), this is a convenience sample.

   (b) Since every fifth part is selected from an assembly line, this is a systematic sample.

   (c) Stratified sample because the population is first stratified and then a sample is collected from each strata.

7. Convenience

8. (a) False. A statistic is a numerical measure that describes a sample characteristic.

   (b) False. Ratio data represents the highest level of measurement.

Descriptive Statistics

CHAPTER 2

2.1 FREQUENCY DISTRIBUTIONS AND THEIR GRAPHS

2.1 Try It Yourself Solutions

1a. The number of classes (8) is stated in the problem.

b. Min = 0   Max = 72   Class width = $\dfrac{(72-0)}{8} = 9$

c.

Lower limit	Upper limit
0	9
10	19
20	29
30	39
40	49
50	59
60	69
70	79

d. See part (e).

e.

Class	Frequency, f
0–9	15
10–19	19
20–29	14
30–39	7
40–49	14
50–59	6
60–69	4
70–79	1

2a. See part (b).

b.

Class	Frequency, f	Midpoint	Relative frequency	Cumulative frequency
0–9	15	4.5	0.1875	15
10–19	19	14.5	0.2375	34
20–29	14	24.5	0.1750	48
30–39	7	34.5	0.0875	55
40–49	14	44.5	0.1750	69
50–59	6	54.5	0.0750	75
60–69	4	64.5	0.0500	79
70–79	1	74.5	0.0125	80
	$N = 80$		$\sum \dfrac{f}{n} = 1$	

c. 42.5% of the population is under 20 years old. 6.25% of the population is over 59 years old.

3a.

Class Boundaries
−0.5–9.5
9.5–19.5
19.5–29.5
29.5–39.5
39.5–49.5
49.5–59.5
59.5–69.5
69.5–79.5

b. Use class midpoints for the horizontal scale and frequency for the vertical scale.

c.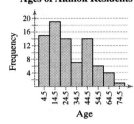
Ages of Akhiok Residents

d. 42.5% of the population is under 20 years old. 6.25% of the population is over 59 years old.

4a. Use class midpoints for the horizontal scale and frequency for the vertical scale.

b. See part (c).

c.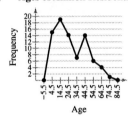
Ages of Ahkiok Residents

d. The population increases up to the age of 14.5 and then decreases. The population increases again between the ages of 34.5 and 44.5, but then after 44.5, the population decreases.

5abc.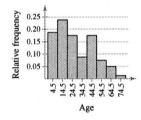
Ages of Akhiok Residents

6a. Use upper class boundaries for the horizontal scale and cumulative frequency for the vertical scale.

b. See part (c).

c.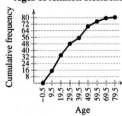
Ages of Akhiok Residents

d. Approximately 69 residents are 49 years old or younger.

7ab.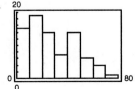

2.1 EXERCISE SOLUTIONS

1. By organizing the data into a frequency distribution, patterns within the data may become more evident.

3. Class limits determine which numbers can belong to that class.

   Class boundaries are the numbers that separate classes without forming gaps between them.

5. False. The midpoint of a class is the sum of the lower and upper limits of the class divided by two.

7. True

9. (a) Class width = 30 − 20 = 10

   (b) and (c)

Class	Frequency, f	Midpoint	Class boundaries
20–29	10	24.5	19.5–29.5
30–39	132	34.5	29.5–39.5
40–49	284	44.5	39.5–49.5
50–59	300	54.5	49.5–59.5
60–69	175	64.5	59.5–69.5
70–79	65	74.5	69.5–79.5
80–89	25	84.5	79.5–89.5
	$\sum f = 991$		

11.

Class	Frequency, f	Midpoint	Relative frequency	Cumulative frequency
20–29	10	24.5	0.01	10
30–39	132	34.5	0.13	142
40–49	284	44.5	0.29	426
50–59	300	54.5	0.30	726
60–69	175	64.5	0.18	901
70–79	65	74.5	0.07	966
80–89	25	84.5	0.03	991
	$\sum f = 991$		$\sum \frac{f}{n} = 1$	

13. (a) Number of classes = 7

    (b) Least frequency ≈ 10

    (c) Greatest frequency ≈ 300

    (d) Class width = 10

15. (a) 50   (b) 12.5–13.5 lb

17. (a) 24   (b) 19.5 lb

19. (a) Class with greatest relative frequency: 8–9 in.
       Class with least relative frequency: 17–18 in.

    (b) Greatest relative frequency ≈ 0.195
        Least relative frequency ≈ 0.005

    (c) Approximately 0.015

21. Class with greatest frequency: 500–550
    Class with least frequency: 250–300 or 700–750

23. Class width $= \dfrac{\text{Max} - \text{Min}}{\text{Number of classes}} = \dfrac{39 - 0}{5} = 7.8 \Rightarrow 8$

Class	Frequency, f	Midpoint	Relative frequency	Cumulative frequency
0–7	8	3.5	0.32	8
8–15	8	11.5	0.32	16
16–23	3	19.5	0.12	19
24–31	3	27.5	0.12	22
32–39	3	35.5	0.12	25
	$\sum f = 25$		$\sum \dfrac{f}{n} = 1$	

25. Class width $= \dfrac{\text{Max} - \text{Min}}{\text{Number of classes}} = \dfrac{7119 - 1000}{6} = 1019.83 \Rightarrow 1020$

Class	Frequency, f	Midpoint	Relative frequency	Cumulative frequency
1000–2019	12	1509.5	0.5455	12
2020–3039	3	2529.5	0.1364	15
3040–4059	2	3549.5	0.0909	17
4060–5079	3	4569.5	0.1364	20
5080–6099	1	5589.5	0.0455	21
6100–7119	1	6609.5	0.0455	22
	$\sum f = 22$		$\sum \dfrac{f}{n} = 1$	

Class with greatest frequency: 1000–2019
Class with least frequency: 5080–6099; 6100–7119

27. Class width $= \dfrac{\text{Max} - \text{Min}}{\text{Number of classes}} = \dfrac{514 - 291}{8} = 27.875 \Rightarrow 28$

Class	Frequency, f	Midpoint	Relative frequency	Cumulative frequency
291–318	5	304.5	0.1667	5
319–346	4	332.5	0.1333	9
347–374	3	360.5	0.1000	12
375–402	5	388.5	0.1667	17
403–430	6	416.5	0.2000	23
431–458	4	444.5	0.1333	27
459–486	1	472.5	0.0455	28
487–514	2	500.5	0.0909	30
	$\sum f = 30$		$\sum \dfrac{f}{n} = 1$	

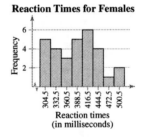

Class with greatest frequency: 403–430
Class with least frequency: 459–486

29. Class width $= \dfrac{\text{Max} - \text{Min}}{\text{Number of classes}} = \dfrac{264 - 146}{5} = 23.6 \Rightarrow 24$

Class	Frequency, f	Midpoint	Relative frequency	Cumulative frequency
146–169	6	157.5	0.2308	6
170–193	9	181.5	0.3462	15
194–217	3	205.5	0.1154	18
218–241	6	229.5	0.2308	24
242–265	2	253.5	0.0769	26
	$\sum f = 26$		$\sum \dfrac{f}{n} = 1$	

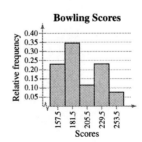

Class with greatest relative frequency: 170–193
Class with least relative frequency: 242–265

31. Class width $= \dfrac{\text{Max} - \text{Min}}{\text{Number of classes}} = \dfrac{52 - 33}{5} = 3.8 \Rightarrow 4$

Class	Frequency, f	Midpoint	Relative frequency	Cumulative frequency
33–36	8	34.5	0.3077	8
37–40	6	38.5	0.2308	14
41–44	5	42.5	0.1923	19
45–48	2	46.5	0.0769	21
49–52	5	50.5	0.1923	26
	$\sum f = 26$		$\sum \dfrac{f}{n} = 1$	

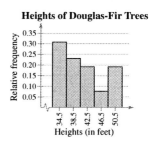

Class with greatest relative frequency: 33–36
Class with least relative frequency: 45–48

33. Class width $= \dfrac{\text{Max} - \text{Min}}{\text{Number of classes}} = \dfrac{73 - 50}{6} = 3.83 \Rightarrow 4$

Class	Frequency, f	Relative frequency	Cumulative frequency
50–53	1	0.0417	1
54–57	0	0.0000	1
58–61	4	0.1667	5
62–65	9	0.3750	14
66–69	7	0.2917	21
70–73	3	0.1250	24
	$\sum f = 24$	$\sum \dfrac{f}{n} \approx 1$	

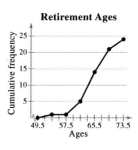

Location of the greatest increase in frequency: 62–65

35. Class width $= \dfrac{\text{Max} - \text{Min}}{\text{Number of classes}} = \dfrac{18 - 2}{6} = 2.67 \Rightarrow 3$

Class	Frequency, f	Relative frequency	Cumulative frequency
2–4	9	0.3214	9
5–7	6	0.2143	15
8–10	7	0.2500	22
11–13	3	0.1071	25
14–16	2	0.0714	27
17–19	1	0.0357	28
	$\sum f = 28$	$\sum \dfrac{f}{n} \approx 1$	

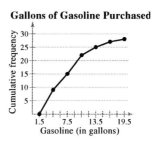

Location of the greatest increase in frequency: 2–4

37. Class width $= \dfrac{\text{Max} - \text{Min}}{\text{Number of classes}} = \dfrac{98 - 47}{5} = 10.2 \Rightarrow 11$

Class	Frequency, f	Midpoint	Relative frequency	Cumulative frequency
47–57	1	52	0.05	1
58–68	1	63	0.05	2
69–79	5	74	0.25	7
80–90	8	85	0.40	15
91–101	5	96	0.25	20
	$\sum f = 20$		$\sum \dfrac{f}{N} = 1$	

Class with greatest frequency: 80–90
Classes with least frequency: 47–57 and 58–68

12 CHAPTER 2 | DESCRIPTIVE STATISTICS

39. (a) Class width = $\dfrac{\text{Max} - \text{Min}}{\text{Number of classes}} = \dfrac{104 - 61}{8} = 5.375 \Rightarrow 6$

Class	Frequency, f	Midpoint	Relative frequency
61–66	1	63.5	0.0333
67–72	3	69.5	0.1000
73–78	6	75.5	0.2000
79–84	10	81.5	0.3333
85–90	5	87.5	0.1667
91–96	2	93.5	0.0667
97–102	2	99.5	0.0667
103–108	1	105.5	0.0333
	$\sum f = 30$		$\sum \dfrac{f}{N} = 1$

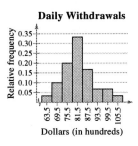

(b) 16.7%, because the sum of the relative frequencies for the last three classes is 0.167.

(c) $9600, because the sum of the relative frequencies for the last two classes is 0.10.

41.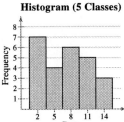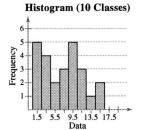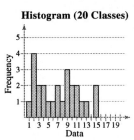

In general, a greater number of classes better-preserves the actual values of the data set, but is not as helpful for observing general trends and making conclusions. When choosing the number of classes, an important consideration is the size of the data set. For instance, you would not want to use 20 classes if your data set contained 20 entries. In this particular example, as the number of classes increases, the histogram shows more fluctuation. The histograms with 10 and 20 classes have classes with zero frequencies. Not much is gained by using more than five classes. Therefore, it appears that five classes would be best.

2.2 MORE GRAPHS AND DISPLAYS

2.2 Try It Yourself Solutions

1a. 0 |
    1 |
    2 |
    3 |
    4 |
    5 |
    6 |
    7 |

b. Key: 3|3 = 33

0 | 5 2 7 1 5 3 1 0 1 3 3 9 0 4 5
1 | 8 2 5 6 3 3 7 3 0 7 8 2 3 8 9 3 6 9 9
2 | 5 4 2 0 3 3 4 0 1 5 9 6 6 6
3 | 9 6 9 7 9 9 3
4 | 4 2 4 7 1 8 0 0 1 9 9 5 1 9
5 | 8 3 1 6 8 9
6 | 0 8 7 8
7 | 2

c. Key: 3|3 = 33

```
0 | 0 0 1 1 1 2 3 3 3 3 4 5 5 5 5 7 9
1 | 0 2 2 3 3 3 3 3 5 6 6 7 7 8 8 8 9 9 9
2 | 0 0 1 2 3 3 4 4 5 5 6 6 6 9
3 | 3 6 7 9 9 9 9
4 | 0 0 1 1 1 2 4 4 5 7 8 9 9 9
5 | 1 3 6 8 8 9
6 | 0 7 8 8
7 | 2
```

d. It appears that the residents of Akhiok are a young population with most of the ages being below 40 years old.

2ab. Key: 3|3 = 33

```
0 | 0 0 1 1 1 2 3 3 3 4
0 | 5 5 5 7 9
1 | 0 2 2 3 3 3 3 3
1 | 5 6 6 7 7 8 8 8 9 9 9
2 | 0 0 1 2 3 3 4 4
2 | 5 5 6 6 6 9
3 | 3
3 | 6 7 9 9 9 9
4 | 0 0 1 1 1 2 4 4
4 | 5 7 8 9 9 9
5 | 1 3
5 | 6 8 8 9
6 | 0
6 | 7 8 8
7 | 2
7 |
```

3a. Use ages for the horizontal axis.

b. **Ages of Akhiok Residents**

(dot plot: Age in years, 0 to 80)

c. It appears that a large percentage of the population is younger than 40 years old.

4a.

Vehicle type	Killed (frequency)	Relative frequency	Central angle
Cars	22,385	0.6556	(0.6555)(360°) ≈ 236°
Trucks	8,457	0.2477	(0.2477)(360°) ≈ 89°
Motorcycles	2,806	0.0822	(0.0822)(360°) ≈ 30°
Other	497	0.0146	(0.0146)(360°) ≈ 5°
	$\sum f = 34{,}145$	$\sum \frac{f}{n} \approx 1$	

b. Motor Vehicle Occupants Killed in 1991

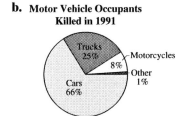

c. It appears the percentage of truck occupants killed has increased by 9% while the percentage of car occupants killed has decreased by 10%. The motorcycle deaths for both years is nearly the same.

14 CHAPTER 2 | DESCRIPTIVE STATISTICS

5a.

Cause	Frequency, f
Auto Dealers	14,668
Auto Repair	9,728
Home Furnishing	7,792
Computer Sales	5,733
Dry Cleaning	4,649

b.

c. It appears that the auto industry (dealers and repair shops) account for the largest portion of complaints filed at the BBB.

6ab.

c. It appears that the longer an employee is with the company, the larger his/her salary will be.

7ab.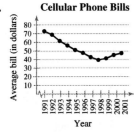

c. It appears that the average monthly bill for cellular telephone subscribers has decreased significantly from 1991 to 1998, then increased from 1998 to 2001.

2.2 EXERCISE SOLUTIONS

1. Quantitative: Stem-and-Leaf Plot, Dot Plot, Histogram, Time Series Chart, Scatter Plot
   Qualitative: Pie Chart, Pareto Chart

3. a

5. b

7. 27, 32, 41, 43, 43, 44, 47, 47, 48, 50, 51, 51, 52, 53, 53, 53, 54, 54, 54, 54, 55, 56, 56, 58, 59, 68, 68, 68, 73, 78, 78, 85

   Max: 85   Min: 27

9. 13, 13, 14, 14, 14, 15, 15, 15, 15, 15, 16, 17, 17, 18, 19

   Max: 19   Min: 13

11. Anheuser-Busch is the top sports advertiser spending approximately $190 million. Honda spends the least. (Answers will vary.)

13. Tailgaters irk drivers the most, while too cautious drivers irk drivers the least. (Answers will vary.)

15. Key: 3|3 = 33

```
3 | 233459
4 | 01134556678
5 | 133
6 | 0069
```

Most elephants tend to drink less than 55 gallons of water per day. (Answers will vary.)

17. Key: 17|5 = 17.5

```
16 | 48
17 | 113455679
18 | 13446669
19 | 0023356
20 | 18
```

It appears that most farmers charge 17 to 19 cents per pound of apples. (Answers will vary.)

19.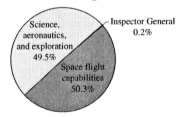

It appears that the lifespan of a fly tends to be between 4 and 14 days. (Answers will vary.)

21.

Category	Budget frequency	Relative frequency	Angle
Science, aeronautics, and exploration	7782	0.5031	$(0.5031)(360°) \approx 181°$
Space flight capabilities	7661	0.4952	$(0.4952)(360°) \approx 178°$
Inspector General	26	0.0017	$(0.0017)(360°) \approx 0.6°$
	$\Sigma f = 15{,}469$	$\sum \frac{f}{N} \approx 1$	

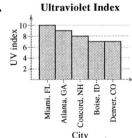

It appears that 50.31% of NASA's budget went to science, aeronautics, and exploration. (Answers will vary.)

23.

Ultraviolet Index

(bar chart: Miami, FL ≈ 10; Atlanta, GA ≈ 9; Concord, NH ≈ 8; Boise, ID ≈ 7; Denver, CO ≈ 7)

It appears that Boise, ID and Denver, CO have the same UV index. (Answers will vary.)

25.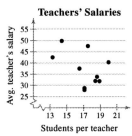

It appears that a teacher's average salary decreases as the number of students per teacher increases. (Answers will vary.)

27.

It appears the price of eggs peaked in 1996. (Answers will vary.)

29. (a) When data is taken at regular intervals over a period of time, a time series chart should be used. (Answers will vary.)

(b)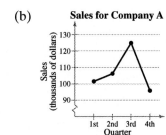

2.3 MEASURES OF CENTRAL TENDENCY

2.3 Try It Yourself Solutions

1a. $\Sigma x = 578$

b. $\bar{x} = \dfrac{\Sigma x}{n} = \dfrac{578}{14} = 41.3$

c. The typical age of an employee in a department store is 41.3 years old.

2a. 0, 0, 1, 1, 1, 2, 3, 3, 4, 5, 5, 5, 9, 10, 12, 12, 13, 13, 13, 13, 13, 15, 16, 16, 17, 17, 18, 18, 18, 19, 19, 19, 20, 20, 21, 22, 23, 23, 24, 24, 25, 25, 26, 26, 26, 29, 36, 39, 39, 39, 39, 40, 40, 41, 41, 41, 42, 44, 44, 45, 47, 48, 49, 49, 49, 51, 53, 56, 58, 58, 60, 67, 68, 68, 72

b. median = middle entry = 23

3a. 0, 0, 1, 1, 1, 2, 3, 3, 3, 4, 5, 5, 5, 7, 9, 10, 12, 12, 13, 13, 13, 13, 13, 15, 16, 16, 17, 17, 18, 18, 18, 19, 19, 19, 20, 20, 21, 22, 23, 23, 24, 24, 25, 25, 26, 26, 26, 29, 33, 36, 37, 39, 39, 39, 39, 40, 40, 41, 41, 41, 42, 44, 44, 45, 47, 48, 49, 49, 49, 51, 53, 56, 58, 58, 59, 60, 67, 68, 68, 72

b. median = mean of two middle entries {23, 24} = 23.5

c. Half of the residents of Akhiok are younger than 23.5 and half are older than 23.5.

4a. 0, 0, 1, 1, 1, 2, 3, 3, 3, 4, 5, 5, 5, 7, 9, 10, 12, 12, 13, 13, 13, 13, 13, 15, 16, 16, 17, 17, 18, 18, 18, 19, 19, 19, 20, 20, 21, 22, 23, 23, 24, 24, 25, 25, 26, 26, 26, 29, 33, 36, 37, 39, 39, 39, 39, 40, 40, 41, 41, 41, 42, 44, 44, 45, 47, 48, 49, 49, 49, 51, 53, 56, 58, 58, 59, 60, 67, 68, 68, 72

b. The age that occurs with the greatest frequency is 13 years old.

c. The mode of the ages of the residents of Akhiok is 13 years old.

5a. "Yes" occurs with the greatest frequency (169).

b. The mode of the responses to the survey is "Yes".

6a. $\bar{x} = \dfrac{\Sigma x}{n} = \dfrac{410}{19} \approx 21.6$

median = 21

mode = 20

b. The mean in Example 6 ($\bar{x} \approx 23.8$) was heavily influenced by the age 65. Neither the median nor the mode was affected as much by the age 65.

7ab.

Source	x	Score, w	Weight, $x \cdot w$
Test Mean	86	0.50	(83)(0.50) = 43
Midterm	96	0.15	(96)(0.15) = 14.4
Final	98	0.20	(98)(0.20) = 19.6
Computer Lab	98	0.10	(98)(0.10) = 9.8
Homework	100	0.05	(100)(0.05) = 5
		$\Sigma w = 1.00$	$\Sigma(x \cdot w) = 91.8$

c. $\bar{x} = \dfrac{\Sigma(x \cdot w)}{\Sigma w} = \dfrac{91.8}{1.00} = 91.8$

d. The weighted mean for the course is 91.8.

8abc.

Class	Midpoint, x	Frequency, f	$x \cdot f$
0–9	4.5	15	67.50
10–19	14.5	19	275.50
20–29	24.5	14	343.00
30–39	34.5	7	241.50
40–49	44.5	14	623.00
50–59	54.5	6	327.00
60–69	64.5	4	258.00
70–79	74.5	1	74.50
		$N = 80$	$\Sigma(x \cdot f) = 2210$

d. $\mu = \dfrac{\Sigma(x \cdot f)}{N} = \dfrac{2210}{80} \approx 27.6$

The average age of a resident of Akhiok is approximately 27.6.

18 CHAPTER 2 | DESCRIPTIVE STATISTICS

2.3 EXERCISE SOLUTIONS

1. False. The mean is the measure of central tendency most likely to be affected by an extreme value (or outlier).

3. False. All quantitative data sets have a median.

5. Answers will vary. A data set with an outlier within it would be an example. For instance, the mean of the prices of existing home sales tends to be "inflated" due to the presence of a few very expensive homes.

7. Skewed right since the "tail" of the distribution extends to the right.

9. Uniform since the bars are approximately the same height.

11. (9), since the distribution values range from 1 to 12 and has (approximately) equal frequencies.

13. (10), since the distribution has a maximum value of 90 and is skewed left due to a few students scoring much lower than the majority of the students.

15. (a) $\bar{x} = \dfrac{\Sigma x}{n} = \dfrac{81}{13} \approx 6.2$

    5 5 5 5 5 5 ⑥ 6 6 7 8 9 9
                 ↑— middle value ⟹ median = 6

    mode = 5 (occurs 6 times)

    (b) Median appears to be the best measure of central tendency since the distribution is skewed.

17. (a) $\bar{x} = \dfrac{\Sigma x}{n} = \dfrac{32}{7} \approx 4.57$

    3.7 4.0 4.8 ④.⑧ 4.8 4.8 5.1
                ↑— middle value ⟹ median = 4.8

    mode = 4.8 (occurs 4 times)

    (b) Median appears to be the best measure of central tendency since there are outliers present.

19. (a) $\bar{x} = \dfrac{\Sigma x}{n} = \dfrac{2720.6}{29} \approx 93.81$

    85.4, 88, 88.7, 89.7, 89.8, 90.1, 90.3, 90.3, 90.7, 91.5,
    91.8, 91.8, 92, 92.8, ⑨2.⑨, 93.3, 94, 94.2, 94.5, 94.8,
    95.3, 96.7, 97.1, 97.2, 98, 98.2, 102.8, 103.5, 105.2
                  — middle value ⟹ median = 92.9

    mode = 90.3, 91.8 (occurs 2 times each)

    (b) Median appears to be the best measure of central tendency since the distribution is skewed.

21. (a) $\bar{x}$ = not possible (nominal data)

    median = not possible (nominal data)

    mode = "Worse"

    (b) Mode appears to be the best measure of central tendency since the data is at the nominal level of measurement.

23. (a) $\bar{x} = \dfrac{\Sigma x}{n} = \dfrac{1194.4}{7} \approx 170.63$

155.7, 158.1, 162.2, (169.3), 180, 181.8, 187.3
⎯⎯ middle value ⇒ median = 169.3

mode = none

(b) Mean appears to be the best measure of central tendency since there are no outliers.

25. (a) $\bar{x} = \dfrac{\Sigma x}{n} = \dfrac{226}{10} = 22.6$

14, 14, 15, 17, 18, 20, 22, 25, 40, 41
⎯⎯ two middle values ⇒ median = $\dfrac{18 + 20}{2} = 19$

mode = 14 (occurs 2 times)

(b) Median appears to be the best measure of central tendency since the distribution is skewed.

27. (a) $\bar{x} = \dfrac{\Sigma x}{n} = \dfrac{197.5}{14} \approx 14.11$

1.5, 2.5, 2.5, 5, 10.5, 11, 13, 15.5, 16.5, 17.5, 20, 26.5, 27, 28.5
⎯⎯ two middle values ⇒ median = $\dfrac{13 + 15.5}{2} = 14.25$

mode = 2.5 (occurs 2 times)

(b) Mean appears to be the best measure of central tendency since there are no outliers.

29. (a) $\bar{x} = \dfrac{\Sigma x}{n} = \dfrac{578}{14} = 41.3$

10, 12, 21, 24, 27, 37, 38, 41, 45, 45, 50, 57, 65, 106
⎯⎯ two middle values ⇒ median = $\dfrac{38 + 41}{2} = 39.5$

mode = 4.5 (occurs 2 times)

(b) Median appears to be the best measure of central tendency since the distribution is skewed.

31. (a) $\bar{x} = \dfrac{\Sigma x}{n} = \dfrac{292}{15} \approx 19.5$

5, 8, 10, 15, 15, 15, 17, (20), 21, 22, 22, 25, 28, 32, 37
⎯⎯ middle value ⇒ median = 20

mode = 15 (occurs 3 times)

(b) Median appears to be the best measure of central tendency since the data is skewed.

33. A = mode (data entry that occurred most often)

B = median (left of mean in skewed right dist.)

C = mean (right of median in skewed right dist.)

35. Mode since the data is nominal.

37. Mean since the data does not contain outliers.

39.

Source	Score, x	Weight, w	$x \cdot w$
Homework	85	0.15	$(85)(0.15) = 12.75$
Quiz	80	0.10	$(80)(0.10) = 8$
Quiz	92	0.10	$(92)(0.10) = 9.2$
Quiz	76	0.10	$(76)(0.10) = 7.6$
Project	100	0.15	$(100)(0.15) = 15$
Speech	90	0.15	$(90)(0.15) = 13.5$
Final Exam	93	0.25	$(93)(0.25) = 23.25$
		$\Sigma w = 1$	$\Sigma(x \cdot w) = 89.3$

$$\bar{x} = \frac{\Sigma(x \cdot w)}{\Sigma w} = \frac{89.3}{1} = 89.3$$

41.

Grade	Points, x	Credits, w	$x \cdot w$
B	3	3	$(3)(3) = 9$
B	3	3	$(3)(3) = 9$
A	4	4	$(4)(4) = 16$
D	1	2	$(1)(2) = 2$
C	2	3	$(2)(3) = 6$
		$\Sigma w = 15$	$\Sigma(x \cdot w) = 42$

$$\bar{x} = \frac{\Sigma(x \cdot w)}{\Sigma w} = \frac{42}{15} = 2.8$$

43.

Midpoint, x	Frequency, f	$x \cdot f$
61	3	$(61)(3) = 183$
64	4	$(64)(4) = 256$
67	7	$(67)(7) = 469$
70	2	$(70)(2) = 140$
	$n = 16$	$\Sigma(x \cdot f) = 1048$

$$\bar{x} = \frac{\Sigma(x \cdot f)}{n} = \frac{1048}{16} \approx 65.5$$

45.

Midpoint, x	Frequency, f	$x \cdot f$
4.5	57	$(4.5)(57) = 256.5$
14.5	68	$(14.5)(68) = 986$
24.5	36	$(24.5)(36) = 882$
34.5	55	$(34.5)(55) = 1897.5$
44.5	71	$(44.5)(71) = 3159.5$
54.5	44	$(54.5)(44) = 2398$
64.5	36	$(64.5)(36) = 2322$
74.5	14	$(74.5)(14) = 1043$
84.5	8	$(84.5)(8) = 676$
	$n = 389$	$\Sigma(x \cdot f) = 13620.5$

$$\bar{x} = \frac{\Sigma(x \cdot f)}{n} = \frac{13{,}620.5}{389} \approx 35.0$$

47. Class width $= \dfrac{\text{Max} - \text{Min}}{\text{Number of classes}} = \dfrac{14 - 3}{6} = 1.83 \Rightarrow 2$

Class	Midpoint, x	Frequency, f
3–4	3.5	3
5–6	5.5	8
7–8	7.5	4
9–10	9.5	2
11–12	11.5	2
13–14	13.5	1
		$\Sigma f = 20$

Shape: Positively skewed

49. Class width = $\frac{\text{Max} - \text{Min}}{\text{Number of classes}} = \frac{76 - 62}{5} = 2.8 \Rightarrow 3$

Class	Midpoint, x	Frequency, f
62–64	63	3
65–67	66	7
68–70	69	9
71–73	72	8
74–76	75	3
		$\Sigma f = 30$

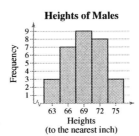

Shape: Symmetric

51. (a) $\bar{x} = \frac{\Sigma x}{n} = \frac{36.03}{6} = 6.005$

5.59, 5.99, 6, 6.02, 6.03, 6.4

two middle values $\Rightarrow$ median $= \frac{6 + 6.02}{2} = 6.01$

(b) $\bar{x} = \frac{\Sigma x}{n} = \frac{35.67}{6} = 5.945$

5.59, 5.99, 6, 6.02, 6.03, 6.4

two middle values $\Rightarrow$ median $= \frac{6 + 6.02}{2} = 6.01$

(c) mean

53. (a) Mean should be used since Car A has the highest mean of the three.

(b) Median should be used since Car B has the highest median of the three.

(c) Mode should be used since Car C has the highest mode of the three.

55. (a) $\bar{x} = 49.2\overline{3}$ (b) median = 46.5

(c) Key: 3 6 = 36   (d) Positively skewed

```
1 | 1 3
2 | 2 8            median
3 | 6 6 6 7 7 7 8
4 | 1 3 4 6 7
5 | 1 1 1 3        mean
6 | 1 2 3 4
7 | 2 2 4 6
8 | 5
9 | 0
```

57. Two different symbols are needed since they describe a measure of central tendency for two different sets of data (sample is a subset of the population).

2.4 MEASURES OF VARIATION

2.4 Try It Yourself Solutions

1a. Min = 23 or $23,000 and Max = 58 or $58,000

b. Range = max − min = 58 − 23 = 35 or $35,000

c. The range of the starting salaries for Corporation B is 35 or $35,000 (much larger than range of Corporation A).

2a. $\mu = \dfrac{\Sigma x}{N} = \dfrac{415}{10} = 41.5$ or $41,500

b.

Salary, x (1000s of dollars)	Deviation, $x - \mu$ (1000s of dollars)
23	23 − 41.5 = −18.5
29	29 − 41.5 = −12.5
32	32 − 41.5 = −9.5
40	40 − 41.5 = −1.5
41	41 − 41.5 = −0.5
41	41 − 41.5 = −0.5
49	49 − 41.5 = 7.5
50	50 − 41.5 = 8.5
52	52 − 41.5 = 10.5
58	58 − 41.5 = 16.5
$\Sigma x = 415$	$\Sigma(x - \mu) = 0$

3ab. $\mu = 41.5$ or $41,500

Salary, x	$x - \mu$	$(x - \mu)^2$
23	−18.5	$(-18.5)^2 = 342.25$
29	−12.5	$(-12.5)^2 = 156.25$
32	−9.5	$(-9.5)^2 = 90.25$
40	−1.5	$(-1.5)^2 = 2.25$
41	−0.5	$(-0.5)^2 = 0.25$
41	−0.5	$(-0.5)^2 = 0.25$
49	7.5	$(7.5)^2 = 56.25$
50	8.5	$(8.5)^2 = 72.25$
52	10.5	$(10.5)^2 = 110.25$
58	16.5	$(16.5)^2 = 272.25$
$\Sigma x = 415$	$\Sigma(x - \mu) = 0$	$\Sigma(x - \mu)^2 = 1102.5$

c. $\sigma^2 = \dfrac{\Sigma(x - \mu)^2}{N} = \dfrac{1102.5}{10}$

d. $\sigma = \sqrt{\sigma^2} = \sqrt{\dfrac{1102.5}{10}} = 10.5$ or $10,500

e. The population variance is 110.3 and the population standard deviation is 10.5 or $10,500.

4a.

Salary, x	$x - \bar{x}$	$(x - \bar{x})^2$
23	−18.5	342.25
29	−12.5	156.25
32	−9.5	90.25
40	−1.5	2.25
41	−0.5	0.25
41	−0.5	0.25
49	7.5	56.25
50	8.5	72.25
52	10.5	110.25
58	16.5	272.25
$\Sigma x = 415$	$\Sigma(x - \bar{x}) = 0$	$\Sigma(x - \bar{x})^2 = 1102.5$

$$SS_x = \Sigma(x - \bar{x})^2 = 1102.5$$

b. $s^2 = \dfrac{\Sigma(x - \bar{x})^2}{(n-1)} = \dfrac{1102.5}{9} = 122.5$

c. $s = \sqrt{s^2} = \sqrt{122.5} \approx 11.1$ or $11,100

5a. (Enter data in computer or calculator)

b. $\bar{x} = 37.89, \quad s = 3.98$

6a. 7, 7, 7, 7, 7, 13, 13, 13, 13, 13

b.

x	$x - \mu$	$(x - \mu)^2$
7	$7 - 10 = -3$	$(-3)^2 = 9$
7	$7 - 10 = -3$	$(-3)^2 = 9$
7	$7 - 10 = -3$	$(-3)^2 = 9$
7	$7 - 10 = -3$	$(-3)^2 = 9$
7	$7 - 10 = -3$	$(-3)^2 = 9$
13	$13 - 10 = 3$	$(3)^2 = 9$
13	$13 - 10 = 3$	$(3)^2 = 9$
13	$13 - 10 = 3$	$(3)^2 = 9$
13	$13 - 10 = 3$	$(3)^2 = 9$
13	$13 - 10 = 3$	$(3)^2 = 9$
$\Sigma x = 100$	$\Sigma(x - \mu) = 0$	$\Sigma(x - \mu)^2 = 90$

$$\mu = \frac{\Sigma x}{N} = \frac{100}{10} = 10$$

$$\sigma = \sqrt{\frac{\Sigma(x-\mu)^2}{N}} = \sqrt{\frac{90}{10}} = \sqrt{9} = 3$$

7a. $64 - 61.25 = 2.75 = 1$ standard deviation

b. 34%

c. The estimated percent of the heights that are between 61.25 and 64 inches is 34%.

8a. $31.6 - 2(19.5) = -7.4 = > 0$

b. $31.6 + 2(19.5) = 70.6$

c. $1 - \dfrac{1}{k^2} = 1 - \dfrac{1}{(2)^2} = 1 - \dfrac{1}{4} = 0.75$

At least 75% of the data lie within 2 standard deviations of the mean. At least 75% of the population of Alaska is between 0 and 70.6 years old.

9a.

x	f	xf
0	10	(0)(10) = 0
1	19	(1)(19) = 19
2	7	(2)(7) = 14
3	7	(3)(7) = 21
4	5	(4)(5) = 20
5	1	(5)(1) = 5
6	1	(6)(1) = 6
	$n = 50$	$\sum xf = 85$

b. $\bar{x} = \dfrac{\sum xf}{n} = \dfrac{85}{50} = 1.7$

c.

$x - \bar{x}$	$(x - \bar{x})^2$	$(x - \bar{x})^2 \cdot f$
0 − 1.7 = −1.70	$(-1.70)^2 = 2.8900$	(2.8900)(10) = 28.90
1 − 1.7 = −0.70	$(-0.70)^2 = 0.4900$	(0.4900)(19) = 9.31
2 − 1.7 = 0.30	$(0.30)^2 = 0.0900$	(0.0900)(7) = 0.63
3 − 1.7 = 1.30	$(1.30)^2 = 1.6900$	(1.6900)(7) = 11.83
4 − 1.7 = 2.30	$(2.30)^2 = 5.2900$	(5.2900)(5) = 26.45
5 − 1.7 = 3.30	$(3.30)^2 = 10.9800$	(10.9800)(1) = 10.89
6 − 1.7 = 4.30	$(4.30)^2 = 18.4900$	(18.4900)(1) = 18.49
		$\sum(x - \bar{x})^2 f = 106.5$

d. $s = \sqrt{\dfrac{\sum(x - \bar{x})^2 f}{(n - 1)}} = \sqrt{\dfrac{106.5}{49}} = \sqrt{2.17} \approx 1.5$

10a.

Class	x	f	xf
0–99	49.5	380	(49.5)(380) = 18,810
100–199	149.5	230	(149.5)(230) = 34,385
200–299	249.5	210	(249.5)(210) = 52,395
300–399	349.5	50	(349.5)(50) = 17,475
400–499	449.5	60	(449.5)(60) = 26,970
500+	650.0	70	(650.0)(70) = 45,500
		$n = 1000$	$\sum xf = 195{,}535$

b. $\bar{x} = \dfrac{\sum xf}{n} = \dfrac{195{,}535}{1000} \approx 195.5$

c.

$x - \bar{x}$	$(x - \bar{x})^2$	$(x - \bar{x})^2 \cdot f$
49.5 − 195.54 = −146.04	$(-146.04)^2 = 21{,}327.68$	(21,327.68)(380) = 8,104,518.4
149.5 − 195.54 = −46.04	$(-46.04)^2 = 2119.68$	(2119.68)(230) = 487,526.4
249.5 − 195.54 = 53.96	$(53.96)^2 = 2911.68$	(2911.68)(210) = 611,452.8
349.5 − 195.54 = 153.96	$(153.96)^2 = 23{,}703.68$	(23,703.68)(50) = 1,185,184
449.5 − 195.54 = 253.96	$(253.96)^2 = 64{,}495.68$	(64,495.68)(60) = 3,869,740.8
650 − 195.54 = 454.96	$(454.96)^2 = 206{,}533.89$	(206,533.89)(70) = 14,457,372.3
		$\sum(x - \bar{x})^2 f = 28{,}715{,}794.7$

d. $s = \sqrt{\dfrac{\sum(x - \bar{x})^2 f}{n - 1}} = \sqrt{\dfrac{28{,}715{,}794.7}{999}} = \sqrt{28{,}744.539} \approx 169.5$

2.4 EXERCISE SOLUTIONS

1. Range = max − min = 11 − 4 = 7

$$\mu = \frac{\Sigma x}{N} = \frac{81}{10} = 8.1$$

x	$x - \mu$	$(x - \mu)^2$
11	11 − 8.1 = 2.9	$(2.9)^2 = 8.41$
10	10 − 8.1 = 1.9	$(1.9)^2 = 3.61$
8	8 − 8.1 = −0.1	$(-0.1)^2 = 0.01$
4	4 − 8.1 = −4.1	$(-4.1)^2 = 16.81$
6	6 − 8.1 = −2.1	$(-2.1)^2 = 4.41$
7	7 − 8.1 = −1.1	$(-1.1)^2 = 1.21$
11	11 − 8.1 = 2.9	$(2.9)^2 = 8.41$
6	6 − 8.1 = −2.1	$(-2.1)^2 = 4.41$
11	11 − 8.1 = 2.9	$(2.9)^2 = 8.41$
7	7 − 8.1 = −1.1	$(-1.1)^2 = 1.21$
$\Sigma x = 81$	$\Sigma(x - \mu) = 0$	$\Sigma(x - \mu)^2 = 56.9$

$$\sigma^2 = \frac{\Sigma(x - \mu)^2}{N} = \frac{56.9}{10} = 5.69 \approx 5.7$$

$$\sigma = \sqrt{\frac{\Sigma(x - \mu)^2}{N}} = \sqrt{\frac{56.9}{10}} = \sqrt{5.69} \approx 2.4$$

3. Range = max − min = 19 − 5 = 14

$$\bar{x} = \frac{\Sigma x}{n} = \frac{100}{9} \approx 11.1$$

x	$x - \mu$	$(x - \mu)^2$
15	15 − 11.11 = 3.89	$(3.89)^2 = 15.13$
8	8 − 11.11 = −3.11	$(-3.11)^2 = 9.67$
12	12 − 11.11 = 0.89	$(0.89)^2 = 0.79$
5	5 − 11.11 = −6.11	$(-6.11)^2 = 37.33$
19	19 − 11.11 = 7.89	$(7.89)^2 = 62.25$
14	14 − 11.11 = 2.89	$(2.89)^2 = 8.35$
8	8 − 11.11 = −3.11	$(-3.11)^2 = 9.67$
6	6 − 11.11 = −5.11	$(-5.11)^2 = 26.11$
13	13 − 11.11 = 1.89	$(1.89)^2 = 3.57$
$\Sigma x = 100$	$\Sigma(x - \bar{x}) = 0$	$\Sigma(x - \bar{x})^2 = 172.89$

$$s^2 = \frac{\Sigma(x - \bar{x})^2}{n - 1} = \frac{172.89}{8} \approx 21.6$$

$$s = \sqrt{\frac{\Sigma(x - \bar{x})^2}{n - 1}} = \sqrt{\frac{172.89}{8}} \approx 4.6$$

5. Range = max − min = 96 − 23 = 73

7. The range is the difference between the maximum and minimum values of a data set. The advantage of the range is that it is easy to calculate. The disadvantage is that it uses only two entries from the data set.

9. The units of variance are squared. Its units are meaningless. (Ex: dollars^2)

11. (a) Range = max − min = 46.6 − 21.5 = 25.1

    (b) Range = max − min = 66.6 − 21.5 = 45.1

    (c) The range has increased substantially.

13. Graph (a) has a standard deviation of 24 and graph (b) has a standard deviation of 16 since graph (a) has more variability.

15. When calculating the population standard deviation, you divide the sum of the squared deviations by n, then take the square root of that value. When calculating the sample standard deviation, you divide the sum of the squared deviations by $n - 1$, then take the square root of that value.

17. Company B. Due to the larger standard deviation in salaries for company B, it would be more likely to be offered a salary of $33,000.

19. (a) LA: range = max − min = 35.9 − 18.3 = 17.6

x	$(x - \bar{x})$	$(x - \bar{x})^2$
20.2	−6.06	36.67
26.1	−0.16	0.02
20.9	−5.36	28.68
32.1	5.84	34.16
35.9	9.64	93.02
23.0	−3.64	10.60
28.2	1.94	3.78
31.6	5.34	28.56
18.3	−7.96	63.29
$\Sigma x = 236.3$		$\Sigma(x - \bar{x})^2 = 298.78$

$$\bar{x} = \frac{\Sigma x}{n} = \frac{236.3}{9} = 26.26$$

$$s^2 = \frac{\Sigma(x - \bar{x})^2}{(n - 1)} = \frac{298.78}{8} \approx 37.35$$

$$s = \sqrt{s^2} \approx 6.11$$

LB: range = max − min = 26.9 − 18.2 = 8.7

x	$(x - \bar{x})$	$(x - \bar{x})^2$
20.9	−1.98	3.91
18.2	−4.68	21.88
20.8	−2.08	4.32
21.1	−1.78	3.16
26.5	3.62	13.12
26.9	4.02	16.18
24.2	1.32	1.75
25.1	2.22	4.94
22.2	−0.68	0.46
$\Sigma x = 205.9$		$\Sigma(x - \bar{x})^2 = 69.72$

$$\bar{x} = \frac{\Sigma x}{n} = \frac{205.9}{9} = 22.88$$

$$s^2 = \frac{\Sigma(x - \bar{x})^2}{(n - 1)} = \frac{69.72}{8} \approx 8.71$$

$$s = \sqrt{s^2} \approx 2.95$$

(b) It appears from the data that the annual salaries in LA are more variable than the salaries in Long Beach.

21. (a) M: range = max − min = 1328 − 923 = 405

x	$(x - \bar{x})$	$(x - \bar{x})^2$
1059	−51.13	2,613.77
1328	217.8	47,469.52
1175	64.88	4,208.77
1123	12.88	165.77
923	−187.13	35,015.77
1017	−93.13	8,672.77
1214	103.88	10,790.02
1042	−68.13	4,641.02
$\Sigma x = 8881$		$\Sigma(x - \bar{x})^2 = 113{,}576.88$

$$\bar{x} = \frac{\Sigma x}{n} = \frac{8881}{8} = 1110.13$$

$$s^2 = \frac{\Sigma(x - \bar{x})^2}{(n - 1)} = \frac{113{,}576.9}{7} \approx 16{,}225.3$$

$$s = \sqrt{s^2} \approx 127.4$$

F: range = max − min = 1393 − 841 = 552

x	$(x - \bar{x})$	$(x - \bar{x})^2$
1226	92.50	8,556.25
965	−168.50	28,392.25
841	−292.50	85,556.25
1053	−80.50	6,480.25
1056	−77.50	6,006.25
1393	259.50	67,340.25
1312	178.50	31,862.25
1222	88.50	7,832.25
$\Sigma x = 9068$		$\Sigma(x - \bar{x})^2 = 242{,}026.00$

$$\bar{x} = \frac{\Sigma x}{n} = \frac{9068}{8} = 1133.50$$

$$s^2 = \frac{\Sigma(x - \bar{x})^2}{(n - 1)} = \frac{242{,}026}{7} \approx 34{,}575.1$$

$$s = \sqrt{s^2} \approx 185.9$$

(b) It appears from the data, the SAT scores for females are more variable than the SAT scores for males.

23. (a) Greatest sample standard deviation: (ii)

Data set (ii) has more entries that are farther away from the mean.

Least same standard deviation: (iii)

Data set (iii) has more entries that are close to the mean.

(b) The three data sets have the same mean, but have different standard deviations.

25. (a) Greatest sample standard deviation: (ii)

Data set (ii) has more entries that are farther away from the mean.

Least sample standard deviation: (iii)

Data set (iii) has more entries that are close to the mean.

(b) The three data sets have the same mean, median, and mode, but have different standard deviations.

27. Similarity: Both estimate proportions of the data contained within k standard deviations of the mean.

Difference: The Empirical Rule assumes the distribution is bell shaped, Chebychev's Theorem makes no such assumption.

29. $(800, 1200) \to (1000 - 1(200), 1000 + 1(200)) \to (\bar{x} - s, \bar{x} + s) \to (\bar{x}, \bar{x} + 2s)$

68% of the farms value between $800 and $1200 per acre.

31. (a) $n = 75$

$68\%(75) = (0.68)(75) = 51$ farm values will be between $800 and $1200 per acre.

(b) $n = 25$

$68\%(25) = (0.68)(25) = 17$ of these farm values will be between $800 and $1200 per acre.

33. $\bar{x} = 1000$
$s = 200$

{1250, 1375, 1450, 550} are outliers. They are more than 2 standard deviations from the mean (800, 1200).

35. $(\bar{x} - 2s, \bar{x} + 2s) \to (1.14, 5.5)$ are 2 standard deviations from the mean.

$1 - \dfrac{1}{k^2} = 1 - \dfrac{1}{(2)^2} = 1 - \dfrac{1}{4} = 0.75 \Rightarrow$ At least 75% of the eruption times lie between 1.14 and 5.5 minutes.

If $n = 32$, at least $(0.75)(32) = 24$ eruptions will lie between 1.14 and 5.5 minutes.

37.

x	f	xf	$x - \bar{x}$	$(x - \bar{x})^2$	$(x - \bar{x})^2 f$
0	5	0	−2.08	4.31	21.53
1	11	11	−1.08	1.16	12.71
2	7	14	−0.08	0.01	0.04
3	10	30	0.93	0.86	8.56
4	7	28	1.93	3.71	25.94
	$n = 40$	$\sum xf = 83$			$\sum(x - \bar{x})^2 f = 68.78$

$\bar{x} = \dfrac{\sum x}{n} = \dfrac{83}{40} \approx 2.1$

$s = \sqrt{\dfrac{\sum(x - \bar{x})^2 f}{n - 1}} = \sqrt{\dfrac{68.78}{39}} = \sqrt{1.76} \approx 1.3$

39. Class width $= \dfrac{\text{Max} - \text{Min}}{5} = \dfrac{14 - 4}{5} = \dfrac{10}{5} = 2$

Class	Midpoint, x	f	xf
4–5	4.5	10	40.5
6–7	6.5	6	39
8–9	8.5	3	25.5
10–11	10.5	7	73.5
12–14	13.0	6	78
		$N = 32$	$\sum xf = 261$

$\mu = \dfrac{\sum xf}{N} = \dfrac{261}{32} \approx 8.2$

$x - \mu$	$(x - \mu)^2$	$(x - \mu)^2 f$
-3.7	13.69	136.90
-1.7	2.89	17.34
0.3	0.09	0.27
2.3	5.29	37.03
4.8	23.04	138.24
		$\sum(x - \mu)^2 f = 329.78$

$$\sigma = \sqrt{\frac{\sum(x - \mu)^2}{N}} = \sqrt{\frac{329.78}{32}} \approx 3.2$$

41.

x	f	xf
70.5	1	70.5
92.5	12	1110
114.5	25	2862.5
136.5	10	1365
158.5	2	317
	$n = 50$	$\sum xf = 5725$

$$\bar{x} = \frac{\sum xf}{n} = \frac{5725}{50} = 114.5$$

$x - \bar{x}$	$(x - \bar{x})^2$	$(x - \bar{x})^2 f$
-44	1936	1936
-22	484	5808
0	0	0
22	484	4840
44	1936	3872
		$\sum(x - \bar{x})^2 f = 16{,}456$

$$s = \sqrt{\frac{\sum(x - \bar{x})^2}{n - 1}} = \sqrt{\frac{16{,}456}{49}} = \sqrt{335.83} \approx 18.33$$

43.

Class	Midpoint, x	f	xf
0–4	2	19.9	39.8
5–13	9	35.2	316.8
14–17	15.5	16.9	261.95
18–24	21	29.8	625.8
25–34	29.5	38.3	1129.85
35–44	39.5	40.0	1580.00
45–64	54.5	78.3	4267.35
65+	70	39.0	2730
		$n = 297.4$	$\sum xf = 10{,}951.55$

$$\bar{x} = \frac{\sum xf}{n} = \frac{10{,}951.55}{297.4} \approx 36.82$$

$x - \bar{x}$	$(x - \bar{x})^2$	$(x - \bar{x})^2 f$
−34.82	1212.43	24,127.36
−27.82	773.95	27,243.04
−21.32	454.54	7,681.73
−15.82	250.27	7,458.05
−7.32	53.58	2,052.11
2.68	7.18	287.20
17.68	312.58	24,475.01
33.18	1100.91	42,935.49
		$\sum(x - \bar{x})^2 f = 136{,}259.99$

$$s = \sqrt{\frac{\sum(x - \bar{x})^2}{n - 1}} = \sqrt{\frac{136{,}259.99}{296.4}} = \sqrt{459.72} \approx 21.44$$

45. $CV_{\text{heights}} = \frac{\sigma}{\mu} \cdot 100\% = \frac{3.44}{72.75} \cdot 100 \approx 4.73$

$CV_{\text{weights}} = \frac{\sigma}{\mu} \cdot 100\% = \frac{18.47}{187.83} \cdot 100 \approx 9.83$

It appears that weight is more variable than height.

47. (a) $\bar{x} = 550$ $s \approx 302.8$

(b) $\bar{x} = 5500$ $s \approx 3028$

(c) $\bar{x} = 55$ $s \approx 30.28$

(d) By multiplying each entry by a constant k, the new sample mean is $k \cdot \bar{x}$ and the new sample standard deviation is $k \cdot s$.

49. $1 - \frac{1}{k^2} = 0.99 \Rightarrow 1 - 0.99 = \frac{1}{k^2} \Rightarrow k^2 = \frac{1}{0.01} \Rightarrow k = \sqrt{\frac{1}{0.01}} = 10$

At least 99% of the data in any data set lie within 10 standard deviations of the mean.

2.5 MEASURES OF POSITION

2.5 Try It Yourself Solutions

1a. 0, 0, 1, 1, 1, 2, 3, 3, 3, 4, 5, 5, 5, 7, 9, 10, 12, 12, 13, 13, 13, 13, 13, 13, 15, 16, 16, 17, 17, 18, 18, 18, 19, 19, 19, 20, 20, 21, 22, 23, 23, 24, 24, 25, 25, 26, 26, 26, 29, 33, 36, 37, 39, 39, 39, 39, 40, 40, 41, 41, 41, 42, 44, 44, 45, 47, 48, 49, 49, 49, 51, 53, 56, 58, 58, 59, 60, 67, 68, 68, 72

b. $Q_2 = 23.5$

c. $Q_1 = 13$ $Q_3 = 41.5$

2a. (Enter the data)

b. $Q_1 = 17$ $Q_2 = 23$ $Q_3 = 28.5$

c. One quarter of the tuition costs is $17,000 or less, one half is $23,000 or less, and three quarters is $28,500 or less.

3a. $Q_1 = 13$ $Q_3 = 41.5$

b. $IQR = Q_3 - Q_1 = 41.5 - 13 = 28.5$

c. The ages in the middle half of the data set vary by 28.5 years.

4a. Min = 0 $Q_1 = 13$ $Q_2 = 23.5$

$Q_3 = 41.5$ Max = 72

bc. Ages of Akhiok Residents

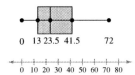

d. It appears that half of the ages are between 13 and 41.5 years.

5a. 80th percentile

b. 80% of the ages are 45 years or younger.

6a. $\mu = 70$, $\sigma = 8$

b. $x = 60$: $z = \dfrac{x - \mu}{\sigma} = \dfrac{60 - 70}{8} = -1.25$

$x = 71$: $z = \dfrac{x - \mu}{\sigma} = \dfrac{71 - 70}{8} = 0.125$

$x = 92$: $z = \dfrac{x - \mu}{\sigma} = \dfrac{92 - 70}{8} = 2.75$

c. From the z-score, the utility bill of $60 is 1.25 standard deviations below the mean, the bill of $71 is 0.125 standard deviations above the mean, and the bill of $92 is 2.75 standard deviations above the mean.

7a. NFL: $\mu = 23.6$, $\alpha = 6.0$

AFL: $\mu = 11.7$, $\sigma = 4.6$

b. Kansas City (NFL): $x = 16 \Rightarrow z = \dfrac{x - \mu}{\sigma} = \dfrac{16 - 23.6}{6.0} = -1.27$

Tampa Bay (AFL): $x = 12 \Rightarrow z = \dfrac{x - \mu}{\sigma} = \dfrac{12 - 11.7}{4.6} = 0.07$

c. The number of field goals scored by Kansas City is 1.27 standard deviations below the mean and the number of field goals scored by Tampa Bay is 0.07 standard deviations above the mean. Comparing the two measures of position indicates Tampa Bay has a higher position within the AFL than Kansas City in the NFL.

2.5 EXERCISE SOLUTIONS

1. (a)   lower half   upper half

1 2 2 4 4 5 5 5 6 6 6 7 7 7 7 8 8 8 9 9
            ↑   ↑   ↑
           Q_1 Q_2 Q_3

$Q_1 = 4.5$ $Q_2 = 6$ $Q_3 = 7.5$

(b)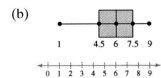

3. The basketball team scored more points per game than 75% of the teams in the league.

5. The student scored above 63% of the students who took the ACT placement test.

7. True

9. False. The 50th percentile is equivalent to Q_2.

11. (a) Min = 10 (b) Max = 20
    (c) $Q_1 = 13$ (d) $Q_2 = 15$
    (e) $Q_3 = 17$ (f) IQR = $Q_3 - Q_1 = 17 - 13 = 4$

13. (a) Min = 900 (b) Max = 2100
    (c) $Q_1 = 1250$ (d) $Q_2 = 1500$
    (e) $Q_3 = 1950$ (f) IQR = $Q_3 - Q_1 = 1950 - 1250 = 700$

15. (a) Min = -1.9 (b) Max = 2.1
    (c) $Q_1 = -0.5$ (d) $Q_2 = 0.1$
    (e) $Q_3 = 0.7$ (f) IQR = $Q_3 - Q_1 = 0.7 - (-0.5) = 1.2$

17. $Q_1 = B$, $Q_2 = A$, $Q_3 = C$

    25% of the entries are below B, 50% are below A, and 75% are below C.

19. (a) $Q_1 = 2$, $Q_2 = 4$, $Q_3 = 5$

    (b) **Watching Television**

21. (a) $Q_1 = 3.2$, $Q_2 = 3.65$, $Q_3 = 3.9$

    (b) **Butterfly Wingspans**

23. (a) 5 (b) 50% (c) 25%

25. A $\Rightarrow z = -1.43$

    B $\Rightarrow z = 0$

    C $\Rightarrow z = 2.14$

    The score 2.14 is unusual because it is so large.

27. (a) Statistics:

$$x = 73 \Rightarrow z = \frac{x - \mu}{\sigma} = \frac{73 - 63}{7} \approx 1.43$$

(b) Biology:

$$x = 26 \Rightarrow z = \frac{x - \mu}{\sigma} = \frac{26 - 23}{3.9} \approx 0.77$$

(c) The student had a better score on the statistics test.

29. (a) Statistics:

$$x = 78 \Rightarrow z = \frac{x - \mu}{\sigma} = \frac{78 - 63}{7} \approx 2.14$$

(b) Biology:

$$x = 29 \Rightarrow z = \frac{x - \mu}{\sigma} = \frac{29 - 23}{3.9} \approx 1.54$$

(c) The student had a better score on the statistics test.

31. (a) $x = 34{,}000 \Rightarrow z = \frac{x - \mu}{\sigma} = \frac{34{,}000 - 35{,}000}{2{,}250} \approx -0.44$

$x = 37{,}000 \Rightarrow z = \frac{x - \mu}{\sigma} = \frac{37{,}000 - 35{,}000}{2{,}250} \approx 0.89$

$x = 31{,}000 \Rightarrow z = \frac{x - \mu}{\sigma} = \frac{31{,}000 - 35{,}000}{2{,}250} \approx -1.78$

None of the selected tires have unusual life spans.

(b) $x = 30{,}500 \Rightarrow z = \frac{x - \mu}{\sigma} = \frac{30{,}500 - 35{,}000}{2{,}250} = -2 \Rightarrow$ 2.5th percentile

$x = 37{,}250 \Rightarrow z = \frac{x - \mu}{\sigma} = \frac{37{,}250 - 35{,}000}{2{,}250} = 1 \Rightarrow$ 84th percentile

$x = 35{,}000 \Rightarrow z = \frac{x - \mu}{\sigma} = \frac{35{,}000 - 35{,}000}{2{,}250} = 0 \Rightarrow$ 50th percentile

33. 67 inches

20% of the heights are below 67 inches.

35. $x = 74$: $z = \frac{x - \mu}{\sigma} = \frac{74 - 69.2}{2.9} \approx 1.66$

$x = 62$: $z = \frac{x - \mu}{\sigma} = \frac{62 - 69.2}{2.9} \approx -2.48$

$x = 80$: $z = \frac{x - \mu}{\sigma} = \frac{80 - 69.2}{2.9} \approx 3.72$

The height of 62 inches is unusual due to a rather small z-score. The height of 80 inches is very unusual due to a rather large z-score.

37. $x = 71.1$: $z = \frac{x - \mu}{\sigma} = \frac{71.1 - 69.2}{2.9} \approx 0.66$

Approximately the 70th percentile.

39. (a) 27 28 31 32 32 33 35 36 36 36 36 37 38 39 39 40 40 40 41 41
41 42 42 42 42 42 42 43 43 43 44 44 45 45 46 47 47 47 47 47
48 48 48 48 48 | 49 49 49 49 49 49 50 50 51 51 51 51 51 51 52
52 52 53 53 54 | 54 54 54 54 54 | 54 54 55 56 56 56 57 57 57 59
59 59 60 60 60 | 61 61 61 62 62 | 63 63 63 63 64 | 65 67 68 74 82

$Q_1 = 42$, $Q_2 = 49$, $Q_3 = 56$

(b) Ages of Executives

(c) Half of the ages are between 42 and 56 years.

(d) About 49 years old ($x = 49.62$ and $Q_2 = 49.00$), since half of the executives are older and half are younger.

41. 22 23 24 32 33 34 36 38 39 40 41 47

$Q_1 = 28$, Q_2, $Q_3 = 39.5$

Midquartile $= \dfrac{Q_1 + Q_3}{2} = \dfrac{28 + 39.5}{2} = 33.75$

43. 13.4 15.2 15.6 16.7 17.2 18.7 19.7 19.8 19.8 20.8 21.4 22.9 28.7 30.1 31.9

Q_1, Q_2, Q_3

Midquartile $= \dfrac{Q_1 + Q_3}{2} = \dfrac{16.7 + 22.9}{2} = 19.8$

CHAPTER 2 REVIEW EXERCISE SOLUTIONS

1.

Class	Midpoint	Boundaries	Frequency	Relative frequency	Cumulative frequency
20–23	21.5	19.5–23.5	1	0.05	1
24–27	25.5	23.5–27.5	2	0.10	3
28–31	29.5	27.5–31.5	6	0.30	9
32–35	33.5	31.5–35.5	7	0.35	16
36–39	37.5	35.5–39.5	4	0.20	20
			$\Sigma f = 20$	$\Sigma \dfrac{f}{n} = 1$	

3.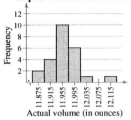

5.

Class	Midpoint, x	Frequency	Cumulative frequency
79–93	86	9	9
94–108	101	12	21
109–123	116	5	26
124–138	131	3	29
139–153	146	2	31
154–168	161	1	32
		$\Sigma f = 32$	

7.
```
1 | 3 7 8 9
2 | 0 1 2 3 3 3 4 4 5 5 5 7 8 8 9
3 | 1 1 2 3 4 5 7 8
4 | 3 4 7
5 | 1
```

9.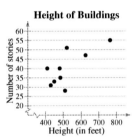

It appears as height increases, the number of stories increases.

11.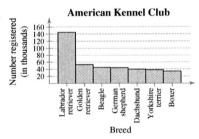

13. $\bar{x} = 8.6$

median = 9

mode = 9

36 CHAPTER 2 | DESCRIPTIVE STATISTICS

15.

Midpoint, x	Frequency, f	xf
21.5	1	21.5
25.5	2	51.0
29.5	6	177.0
33.5	7	234.5
37.5	4	150.0
	$n = 20$	$\Sigma xf = 634$

$$\bar{x} = \frac{\Sigma xf}{n} = \frac{634}{20} = 31.7$$

17. $\bar{x} = \dfrac{\Sigma xw}{w} = \dfrac{(65)(0.15) + (72)(0.15) + (84)(0.15) + (89)(0.15) + (70)(0.15) + (90)(0.25)}{0.15 + 0.15 + 0.15 + 0.15 + 0.15 + 0.25}$

$= \dfrac{79.5}{1} = 79.5$

19. Skewed   **21.** Skewed left   **23.** Median

25. Range = max − min = 8.26 − 5.46 = 2.8

27. $\mu = \dfrac{\Sigma x}{N} = \dfrac{108}{12} = 9$

$\sigma = \sqrt{\dfrac{\Sigma(x - \mu)^2}{N}} = \sqrt{\dfrac{(6 - 9)^2 + (14 - 9)^2 + \cdots + (12 - 9)^2 + (10 - 9)^2}{12}}$

$= \sqrt{\dfrac{122}{12}} \approx \sqrt{10.17} \approx 3.2$

29. $\bar{x} = \dfrac{\Sigma x}{n} = \dfrac{36{,}801}{15} = 2453.4$

$s = \sqrt{\dfrac{\Sigma(x - \bar{x})^2}{n - 1}} = \sqrt{\dfrac{(2445 - 2453.4)^2 + \cdots + (2.377 - 2453.4)^2}{14}}$

$= \sqrt{\dfrac{1{,}311{,}783.6}{14}} \approx \sqrt{93{,}698.8} \approx 306.1$

31. 99.7% of the distribution lies within 3 standard deviations of the mean.

$\mu + 3\sigma = 29 + (3)(2.50) = 36.5$

$\mu - 3\sigma = 29 - (3)(2.50) = 21.5$

99.7% of the distribution lies between $21.50 and $36.50.

33. $n = 40 \qquad \mu = 23 \qquad \sigma = 6$

$(11, 35) \rightarrow (23 - 2(6), 23 + 2(6)) \Rightarrow (\mu - 2\sigma, \mu + 2\sigma) \Rightarrow k = 2$

$1 = \dfrac{1}{k^2} = 1 - \dfrac{1}{(2)^2} = 1 - \dfrac{1}{4} = 0.75$

At least $(40)(0.75) = 30$ customers have a mean sale between $11 and $35.

35. $\bar{x} = \dfrac{\Sigma xf}{n} = \dfrac{99}{40} \approx 2.5$

$s = \sqrt{\dfrac{\Sigma (x - \bar{x})^2 f}{n - 1}} = \sqrt{\dfrac{(0 - 1.24)^2(1) + (1 - 1.24)^2(8) + \cdots + (5 - 1.24)^2(3)}{39}}$

$ = \sqrt{\dfrac{59.975}{39}} \approx 1.2$

37. $Q_1 = 56$

39. IQR $= Q_3 - Q_1 = 70 - 56 = 14$

41. IQR $= Q_3 - Q_1 = 33 - 29 = 4$

43. 23% of the students scored higher than 68.

45. $x = 248 \Rightarrow z = \dfrac{x - \mu}{\sigma} = \dfrac{248 - 192}{24} \approx 2.33$

This is an unusually heavy player.

47. $x = 222 \Rightarrow z = \dfrac{x - \mu}{\sigma} = \dfrac{222 - 192}{24} = 1.25$

This player is not unusual.

CHAPTER 2 QUIZ SOLUTIONS

1. (a)

Class limits	Midpoint	Class boundaries	Frequency	Relative frequency	Cumulative frequency
101–112	106.5	100.5–112.5	3	0.12	3
113–124	118.5	112.5–124.5	11	0.44	14
125–136	130.5	124.5–136.5	7	0.28	21
137–148	142.5	136.5–148.5	2	0.08	23
149–160	154.5	148.5–160.5	2	0.08	25

(b) Frequency Histogram and Polygon

(c) Relative Frequency Histogram

(d) Skewed

(e)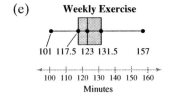

(f)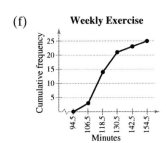

38 CHAPTER 2 | DESCRIPTIVE STATISTICS

2. $\bar{x} = \dfrac{\Sigma xf}{n} = \dfrac{3130.5}{25} \approx 125.22$

$s = \sqrt{\dfrac{\Sigma(x-\bar{x})^2 f}{n-1}} = \sqrt{\dfrac{4055.04}{24}} \approx 13.00$

3. (a)  (b)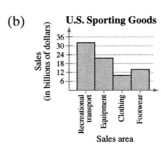

4. (a) $\bar{x} = \dfrac{\Sigma x}{n} = 751.6$

   median = 784.5

   mode = (none)

   The mean best describes a typical salary because there are no outliers.

   (b) range = max − min = 575

   $s^2 = \dfrac{\Sigma(x-\bar{x})^2}{n-1} = 48{,}135.1$

   $s = \sqrt{\dfrac{\Sigma(x-\bar{x})^2}{n-1}} = 219.4$

5. $\bar{x} - 2s = 155{,}000 - 2 \cdot 15{,}000 = \$125{,}000$

   $\bar{x} + 2s = 155{,}000 + 2 \cdot 15{,}000 = \$185{,}000$

   95% of the new home prices fall between $125,000 and $185,000.

6. (a) $x = 200{,}000 \quad z = \dfrac{x-\mu}{\sigma} = \dfrac{200{,}000 - 155{,}000}{15{,}000} = 3 \Rightarrow$ unusual price

   (b) $x = 55{,}000 \quad z = \dfrac{x-\mu}{\sigma} = \dfrac{55{,}000 - 155{,}000}{15{,}000} \approx -6.67 \Rightarrow$ very unusual price

   (c) $x = 175{,}000 \quad z = \dfrac{x-\mu}{\sigma} = \dfrac{175{,}000 - 155{,}000}{15{,}000} \approx 1.33 \Rightarrow$ not unusual

   (d) $x = 122{,}000 \quad z = \dfrac{x-\mu}{\sigma} = \dfrac{122{,}000 - 155{,}000}{15{,}000} = -2.2 \Rightarrow$ unusual price

7. (a) $Q_1 = 71 \quad Q_2 = 84.5 \quad Q_3 = 90$

   (b) IQR = $Q_3 - Q_1 = 90 - 71 = 19$

   (c)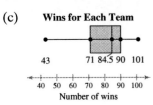

Probability

3.1 BASIC CONCEPTS OF PROBABILITY

3.1 Try It Yourself Solutions

1ab.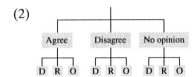

  c. (1) 6 outcomes (2) 9 outcomes

  d. (1) Let A = Agree, D = Disagree, N = No Opinion, M = Male and F = Female
  Sample space = $\{AM, AF, DM, DF, NM, NF\}$

  (2) Let A = Agree, D = Disagree, N = No Opinion, R = Republican, De = Democrat, O = Other
  Sample space = $\{AR, ADe, AO, DR, DDe, DO, NR, NDe, NO\}$

2a. (1) 6 outcomes (2) 1 outcome

  b. (1) Not a simple event (2) Simple event

3a. (1) 52 (2) 1 (3) $P(\text{7 of diamonds}) = \frac{1}{52} \approx 0.0192$

  b. (1) 52 (2) 13 (3) $P(\text{diamond}) = \frac{13}{52} = 0.25$

  c. (1) 52 (2) 52 (3) $P(\text{diamond, heart, club, or spade}) = \frac{52}{52} = 1$

4a. Event = the next claim processed is fraudulent (Freq = 4)

  b. Total Frequency = 100

  c. $P(\text{fraudulent claim}) = \frac{4}{100} = 0.04$

5a. Frequency = 54

  b. Total of the Frequencies = 1000

  c. $P(\text{age 15 to 24}) = \frac{54}{1000} = 0.054$

6a. Event = salmon successfully passing through a dam on the Columbia River.

  b. Estimated from the results of an experiment.

  c. Empirical probability

7a. $P(\text{red gill}) = \frac{17}{40} = 0.425$

  b. $P(\text{not red gill}) = 1 - \frac{17}{40} = \frac{23}{40} = 0.575$

  c. $\frac{23}{40}$ or 0.575

3.1 EXERCISE SOLUTIONS

1. (a) Yes, the probability of an event occurring must be contained in the interval [0, 1] or [0%, 100%].

   (b) No, the probability of an event occurring cannot be greater than 1.

   (c) No, the probability of an event occurring cannot be less than 0.

   (d) Yes, the probability of an event occurring must be contained in the interval [0, 1] or [0%, 100%].

   (e) Yes, the probability of an event occurring must be contained in the interval [0, 1] or [0%, 100%].

3. $\{0, 1, 2, 3, 4, 5, 6, 7, 8, 9\}$

5. $\{(P, A), (P, B), (P, AB), (P, O), (N, A), (N, B), (N, AB), (N, O)\}$ where (P, A) represents positive Rh-factor with A- blood type and (N, A) represents negative Rh-factor with A- blood type.

7. Simple event because it is an event that consists of a single outcome.

9. Not a simple event because it is an event that consists of more than a single outcome.

   king = {king of hearts, king of spades, king of clubs, king of diamonds}

11. Empirical probability since company records were probably used to calculate the frequency of a washing machine breaking down.

13. $P(\text{less than 1000}) = \dfrac{999}{6296} \approx 0.159$

15. $P(\text{number divisible by 1000}) = \dfrac{6}{6296} \approx 0.000953$

17. $P(\text{voted in 2002 Gubernatorial election}) = \dfrac{3{,}219{,}864}{6{,}797{,}293} \approx 0.474$

19. $P(\text{between 21 and 24}) = \dfrac{7.3}{129.5} \approx 0.056$

21. $P(\text{not between 18 and 20}) = 1 - \dfrac{4.8}{129.5} \approx 1 - 0.037 \approx 0.963$

23. $P(\text{Phd}) = \dfrac{8}{89} \approx 0.090$

25. $P(\text{master's}) = \dfrac{21}{89} \approx 0.236$

27. (a) $P(\text{pink}) = \dfrac{2}{4} = 0.5$ (b) $P(\text{red}) = \dfrac{1}{4} = 0.25$ (c) $P(\text{white}) = \dfrac{1}{4} = 0.25$

29. $P(\text{service industry}) = \dfrac{102{,}811}{135{,}073} \approx 0.761$

31. $P(\text{not in service industry}) = 1 - P(\text{service industry}) = 1 - 0.761 = 0.239$

33. The probability of choosing a tea drinker who does not have a college degree.

35. (a) $P(\text{at least } 21) = \dfrac{44}{80} \approx 0.55$ (b) $P(\text{between 40 and 50 inclusive}) = \dfrac{14}{80} \approx 0.175$

(c) $P(\text{older than } 65) = \dfrac{4}{80} = 0.05$

37. (a)

Sum	P(sum)	Probability
2	1/36	0.028
3	2/36	0.056
4	3/36	0.083
5	4/36	0.111
6	5/36	0.139
7	6/36	0.167
8	5/36	0.139
9	4/36	0.111
10	3/36	0.083
11	2/36	0.056
12	1/36	0.028

(b) Answers will vary.

(c) The answers in part (a) and (b) will be similar.

39. Let S = Sunny Day and R = Rainy Day

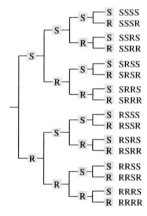

41. No, the odds of winning a prize are 1:6. (One winning cap and 6 losing caps) Thus, the statement should read, "one in seven game pieces win a prize."

43. $13:39 = 1:3$

3.2 CONDITIONAL PROBABILITY AND THE MULTIPLICATION RULE

3.2 Try It Yourself Solutions

1a. (1) 30 and 102   (2) 11 and 50

 b. (1) $P(\text{not have gene}) = \dfrac{30}{102} \approx 0.294$ (2) $P(\text{not have gene}|\text{normal IQ}) = \dfrac{11}{50} = 0.22$

2a. (1) No   (2) Yes

 b. (1) Independent   (2) Dependent

42 CHAPTER 3 | PROBABILITY

3a. (1) Independent (2) Dependent

b. (1) Let A = {swimming through 1st dam}

B = {swimming through 2nd dam}

$P(A \text{ and } B) = P(A) \cdot P(B|A) = (0.85) \cdot (0.85) = 0.723$

(2) Let A = {selecting a heart}

B = {selecting a second heart}

$P(A \text{ and } B) = P(A) \cdot P(B|A) = \left(\dfrac{13}{52}\right) \cdot \left(\dfrac{12}{51}\right) \approx 0.059$

4a. (1) Find probability of the event (2) Find probability of the complement of the event

b. (1) $P(3 \text{ salmon successful}) = (0.90) \cdot (0.90) \cdot (0.90) = 0.729$

(2) $P(\text{at least one salmon successful}) = 1 - P(\text{none are successful})$

$= 1 - (0.10) \cdot (0.10) \cdot (0.10) = 0.999$

3.2 EXERCISE SOLUTIONS

1. Two events are independent if the occurrence of one of the events does not affect the probability of the occurrence of the other event.

   If $P(B|A) = P(B)$ or $P(A|B) = P(A)$, then Events A and B are independent.

3. False. If two events are independent, $P(A|B) = P(A)$.

5. These events are independent since the outcome of the 1st card drawn does not affect the outcome of the 2nd card drawn.

7. These events are dependent since the outcome of the 1st ball drawn affects the outcome of the 2nd ball drawn.

9. Let A = {have mutated BRCA gene} and B = {develop breast cancer}. Thus

   $P(B) = \dfrac{1}{8}$, $P(A) = \dfrac{1}{250}$, and $P(B|A) = \dfrac{8}{10}$.

   (a) $P(B|A) = \dfrac{8}{10} = 0.8$

   (b) $P(A \text{ and } B) = P(A) \cdot P(B|A) = \left(\dfrac{1}{250}\right) \cdot \left(\dfrac{8}{10}\right) = 0.0032$

   (c) Dependent since $P(B|A) \neq P(B)$.

11. Let A = {own a computer} and B = {summer vacation this year}.

    (a) $P(B|A) = \dfrac{46}{57} \approx 0.807$ (b) $P(A \text{ and } B) = P(A)P(B|A) = \left(\dfrac{57}{146}\right)\left(\dfrac{46}{57}\right) = 0.315$

    (c) Dependent, since $P(B|A) \neq P(B)$. The probability that the family takes a summer vacation depends on whether or not they own a computer.

13. Let A = {pregnant} and B = {multiple births}. Thus $P(A) = 0.24$ and $P(B|A) = 0.07$.

  (a) $P(A \text{ and } B) = P(A) \cdot P(B|A) = (0.24) \cdot (0.07) = 0.0168$

  (b) $P(B'|A) = 1 - P(B|A) = 1 - 0.07 = 0.93$

  (c) It is not unusual since the probability of a pregnancy and multiple births is 0.0168.

15. Let A = {household in U.S. has a computer} and B = {has Internet access}

  $P(A \text{ and } B) = P(A)P(B|A) = (0.57)(0.51) = 0.291$

17. Let A = {1st part drawn is defective} and B = {2nd part drawn is defective}

  (a) $P(A \text{ and } B) = P(A) \cdot P(A|B) = \left(\frac{4}{11}\right) \cdot \left(\frac{3}{10}\right) \approx 0.109$

  (b) $P(A' \text{ and } B') = P(A') \cdot P(B'|A') = \left(\frac{7}{11}\right) \cdot \left(\frac{6}{10}\right) \approx 0.382$

  (c) $P(\text{at least one part defective})$

19. Let A = {have one month's income or more} and B = {man}

  (a) $P(A) = \frac{138}{287} \approx 0.481$

  (b) $P(A'|B) = \frac{66}{142} \approx 0.465$

  (c) $P(B'|A) = \frac{62}{138} \approx 0.449$

  (d) Dependent since $P(A') \approx 0.519 \neq 0.465 \approx P(A'|B)$ Whether a person has at least one month's income saved depends on whether or not the person is male.

21. (a) $P(\text{all five have AB+}) = (0.03) \cdot (0.03) \cdot (0.03) \cdot (0.03) \cdot (0.03) = 0.0000000243$

  (b) $P(\text{none have AB+}) = (0.97) \cdot (0.97) \cdot (0.97) \cdot (0.97) \cdot (0.97) \approx 0.859$

  (c) $P(\text{at least one has AB+}) = 1 - P(\text{none have AB+}) = 1 - 0.859 = 0.141$

23. (a) $P(\text{first question correct}) = 0.2$

  (b) $P(\text{first two questions correct}) = (0.2) \cdot (0.2) = 0.04$

  (c) $P(\text{first three questions correct}) = (0.2)^3 = 0.008$

  (d) $P(\text{none correct}) = (0.8)^3 = 0.512$

  (e) $P(\text{at least one correct}) = 1 - P(\text{none correct}) = 1 - 0.512 = 0.488$

25. $P(\text{all three products came from the third factory}) = \frac{25}{110} \cdot \frac{24}{109} \cdot \frac{23}{108} \approx 0.011$

27. $P(A|B) = \dfrac{P(A) \cdot P(B|A)}{P(A) \cdot P(B|A) + P(A') \cdot P(B|A')} = \dfrac{\left(\frac{2}{3}\right) \cdot \left(\frac{1}{5}\right)}{\left(\frac{2}{3}\right) \cdot \left(\frac{1}{5}\right) + \left(\frac{1}{3}\right) \cdot \left(\frac{1}{2}\right)}$

  $= \dfrac{0.133}{0.133 + 0.167} = 0.444$

44 CHAPTER 3 | PROBABILITY

29. $P(A|B) = \dfrac{P(A)P(B|A)}{(A)P(B|A) + P(A')P(B|A')}$

$= \dfrac{(0.25)(0.3)}{(0.25)(0.3) + (0.75)(0.5)} = \dfrac{0.075}{0.075 + 0.375} = 0.167$

31. $P(A) = \dfrac{1}{200} = 0.005$

$P(B|A) = 0.80$

$P(B|A') = 0.05$

(a) $P(A|B) = \dfrac{P(A) \cdot P(B|A)}{P(A) \cdot P(B|A) + P(A') \cdot P(B|A')}$

$= \dfrac{(0.005) \cdot (0.8)}{(0.005) \cdot (0.8) + (0.995) \cdot (0.05)} = \dfrac{0.004}{0.004 + 0.04975} \approx 0.074$

(b) $P(A'|B') = \dfrac{P(A') \cdot P(B'|A')}{P(A') \cdot P(B'|A') + P(A) \cdot P(B'|A)}$

$= \dfrac{(0.995) \cdot (0.95)}{(0.995) \cdot (0.95) + (0.005) \cdot (0.2)} = \dfrac{0.94525}{0.94525 + 0.001} \approx 0.999$

33. Let A = {flight departs on time} and B = {flight arrives on time}

$P(A|B) = \dfrac{P(A \text{ and } B)}{P(B)} = \dfrac{(0.83)}{(0.87)} \approx 0.954$

3.3 THE ADDITION RULE

3.3 Try It Yourself Solutions

1a. (1) None of the statements are true.

   (2) None of the statements are true.

   (3) Events A and B cannot occur at the same time.

b. (1) A and B are not mutually exclusive.

   (2) A and B are not mutually exclusive.

   (3) A and B are mutually exclusive.

2a. (1) Mutually exclusive   (2) Not mutually exclusive

b. (1) Let A = {6} and B = {odd}

$P(A) = \dfrac{1}{6}$ and $P(B) = \dfrac{3}{6} = \dfrac{1}{2}$

(2) Let A = {face card} and B = {heart}

$P(A) = \dfrac{12}{52}, P(B) = \dfrac{13}{52}$, and $P(A \text{ and } B) = \dfrac{3}{52}$

c. (1) $P(A \text{ or } B) = P(A) + P(B) = \frac{1}{6} + \frac{1}{2} \approx 0.667$

(2) $P(A \text{ or } B) = P(A) + P(B) - P(A \text{ and } B) = \frac{12}{52} + \frac{13}{52} - \frac{3}{52} \approx 0.423$

3a. $A = \{\text{sales between } \$0 \text{ and } \$24{,}999\}$

$B = \{\text{sales between } \$25{,}000 \text{ and } \$49{,}000\}$

b. A and B cannot occur at the same time $\rightarrow A$ and B are mutually exclusive

c. $P(A) = \frac{3}{36}$ and $P(B) = \frac{5}{36}$

d. $P(A \text{ or } B) = P(A) + P(B) = \frac{3}{36} + \frac{5}{36} \approx 0.222$

4a. (1) Mutually exclusive (2) Not mutually exclusive

b. (1) $P(B \text{ or } AB) = P(B) + P(AB) = \frac{45}{409} + \frac{16}{409} \approx 0.149$

(2) $P(O \text{ or } Rh+) = P(O) + P(Rh+) - P(O \text{ and } Rh+) = \frac{184}{409} + \frac{344}{409} - \frac{156}{409} \approx 0.910$

5a. Let $A = \{\text{linebacker}\}$ and $B = \{\text{quarterback}\}$

$P(A \text{ or } B) = P(A) + P(B) = \frac{31}{255} + \frac{17}{255} \approx 0.188$

b. $P(\text{not a linebacker or quarterback}) = 1 - P(A \text{ or } B) \approx 1 - 0.188 \approx 0.812$

3.3 EXERCISE SOLUTIONS

1. $P(A \text{ and } B) = 0$ because A and B cannot occur at the same time.

3. True

5. False, $P(A \text{ or } B) = P(A) + P(B) - P(A \text{ and } B)$

7. Not mutually exclusive since the two events can occur at the same time.

9. Not mutually exclusive since the two events can occur at the same time. The worker can be female and have a college degree.

11. Mutually exclusive since the two events cannot occur at the same time. The person cannot be in both age classes.

13. (a) No, it is possible for the events {overtime} and {temporary help} to occur at the same time.

(b) $P(\text{OT or temp}) = P(\text{OT}) + P(\text{temp}) - P(\text{OT and temp}) = \frac{18}{52} + \frac{9}{52} - \frac{5}{52} \approx 0.423$

15. (a) Not mutually exclusive since the two events can occur at the same time. A carton can have a puncture and a smashed corner.

(b) $P(\text{puncture or corner}) = P(\text{puncture}) + P(\text{corner}) - P(\text{puncture and corner})$
$= 0.05 + 0.08 - 0.004 = 0.126$

17. (a) $P(\text{heart or 3}) = P(\text{heart}) + P(3) \cdot P(\text{heart and 3})$

$$= \frac{13}{52} + \frac{4}{52} - \frac{1}{52} \approx 0.308$$

(b) $P(\text{black or king}) = P(\text{black}) + P(\text{king}) - P(\text{black and king})$

$$= \frac{26}{52} + \frac{4}{52} - \frac{2}{52} \approx 0.538$$

(c) $P(\text{5 or face card}) = P(5) + P(\text{face card}) - P(\text{5 and face card})$

$$= \frac{4}{52} + \frac{12}{52} - 0 \approx 0.307$$

19. (a) $P(\text{under 5}) = 0.067$

(b) $P(\text{not 65+}) = 1 - P(65+) = 1 - 0.132 = 0.868$

(c) $P(\text{between 18 and 34}) = P(\text{between 18 and 24 or between 25 and 34}) = P(\text{between 18 and 24}) + P(\text{between 25 and 34}) = 0.101 + 0.130 = 0.231$

21. (a) $P(\text{not confident}) = \frac{112}{1018} \approx 0.11$

(b) $P(\text{somewhat confident or very confident}) = P(\text{somewhat confident}) + P(\text{very confident})$

$$= \frac{397}{1018} + \frac{336}{1018} = \frac{733}{1018} \approx 0.72$$

23. (a) $P(\text{1st LH and 2nd LH}) = \left(\frac{120}{1000}\right) \cdot \left(\frac{119}{999}\right) \approx 0.014$

(b) $P(\text{neither are LH}) = \left(\frac{880}{1000}\right) \cdot \left(\frac{879}{999}\right) \approx 0.774$

(c) $P(\text{at least one is LH}) = 1 - P(\text{neither are LH}) = 1 - \left(\frac{880}{1000}\right) \cdot \left(\frac{879}{999}\right) \approx 0.226$

25. Answers will vary.

Conclusion: If two events, {A} and {B}, are independent, $P(A \text{ and } B) = P(A) \cdot P(B)$. If two events are mutually exclusive, $P(A \text{ and } B) = 0$. The only scenario when two events can be independent and mutually exclusive is if $P(A) = 0$ or $P(B) = 0$.

27. $P(A \text{ or } B \text{ or } C) = P(A) + P(B) + P(C) - P(A \text{ and } B) - P(A \text{ and } C) - P(B \text{ and } C)$

$$+ P(A \text{ and } B \text{ and } C)$$

$$= 0.35 + 0.35 + 0.30 - 0.15 - 0.25 - 0.19 + 0.13$$

$$= 0.54$$

3.4 COUNTING PRINCIPLES

3.4 Try It Yourself Solutions

1a. Manufacturer: 4
Size: 3
Color: 6

b. $4 \cdot 3 \cdot 6 = 72$ ways

2a. (1) Each letter is an event (26 choices)

(2) Each letter is an event (26, 25, 24, 23, 22, and 21 choices)

b. (1) $26 \cdot 26 \cdot 26 \cdot 26 \cdot 26 \cdot 26 = 26^6 = 308{,}915{,}776$

(2)

3a. $n =$

4a. $_8P \cdots \cdot 7 \cdot 6 = 336$

b. ... sh in first, second, and third place.

5a. $\cdots 11 \cdot 10 \cdot 9 = 11{,}880$

6a. $= \dfrac{20!}{6! \cdot 9! \cdot 5!} = 77{,}597{,}520$

7a.

b.

c. ...tees that can be selected from 16 employees.

8. ...istinguishable permutations.

c. $\dfrac{10}{220} \approx 0.045$

...more events can occur in sequence.

...ent of objects.

5. $7! = (7)(6)(5)(4)(3)(2)(1) = 5040$

7. $_7C_4 = \dfrac{7!}{4!(7-4)!} = \dfrac{(7)(6)(5)(4)(3)(2)(1)}{[(4)(3)(2)(1)][(3)(2)(1)]} = \dfrac{540}{(24)(6)} = 35$

9. Permutation, since order of the fifteen people in line matters.

11. $10 \cdot 8 \cdot {}_{13}C_2 = 6240$

13. $9 \cdot 10 \cdot 10 \cdot 5 = 4500$

15. $8! = 40{,}320$

17. $10! = 3{,}628{,}800$

19. $_{12}P_3 = 1320$

21. $_{14}P_4 = 24{,}024$

23. $\dfrac{18!}{4! \cdot 8! \cdot 6!} = 9{,}189{,}180$

48 CHAPTER 3 | PROBABILITY

25. $_{40}C_{12} = 5{,}586{,}853{,}480$

27. $_{9}C_{3} = 84$

29. (a) $_{8}C_{4} = 70$ (b) $2 \cdot 2 \cdot 2 \cdot 2 = 16$ (c) $_{4}C_{2}\left[\dfrac{(_{2}C_{0}) \cdot (_{2}C_{0}) \cdot (_{2}C_{2}) \cdot (_{2}C_{2})}{_{8}C_{4}}\right] \approx 0.086$

31. (a) $(26)(26)(10)(10)(10)(10)(10) = 67{,}600{,}000$

  (b) $(26)(25)(10)(9)(8)(7)(6) = 19{,}656{,}000$

  (c) $\dfrac{1}{67{,}600{,}000}$

33. (a) $5! = 120$ (b) $2! \cdot 3! = 12$ (c) $3! \cdot 2! = 12$ (d) 0.4

35. $(6\%)(1200) = (0.06)(1200) = 72$ of the 1200 rate financial shape as excellent.

  $P(\text{all four rate excellent}) = \dfrac{_{72}C_{4}}{_{1200}C_{4}} = \dfrac{1{,}028{,}790}{85{,}968{,}659{,}700} \approx 0.0000120$

37. $(39\%)(500) = (0.39)(500) = 195$ of the 1500 rate financial shape as fair $\Rightarrow 500 - 195 = 305$ rate shape as not fair.

  $P(\text{none of 80 selected rate fair}) = \dfrac{_{305}C_{80}}{_{500}C_{80}} \approx 6.00 \times 10^{-20}$

39. (a) $_{40}C_{5} = 658{,}008$ (b) $P(\text{win}) = \dfrac{1}{658{,}008} \approx 0.00000152$

41. $_{14}C_{4} = 1001$ possible 4 digit arrangements if order is not important.

  Assign 1000 of the 4 digit arrangements to the 13 teams since 1 arrangement is excluded.

43. $P(\text{1st}) = \dfrac{250}{1000} = 0.250$ $P(\text{8th}) = \dfrac{29}{1000} = 0.029$

  $P(\text{2nd}) = \dfrac{200}{1000} = 0.200$ $P(\text{9th}) = \dfrac{18}{1000} = 0.018$

  $P(\text{3rd}) = \dfrac{157}{1000} = 0.157$ $P(\text{10th}) = \dfrac{11}{1000} = 0.011$

  $P(\text{4th}) = \dfrac{120}{1000} = 0.120$ $P(\text{11th}) = \dfrac{7}{1000} = 0.007$

  $P(\text{5th}) = \dfrac{89}{1000} = 0.089$ $P(\text{12th}) = \dfrac{6}{1000} = 0.006$

  $P(\text{6th}) = \dfrac{64}{1000} = 0.064$ $P(\text{13th}) = \dfrac{5}{1000} = 0.005$

  $P(\text{7th}) = \dfrac{44}{1000} = 0.044$

45. Let $A = \{\text{team with the worst record wins third pick}\}$ and

  $B = \{\text{team with the best record, ranked 13th, wins first pick}\}$ and

  $C = \{\text{team ranked 2nd wins the second pick}\}$.

  $P(A|B \text{ and } C) = \dfrac{250}{995} + \dfrac{250}{(1000 - (200 + 5))} = \dfrac{250}{995} + \dfrac{250}{795} \approx 0.2513 + 0.3144 \approx 0.566$

CHAPTER 3 REVIEW EXERCISE SOLUTIONS

1. Sample space:
   {HHHH, HHHT, HHTH, HHTT, HTHH, HTHT, HTTH, HTTT, THHH, THHT, THTH, THTT, TTHH, TTHT, TTTH, TTTT}

   Event: Getting three heads
   {HHHT, HHTH, HTHH, THHH}

3. Sample space: {0, 1, 2, 3, 4, 5, 6, 7, 8, 9}

   Event: Choosing a number less than 2
   {0, 1}

5. Empirical probability

7. Subjective probability

9. Classical probability

11. $P(\text{at least } 20) = 1 - P(\text{less than } 19) = 1 - 0.28 = 0.72$

13. $P(\text{undergrad} \mid +) = 0.92$

15. Independent, the first event does not affect the outcome of the second event.

17. $P(\text{correct toothpaste and correct dental rinse}) = P(\text{correct toothpaste}) \cdot P(\text{correct dental rinse})$
    $= \left(\frac{1}{8}\right) \cdot \left(\frac{1}{5}\right) \approx 0.025$

19. Mutually exclusive since both events cannot occur at the same time.

21. $P(\text{home or work}) = P(\text{home}) + P(\text{work}) - P(\text{home and work}) = 0.44 + 0.37 - 0.21 = 0.60$

23. $P(4\text{–}8 \text{ or club}) = P(4\text{–}8) + P(\text{club}) - P(4\text{–}8 \text{ and club}) = \frac{20}{52} + \frac{13}{52} - \frac{5}{52} \approx 0.538$

25. $P(\text{odd or less than } 4) = P(\text{odd}) + P(\text{less than } 4) - P(\text{odd and less than } 4)$
    $= \frac{6}{12} + \frac{3}{12} - \frac{2}{12} \approx 0.583$

27. $P(600 \text{ or more}) = P(600 - 999) + P(1000 \text{ or more})$
    $= 0.192 + 0.271 = 0.463$

29. $P(\text{poor taste or hard to find}) = P(\text{poor taste}) + P(\text{hard to find})$
    $= \frac{60}{500} + \frac{55}{500} = \frac{115}{500} = 0.23$

31. $8 \cdot 2 \cdot 9 = 144$

33. Order is important: $_{15}P_3 = 2730$

35. Order is not important: $_{17}C_4 = 2380$

37. $P(3 \text{ kings and } 2 \text{ queens}) = \frac{_4C_3 \cdot {_4C_2}}{_{52}C_5} = \frac{4 \cdot 6}{2{,}598{,}960} \approx 0.00000923$

39. (a) $P(\text{no defectives}) = \frac{_{197}C_3}{_{200}C_3} = \frac{1{,}254{,}890}{1{,}313{,}400} \approx 0.955$

(b) $P(\text{all defective}) = \dfrac{_3C_3}{_{200}C_3} = \dfrac{1}{1{,}313{,}400} = \approx 0.000000761$

(c) $P(\text{at least one defective}) = 1 - P(\text{no defective}) = 1 - 0.955 = 0.045$

(d) $P(\text{at least one non-defective}) = 1 - P(\text{all defective}) = 1 - 0.000000761 \approx 0.999999239$

CHAPTER 3 QUIZ SOLUTIONS

1. (a) $P(\text{bachelor}) = \dfrac{1322}{2466} \approx 0.536$

   (b) $P(\text{bachelor}|F) = \dfrac{769}{1460} \approx 0.527$

   (c) $P(\text{bachelor}|M) = \dfrac{553}{1006} \approx 0.550$

   (d) $P(\text{associate or bachelor}) = P(\text{associate}) + P(\text{bachelor})$
   $$= \dfrac{632}{2466} + \dfrac{1322}{2466} \approx 0.792$$

   (e) $P(\text{doctorate}|M) = \dfrac{25}{1006} \approx 0.025$

   (f) $P(\text{master or female}) = P(\text{master}) + P(\text{female}) - P(\text{master and female})$
   $$= \dfrac{467}{2466} + \dfrac{1460}{2466} - \dfrac{270}{2466} = 0.672$$

   (g) $P(\text{associate and male}) = P(\text{associate}) \cdot P(\text{male}|\text{associate})$
   $$= \dfrac{632}{2466} \cdot \dfrac{231}{632} = \dfrac{145{,}992}{1{,}558{,}512} \approx 0.094$$

   (h) $P(F|\text{bachelor}) = \dfrac{769}{1322} \approx 0.582$

2. Not mutually exclusive since both events can occur at the same time.

   Dependent since one event can affect the occurrence of the second event.

3. (a) $_{147}C_3 = 518{,}665$ (b) $_3C_3 = 1$ (c) $_{150}C_3 - {_3C_3} = 551{,}300 - 1 = 551{,}299$

4. (a) $\dfrac{_{147}C_3}{_{150}C_3} = \dfrac{518{,}665}{551{,}300} \approx 0.94$

   (b) $\dfrac{_3C_3}{_{150}C_3} = \dfrac{1}{551{,}300} \approx 0.00000181$

   (c) $\dfrac{_{150}C_3 - {_3C_3}}{_{150}C_3} = \dfrac{551{,}299}{551{,}300} \approx 0.999998$

5. $9 \cdot 10 \cdot 10 \cdot 5 = 4500$

6. $_{25}P_4 = 303{,}600$

Discrete Probability Distributions

4.1 PROBABILITY DISTRIBUTIONS

4.1 Try It Yourself Solutions

1a. (1) measured   (2) counted

b. (1) Random variable is continuous since x can be any amount of time needed to complete a test.

(2) Random variable is discrete since x can be counted.

2ab.

x	f	P(x)
0	16	0.16
1	19	0.19
2	15	0.15
3	21	0.21
4	9	0.09
5	10	0.10
6	8	0.08
7	2	0.02
	$n = 100$	$\Sigma P(x) = 1$

New Employee Sales

3a. Each $P(x)$ is between 0 and 1.

b. $\Sigma P(x) = 1$

c. Since both conditions are met, the distribution is a probability distribution.

4a. (1) Yes, each outcome is between 0 and 1.   (2) Yes, each outcome is between 0 and 1.

b. (1) No, $\Sigma P(x) = \frac{18}{16} \neq 1$.   (2) Yes, $\Sigma P(x) = 1$.

c. (1) Not a probability distribution   (2) Is a probability distribution

5ab.

x	P(x)	xP(x)
0	0.16	(0)(0.16) = 0.00
1	0.19	(1)(0.19) = 0.19
2	0.15	(2)(0.15) = 0.30
3	0.21	(3)(0.21) = 0.63
4	0.09	(4)(0.09) = 0.36
5	0.10	(5)(0.10) = 0.50
6	0.08	(6)(0.08) = 0.48
7	0.02	(7)(0.02) = 0.14
	$\Sigma P(x) = 1$	$\Sigma xP(x) = 2.60$

c. $\mu = \Sigma xP(x) = 2.6$

On average, 2.6 sales are made per day.

6ab.

x	$P(x)$	$x - \mu$	$(x - \mu)^2$	$P(x)(x - \mu)^2$
0	0.16	−2.6	6.76	(0.16)(6.76) = 1.0816
1	0.19	−1.6	2.56	(0.19)(2.56) = 0.4864
2	0.15	−0.6	0.36	(0.15)(0.36) = 0.054
3	0.21	0.4	0.16	(0.21)(0.16) = 0.0336
4	0.09	1.4	1.96	(0.09)(1.96) = 0.1764
5	0.10	2.4	5.76	(0.10)(5.76) = 0.576
6	0.08	3.4	11.56	(0.08)(11.56) = 0.9248
7	0.02	4.4	19.36	(0.02)(19.36) = 0.3872
	$\sum P(x) = 1$			$\sum P(x)(x - \mu)^2 = 3.72$

c. $\sigma = \sqrt{\sigma^2} = \sqrt{3.720} \approx 1.9$

d. A typical distance or deviation of the random variable from the mean is 1.9 sales per day.

7ab.

x	f	$P(x)$	$xP(x)$
0	25	0.111	(0)(0.111) = 0.000
1	48	0.213	(1)(0.213) = 0.213
2	60	0.267	(2)(0.267) = 0.533
3	45	0.200	(3)(0.200) = 0.600
4	20	0.089	(4)(0.089) = 0.356
5	10	0.044	(5)(0.044) = 0.222
6	8	0.036	(6)(0.036) = 0.213
7	5	0.022	(7)(0.022) = 0.156
8	3	0.013	(8)(0.013) = 0.107
9	1	0.004	(9)(0.004) = 0.040
	$n = 225$	$\sum P(x) \approx 1$	$\sum xP(x) = 2.440$

c. $E(x) = \sum xP(x) = 2.4$

d. You can expect an average of 2.4 sales per day.

4.1 EXERCISE SOLUTIONS

1. A random variable represents a numerical value assigned to an outcome of a probability experiment.

   Examples: Answers will vary.

3. False. In most applications, discrete random variables represent counted data, while continuous random variables represent measured data.

5. True

7. Discrete, because home attendance is a random variable that is countable.

9. Continuous, because annual vehicle-miles driven is a random variable that cannot be counted.

11. Discrete, because the random variable is countable.

13. Continuous, because the random variable has an infinite number of possible outcomes and cannot be counted.

15. Discrete, because the random variable is countable.

17. Continuous, because the random variable has an infinite number of possible outcomes and cannot be counted.

19. (a) $P(x > 2) = 0.25 + 0.10 = 0.35$

    (b) $P(x < 4) = 1 - P(4) = 1 - 0.10 = 0.90$

21. $\Sigma P(x) = 1 \to P(3) = 0.22$

23. Yes

25. No, $\Sigma P(x) = 0.95$ and $P(5) < 0$.

27. (a)

x	f	$P(x)$	$xP(x)$	$(x - \mu)$	$(x - \mu)^2$	$(x - \mu)^2 P(x)$
0	1491	0.686	(0)(0.686) = 0	−0.501	0.251	(0.251)(0.686) = 0.169
1	425	0.195	(1)(0.195) = 0.195	0.499	0.249	(0.249)(0.195) = 0.049
2	168	0.077	(2)(0.077) = 0.155	1.499	2.246	(2.246)(0.077) = 0.173
3	48	0.022	(3)(0.022) = 0.066	2.499	6.244	(6.244)(0.022) = 0.138
4	29	0.013	(4)(0.013) = 0.053	3.499	12.241	(12.241)(0.013) = 0.160
5	14	0.006	(5)(0.006) = 0.030	4.499	20.238	(20.238)(0.006) = 0.122
	$n = 2175$	$\Sigma P(x) \approx 1$	$\Sigma xP(x) = 0.497$			$\Sigma (x - \mu)^2 P(x) = 0.811$

(b) $\mu = \Sigma xP(x) \approx 0.5$

(c) $\sigma^2 = \Sigma(x - \mu)^2 P(x) \approx 0.8$

(d) $\sigma = \sqrt{\sigma^2} \approx \sqrt{0.8246} \approx 0.9$

(e) A household on average has 0.5 dogs with a standard deviation of 0.9.

29. (a)

x	f	$P(x)$	$xP(x)$	$(x - \mu)$	$(x - \mu)^2$	$(x - \mu)^2 P(x)$
0	300	0.432	0.000	−0.764	0.584	(0.584)(0.432) = 0.252
1	280	0.403	0.403	0.236	0.056	(0.056)(0.403) = 0.022
2	95	0.137	0.274	1.236	1.528	(1.528)(0.137) = 0.209
3	20	0.029	0.087	2.236	5.000	(5.000)(0.029) = 0.145
	$n = 695$	$\Sigma P(x) \approx 1$	$\Sigma xP(x) = 0.764$			$\Sigma (x - \mu)^2 P(x) = 0.629$

(b) $\mu = \Sigma xP(x) \approx 0.8$

(c) $\sigma^2 = \Sigma(x - \mu)^2 P(x) \approx 0.6$

(d) $\sigma = \sqrt{\sigma^2} \approx 0.8$

(e) A household on average has 0.8 computers with a standard deviation of 0.8 computers.

31. (a)

x	f	$P(x)$	$xP(x)$	$(x - \mu)$	$(x - \mu)^2$	$(x - \mu)^2 P(x)$
0	6	0.031	0.000	−3.411	11.638	(11.638)(0.031) = 0.360
1	12	0.063	0.063	−2.411	5.815	(5.815)(0.063) = 0.366
2	29	0.151	0.302	−1.411	1.992	(1.992)(0.151) = 0.300
3	57	0.297	0.891	−0.411	0.169	(0.169)(0.297) = 0.050
4	42	0.219	0.876	0.589	0.346	(0.346)(0.219) = 0.076
5	30	0.156	0.780	1.589	2.523	(2.523)(0.156) = 0.394
6	16	0.083	0.498	2.589	6.701	(6.701)(0.083) = 0.557
	$n = 192$	$\Sigma P(x) = 1$	$\Sigma xP(x) = 3.410$			$\Sigma (x - \mu)^2 P(x) = 2.103$

(b) $\mu = \Sigma xP(x) = 3.4$

(c) $\sigma^2 = \Sigma(x - \mu)^2 P(x) = 2.1$

(d) $\sigma = \sqrt{\sigma^2} = 1.4$

(e) An employee works an average of 3.4 overtime hours per week with a standard deviation of 1.4 hours.

33.

x	$P(x)$	$xP(x)$	$(x - \mu)$	$(x - \mu)^2$	$(x - \mu)^2 P(x)$
0	0.02	0.00	−5.30	28.09	0.562
1	0.02	0.02	−4.30	18.49	0.372
2	0.06	0.12	−3.30	10.89	0.653
3	0.06	0.18	−2.30	5.29	0.317
4	0.08	0.32	−1.30	1.69	0.135
5	0.22	1.10	−0.30	0.09	0.020
6	0.30	1.80	0.70	0.49	0.147
7	0.16	1.12	1.70	2.89	0.462
8	0.08	0.64	2.70	7.29	0.583
	$\sum P(x) = 1$	$\sum xP(x) = 5.30$			$\sum (x - \mu)^2 P(x) = 3.250$

(a) $\mu = \Sigma xP(x) = 5.3$

(b) $\sigma^2 = \Sigma(x - \mu)^2 P(x) = 3.3$

(c) $\sigma = \sqrt{\sigma^2} = 1.8$

(d) $E[x] = \mu = \Sigma xP(x) = 5.3$

(e) The expected number of correctly answered questions is 5.3 with a standard deviation of 1.8.

35. (a) $\mu = \Sigma xP(x) = 2.150$

(b) $\sigma^2 = \Sigma(x - \mu)^2 P(x) \approx 1.142$

(c) $\sigma = \sqrt{\sigma^2} \approx 1.069$

(d) $E[x] = \mu = \Sigma xP(x) = 2.150$

(e) The average number of hurricanes that hit the U.S. is 2.150 with a standard deviation of 1.069.

37. (a) $\mu = \Sigma xP(x) = 2.55$

(b) $\sigma^2 = \Sigma(x - \mu)^2 P(x) \approx 1.90$

(c) $\sigma = \sqrt{\sigma^2} \approx 1.38$

(d) $E[x] = \mu = \Sigma xP(x) = 2.55$

(e) The average household size is 2.55 with a standard deviation of 1.38.

39. (a) $P(x < 2) = 0.6855 + 0.1954 = 0.8809$

(b) $P(x \geq 2) = 1 - P(x < 2) = 1 - 0.8809 - 0.1191$

(c) $P(2 \leq x \leq 4) = 0.0772 + 0.0221 + 0.0133 = 0.1126$

41. A household with three dogs is unusual since the probability is only 0.0221.

43. $E(x) = \mu = \Sigma xP(x) = (-1) \cdot \left(\frac{37}{38}\right) + (35) \cdot \left(\frac{1}{38}\right) \approx -\$0.05$

4.2 BINOMIAL DISTRIBUTIONS

4.2 Try It Yourself Solutions

1a. Trial: answering a question (10 trials)
Success: question answered correctly

  b. Yes, the experiment satisfies the four conditions of a binomial experiment.

  c. $n = 10, p = 0.25, q = 0.75, x = 0, 1, 2, \ldots, 9, 10$

2a. Trial: drawing a card with replacement (5 trials)
Success: card drawn is a club
Failure: card drawn is not a club

  b. $n = 5, p = 0.25, q = 0.75, x = 3$

  c. $P(3) = \dfrac{5!}{2!3!}(0.25)^3(0.75)^2 \approx 0.088$

3a. Trial: selecting a worker and asking a question (7 trials)
Success: Selecting a worker who will rely on pension
Failure: Selecting a worker who will not rely on pension

  b. $n = 7, p = 0.34, q = 0.66, x = 0, 1, 2, \ldots, 6, 7$

  c. $P(0) = {}_7C_0(0.34)^0(0.66)^7 = 0.0546$
  $P(1) = {}_7C_1(0.34)^1(0.66)^6 = 0.197$
  $P(2) = {}_7C_2(0.34)^2(0.66)^5 = 0.304$
  $P(3) = {}_7C_3(0.34)^3(0.66)^4 = 0.261$
  $P(4) = {}_7C_4(0.34)^4(0.66)^3 = 0.134$
  $P(5) = {}_7C_5(0.34)^5(0.66)^2 = 0.0416$
  $P(6) = {}_7C_6(0.34)^6(0.66)^1 = 0.00714$
  $P(7) = {}_7C_7(0.34)^7(0.66)^0 = 0.000525$

  d.

x	$P(x)$
0	0.0546
1	0.197
2	0.304
3	0.261
4	0.134
5	0.0416
6	0.00714
7	0.000525

4a. $n = 250, p = 0.71, x = 178$ **b.** $P(178) \approx 0.056$

  c. The probability that exactly 178 people in the United States will use more than one topping on their hot dog is about 0.056.

5a. (1) $x = 2$ (2) $x = 2, 3, 4,$ or 5 (3) $x = 0$ or 1

  b. (1) $P(2) \approx 0.217$

  (2) $P(x \geq 2) = 1 - P(0) - P(1) = 1 - 0.308 - 0.409 = 0.283$

  or

  $P(x \geq 2) = P(2) + P(3) + P(4) + P(5)$

  $= 0.217 + 0.058 + 0.008 + 0.0004$

  $= 0.283$

  (3) $P(x < 2) = P(0) + P(1) = 0.308 + 0.409 = 0.717$

  c. (1) The probability that exactly two men consider fishing their favorite leisure-time activity is about 0.217.

  (2) The probability that at least two men consider fishing their favorite leisure-time activity is about 0.283.

  (3) The probability that fewer than two men consider fishing their favorite leisure-time activity is about 0.717.

56 CHAPTER 4 | PROBABILITY DISTRIBUTIONS

6a. Trial: selecting a business and asking a question (10 trials)
Success: Selecting a business with a Web site
Failure: Selecting a business without a site

b. $n = 10, p = 0.30, x = 4$ **c.** $P(4) \approx 0.200$

d. The probability of randomly selecting 10 small businesses and finding exactly four that have a website is 0.200.

7a. $P(0) = {}_6C_0(0.57)^0(0.43)^6 = 1(0.57)^0(0.43)^6 = 0.006$
$P(1) = {}_6C_1(0.57)^1(0.43)^5 = 6(0.57)^1(0.43)^5 = 0.050$
$P(2) = {}_6C_2(0.57)^2(0.43)^4 = 15(0.57)^2(0.43)^4 = 0.167$
$P(3) = {}_6C_3(0.57)^3(0.43)^3 = 20(0.57)^3(0.43)^3 = 0.294$
$P(4) = {}_6C_4(0.57)^4(0.43)^2 = 15(0.57)^4(0.43)^2 = 0.293$
$P(5) = {}_6C_5(0.57)^5(0.43)^1 = 6(0.57)^5(0.43)^1 = 0.155$
$P(6) = {}_6C_6(0.57)^6(0.43)^0 = 1(0.57)^6(0.43)^0 = 0.034$

b.

x	P(x)
0	0.006
1	0.050
2	0.167
3	0.294
4	0.293
5	0.155
6	0.340

c.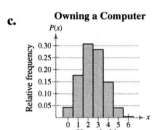
Owning a Computer

8a. Success: Selecting a clear day
$n = 31, p = 0.44, q = 0.56$

b. $\mu = np = (31)(0.44) = 13.6$

c. $\sigma^2 = npq = (31)(0.44)(0.56) = 7.6$

d. $\sigma = \sqrt{\sigma^2} = 2.8$

e. On average, there are about 14 clear days during the year. The standard deviation is about 3 days.

4.2 EXERCISE SOLUTIONS

1. (a) $p = 0.50$ (graph is symmetric)

(b) $p = 0.20$ (graph is skewed right $\rightarrow p < 0.5$)

(c) $p = 0.80$ (graph is skewed left $\rightarrow p > 0.5$)

3. (a) $n = 12, (x = 0, 1, 2, \ldots, 12)$

(b) $n = 4, (x = 0, 1, 2, 3, 4)$

(c) $n = 8, (x = 0, 1, 2, \ldots, 8)$

As n increases, the probability distribution becomes more symmetric.

5. Is a binomial experiment.
Success: baby recovers
$n = 5, p = 0.80, q = 0.20, x = 0, 1, 2, \ldots, 5$

7. Is not a binomial experiment because there are more than 2 possible outcomes for each trial.

9. $\mu = np = (100)(0.4) = 40$

   $\sigma^2 = npq = (100)(0.4)(0.6) = 24$

   $\sigma = \sqrt{\sigma^2} = 4.9$

11. $\mu = np = (138)(0.16) = 22.08$

    $\sigma^2 = npq = (138)(0.16)(0.84) = 18.547$

    $\sigma = \sqrt{\sigma^2} = 4.307$

13. $n = 5, p = .25$

    (a) $P(3) \approx 0.088$

    (b) $P(x \geq 3) = P(3) + P(4) + P(5) = 0.088 + 0.015 + .001 = 0.104$

    (c) $P(x < 3) = 1 - P(x \geq 3) = 1 - 0.104 = 0.896$

15. $n = 10, p = 0.54$ (using binomial formula)

    (a) $P(8) = 0.069$

    (b) $P(x \geq 8) = P(8) + P(9) + P(10) = 0.069 + 0.018 + 0.002 = 0.089$

    (c) $P(x < 8) = 1 - P(x \geq 8) = 1 - 0.089 = 0.911$

17. $n = 10, p = 0.21$ (using binomial formula)

    (a) $P(3) \approx 0.213$

    (b) $P(x > 3) = 1 - P(0) - P(1) - P(2) - P(3)$

    $= 1 - 0.095 - 0.252 - 3.01 - 0.213 = 0.139$

    (c) $P(x \leq 3) = 1 - P(x > 3) = 1 - 0.139 = 0.861$

19. (a) $n = 6, p = 0.36$ (b) **Basketball Fans**   (c) Symmetric

x	P(x)
0	0.069
1	0.232
2	0.326
3	0.245
4	0.103
5	0.023
6	0.002

    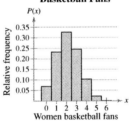

    (d) $\mu = np = (6)(0.36) = 2.2$

    (e) $\sigma^2 = npq = (6)(0.36)(0.64) \approx 1.4$

    (f) $\sigma = \sqrt{\sigma^2} \approx 1.2$

    (g) On average 2.2, out of 6, women would consider themselves basketball fans. The standard deviation is 1.2 women.

    $x = 0, 5,$ or 6 would be uncommon due to their low probabilities.

21. (a) $n = 4, p = 0.05$ (b) 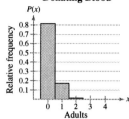 (c) Skewed right

x	P(x)
0	0.814506
1	0.171475
2	0.013537
3	0.000475
4	0.000006

Donating Blood

(d) $\mu = np = (4)(0.05) = 0.2$

(e) $\sigma^2 = npq = (4)(0.05)(0.95) = 0.2$

(f) $\sigma = \sqrt{\sigma^2} \approx 0.4$

(g) On average 0.2 eligible adults, out of every 4, give blood. The standard deviation is 0.4 adults.

$x = 2, 3,$ or 4 would be uncommon due to their low probabilities.

23. (a) $n = 6, p = 0.61$ (b) $P(2) = 0.129$

x	P(x)
0	0.004
1	0.033
2	0.129
3	0.269
4	0.316
5	0.198
6	0.052

(c) $P(\text{at least } 5) = P(5) + P(6) = 0.197 + 0.052 = 0.249$

25. $\mu = np = (6)(0.61) = 3.7$

$\sigma^2 = npq = (16)(0.61)(0.39) = 1.4$

$\sigma = \sqrt{\sigma^2} = 1.2$

If six drivers are randomly selected on average 3.7 drivers will name talking on cell phones as the most annoying habit of other drivers.

None of six randomly selected drivers naming talking on cell phones as the most annoying habit of other drivers is rare since $P(0) = 0.004$.

27. $P(5, 2, 2, 1) = \dfrac{10!}{5!2!2!1!}\left(\dfrac{9}{16}\right)^5\left(\dfrac{3}{16}\right)^2\left(\dfrac{3}{16}\right)^2\left(\dfrac{1}{16}\right)^1 \approx 0.033$

4.3 MORE DISCRETE PROBABILITY DISTRIBUTIONS

4.3 Try It Yourself Solutions

1a. $P(1) = (0.23)(0.77)^0 = 0.23$

$P(2) = (0.23)(0.77)^1 = 0.177$

$P(3) = (0.23)(0.77)^2 = 0.136$

b. $P(x < 4) = P(1) + P(2) + P(3) = 0.543$

c. The probability that your first sale will occur before your fourth sales call is 0.543.

2a. $P(0) = \dfrac{3^0(2.71828)^{-3}}{0!} \approx 0.050$

$P(1) = \dfrac{3^1(2.71828)^{-3}}{1!} \approx 0.149$

$P(2) = \dfrac{3^2(2.71828)^{-3}}{2!} \approx 0.224$

$P(3) = \dfrac{3^3(2.71828)^{-3}}{3!} \approx 0.224$

$P(4) = \dfrac{3^4(2.71828)^{-3}}{4!} \approx 0.168$

b. $P(0) + P(1) + P(2) + P(3) + P(4) \approx 0.050 + 0.149 + 0.224 + 0.224 + 0.168 \approx 0.815$

c. $1 - 0.815 \approx 0.185$

d. The probability that more than four accidents will occur in any given month at the intersection is 0.185.

3a. $\mu = \dfrac{2000}{20,000} = 0.10$

b. $\mu = 0.10, x = 3$

c. $P(3) = 0.0002$

d. The probability of finding three brown trout in any given cubic meter of the lake is 0.0002.

4.3 EXERCISE SOLUTIONS

1. $P(3) = (0.8)(0.2)^2 = 0.032$

3. $P(6) = (0.09)(0.91)^5 = 0.056$

5. $P(4) = \dfrac{(2)^4(e^{-2})}{4!} = 0.090$

7. $P(2) = \dfrac{(1.5)^2(e^{-1.5})}{2!} = 0.251$

9. Geometric. We are interested in counting the number of trials until the first success.

11. Poisson. We are interested in counting the number of occurrences that take place within a given unit of space.

13. Binomial. We are interested in counting the number of successes out of *n* trials.

15. $p = 0.19$

(a) $P(5) = (0.19)(0.81)^4 \approx 0.082$

(b) $P(\text{sale on 1st, 2nd, or 3rd call})$
$= P(1) + P(2) + P(3) = (0.19)(0.81)^0 + (0.19)(0.81)^1 + (0.19)(0.81)^2 \approx 0.469$

(c) $P(x > 3) = 1 - P(x \leq 3) = 1 - 0.469 = 0.531$

17. $p = 0.01$

(a) $P(10) = (0.01)(0.99)^9 \approx 0.009$

(b) $P(\text{1st, 2nd, or 3rd part is defective}) = P(1) + P(2) + P(3)$
$$= (0.01)(0.99)^0 + (0.01)(0.99)^1 + (0.01)(0.99)^2 \approx 0.030$$

(c) $P(x > 10) = 1 - P(x \leq 10) = 1 - [P(1) + P(2) + \cdots + P(10)] = 1 - [0.096] \approx 0.904$

19. $\mu = 8$

(a) $P(4) = \dfrac{8^4 e^{-8}}{4!} \approx 0.057$

(b) $P(x \geq 4) = 1 - (P(0) + P(1) + P(2) + P(3))$
$$\approx 1 - (0.0003 + 0.0027 + 0.0107 + 0.0286)$$
$$= 0.958$$

(c) $P(x > 4) = 1 - (P(0) + P(1) + P(2) + P(3) + P(4))$
$$\approx 1 - (0.0003 + 0.0027 + 0.0107 + 0.0286 + 0.0573)$$
$$= 0.900$$

21. $\mu = 0.7$

(a) $P(1) = 0.348$

(b) $P(X \leq 1) = P(0) + P(1) = 0.4966 + 0.3476 = 0.8442$

(c) $P(X > 1) = 1 - P(X \leq 1) = 1 - 0.8442 = 0.156$

23. (a) $n = 6500$, $p = 0.001$
$$P(5) = \dfrac{6500!}{6495!5!}(0.001)^5(0.999)^{6495} \approx 0.1453996179$$

(b) $\mu = \dfrac{6500}{1000} = 6.5$ warped glass items per 6500.

$P(5) = 0.1453688667$

The results are approximately the same.

25. $p = 0.001$

(a) $\mu = \dfrac{1}{p} = \dfrac{1}{0.001} = 1000$

$\sigma^2 = \dfrac{q}{p^2} = \dfrac{0.999}{(0.001)^2} = 999{,}000$

$\sigma = \sqrt{\sigma^2} \approx 999.5$

On average you would have to play 1000 times until you won the lottery. The standard deviation is 999.5.

(b) 1000 times

Lose money. On average you would win $500 every 1000 times you play the lottery. Hence, the net gain would be −$500.

27. $\mu = 3.9$

(a) $\sigma^2 = 3.9$

$\sigma = \sqrt{\sigma^2} \approx 2.0$

The standard deviation is 2.0 strokes.

(b) $P(X > 4) = 1 - P(X \leq 4) = 1 - 0.648 = 0.352$

CHAPTER 4 REVIEW EXERCISE SOLUTIONS

1. Discrete

3. Continuous

5. No, $\Sigma P(x) \neq 1$.

7. Yes

9. Yes

11. (a)

x	Frequency	$P(x)$	$xP(x)$	$x - \mu$	$(x - \mu)^2$	$(x - \mu)^2 P(x)$
2	3	0.005	0.009	−4.371	19.104	0.088
3	12	0.018	0.055	−3.371	11.362	0.210
4	72	0.111	0.443	−2.371	5.621	0.623
5	115	0.177	0.885	−1.371	1.879	0.332
6	169	0.260	1.560	−0.371	0.137	0.036
7	120	0.185	1.292	0.629	0.396	0.073
8	83	0.128	1.022	1.629	2.654	0.339
9	48	0.074	0.665	2.629	6.913	0.510
10	22	0.034	0.338	3.629	13.171	0.446
11	6	0.009	0.102	4.629	21.430	0.198
	$n = 650$	$\Sigma P(x) = 1$	$\Sigma xP(x) = 6.371$			$\Sigma(x - \mu)^2 P(x) = 2.855$

(b)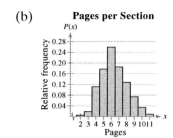
Pages per Section

(c) $\mu = \Sigma xP(x) \approx 6.4$

$\sigma^2 = \Sigma(x - \mu)^2 P(x) \approx 2.9$

$\sigma = \sqrt{\sigma^2} \approx 1.7$

13. (a)

x	Frequency	$P(x)$	$xP(x)$	$x - \mu$	$(x - \mu)^2$	$(x - \mu)^2 P(x)$
0	3	0.015	0.000	−2.315	5.359	0.080
1	38	0.190	0.190	−1.315	1.729	0.329
2	83	0.415	0.830	−0.315	0.099	0.041
3	52	0.260	0.780	0.685	0.469	0.122
4	18	0.090	0.360	1.685	2.839	0.256
5	5	0.025	0.125	2.685	7.209	0.180
6	1	0.005	0.030	3.685	13.579	0.068
	$n = 200$	$\Sigma P(x) = 1$	$\Sigma xP(x) = 2.315$			$\Sigma(x - \mu)^2 P(x) = 1.076$

(b)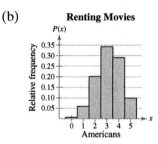

Televisions per Household

(c) $\mu = \Sigma x P(x) \approx 2.3$

$\sigma^2 = \Sigma(x - \mu)^2 P(x) \approx 1.1$

$\sigma = \sqrt{\sigma^2} \approx 1.0$

15. $E(x) = \mu = \Sigma x P(x) = 3.4$

17. Yes, $n = 12, p = 0.24, q = 0.76, x = 0, 1, \ldots, 12$.

19. $n = 8, p = 0.25$

(a) $P(3) = 0.208$

(b) $P(x \geq 3) = 1 - P(x < 3) = 1 - [P(0) + P(1) + P(2)] \approx 1 - [0.100 + 0.267 + 0.311] = 0.322$

(c) $P(x > 3) = 1 - P(x \leq 3) = 1 - [P(0) + P(1) + P(2) + P(3)]$

$= 1 - [0.100 + 0.267 + 0.311 + 0.208] = 0.114$

21. $n = 7, p = 0.43$ (use binomial formula)

(a) $P(3) = 0.294$

(b) $P(x \geq 3) = 1 - P(x < 3) = 1 - [P(0) + P(1) + P(2)] \approx 1 - [0.020 + 0.103 + 0.234] = 0.643$

(c) $P(x > 3) = 1 - P(x \leq 3) \approx 1 - [P(0) + P(1) + P(2) + P(3)]$

$\approx 1 - (0.020 + 0.103 + 0.234 + 0.294) = 0.349$

23. (a)

x	$P(x)$
0	0.007
1	0.059
2	0.201
3	0.342
4	0.291
5	0.099

(b)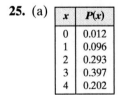

Renting Movies

(c) $\mu = np = (5)(0.63) \approx 3.2$

$\sigma^2 = npq = (5)(0.63)(0.37) = 1.2$

$\sigma = \sqrt{\sigma^2} \approx 1.1$

25. (a)

x	$P(x)$
0	0.012
1	0.096
2	0.293
3	0.397
4	0.202

(b)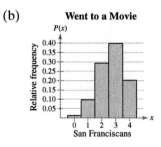

Went to a Movie

(c) $\mu = np = (4)(0.67) \approx 2.7$

$\sigma^2 = npq = (4)(0.67)(0.33) = 0.9$

$\sigma = \sqrt{\sigma^2} \approx 0.9$

27. $p = 0.167$

(a) $P(4) \approx 0.096$

(b) $P(x \le 4) = P(1) + P(2) + P(3) + P(4) \approx 0.518$

(c) $P(x > 3) = 1 - P(x \le 3) = 1 - [P(1) + P(2) + P(3)] \approx 0.579$

29. $\mu = \dfrac{2768}{36} \approx 76.89$ lightning deaths/year $\to \mu = \dfrac{76.89}{365} \approx 0.211$ deaths/day

(a) $P(0) = \dfrac{0.211^0 e^{-0.211}}{0!} \approx 0.810$

(b) $P(1) = \dfrac{0.211^1 e^{-0.211}}{1!} \approx 0.171$

(c) $P(x > 1) = 1 - [P(0) + P(1)] = 1 - [0.810 + 0.171] = 0.019$

CHAPTER 4 QUIZ SOLUTIONS

1. (a) Discrete because the random variable is countable.

(b) Continuous because the random variable has an infinite number of possible outcomes and cannot be counted.

2. (a)

x	Frequency	$P(x)$	$xP(x)$	$x - \mu$	$(x - \mu)^2$	$(x - \mu)^2 P(x)$
1	61	0.370	0.370	−1.145	1.312	0.485
2	39	0.236	0.473	−0.145	0.021	0.005
3	48	0.291	0.873	0.855	0.730	0.212
4	14	0.085	0.339	1.855	3.439	0.292
5	3	0.018	0.091	2.855	8.148	0.148
	$n = 165$	$\sum P(x) = 1$	$\sum xP(x) = 2.145$			$\sum (x - \mu)^2 P(x) = 1.142$

(b)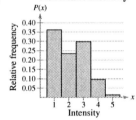

Hurricane Intensity

(c) $\mu = \Sigma xP(x) \approx 2.1$

$\sigma^2 = \Sigma(x - \mu)^2 P(x) \approx 1.1$

$\sigma = \sqrt{\sigma^2} \approx 1.1$

On average the intensity of a hurricane will be 2.1. The standard deviation is 1.1.

(d) $P(x \ge 4) = P(4) + P(5) = 0.085 + 0.018 = 0.103$

3. $n = 8, p = 0.80$

(a)

x	$P(x)$
0	0.000003
1	0.000082
2	0.001147
3	0.009175
4	0.045875
5	0.146801
6	0.293601
7	0.335544
8	0.167772

(b)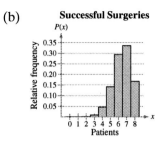

(c) $\mu = np = (8)(0.80) = 6.4$

$\sigma^2 = npq = (8)(0.80)(0.20) = 1.3$

$\sigma = \sqrt{\sigma^2} \approx 1.1$

(d) $P(2) = 0.001$

(e) $P(x < 2) = P(0) + P(1) = 0.000003 + 0.000082 = 0.000085$

4. $\mu = 5$

(a) $P(5) = 0.1755$

(b) $P(x < 5) = P(0) + P(1) + P(2) + P(3) + P(4)$
$= 0.0067 + 0.0337 + 0.0842 + 0.1404 + 0.1755 = 0.4405$

(c) $P(0) = 0.0067$

Normal Probability Distributions

CHAPTER 5

5.1 INTRODUCTION TO NORMAL DISTRIBUTIONS AND THE STANDARD NORMAL DISTRIBUTION

5.1 Try It Yourself Solutions

1a. A: 45, B: 60, C: 45 (B has the greatest mean)

 b. Curve C is more spread out, so curve C has the greatest standard deviation.

2a. Mean = 3.5 feet

 b. Inflection points: 3.3 and 3.7
 Standard deviation = 0.2 feet

3a. (1) 0.0143 (2) 0.985

4a.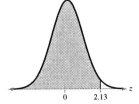

5a.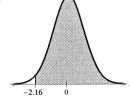

 b. 0.9834

 b. 0.0154

 c. Area = 1 − 0.0154 = 0.9846

6a. 0.0885 **b.** 0.0154

 c. Area = 0.0885 − 0.0154 = 0.0731

5.1 EXERCISE SOLUTIONS

1. Answers will vary.

3. Answers will vary.
 Similarities: Both curves will have the same line of symmetry.
 Differences: One curve will be more spread out than the other.

5. $\mu = 0, \sigma = 1$

7. "The" standard normal distribution is used to describe one specific normal distribution ($\mu = 0, \sigma = 1$). "A" normal distribution is used to describe a normal distribution with any mean and standard deviation.

9. No, the graph crosses the x-axis.

11. Yes, the graph fulfills the properties of the normal distribution.

13. No, the graph is skewed to the right.

15. The histogram represents data from a normal distribution since it's bell-shaped.

17. (Area left of $z = 1.2$) − (Area left of $z = 0$) = 0.8849 − 0.5 = 0.3849

19. (Area left of $z = 1.5$) − (Area left of $z = -0.5$) = $0.9332 - 0.3085 = 0.6247$

21. 0.9382   **23.** 0.975

25. $1 - 0.1711 = 0.8289$ **27.** $1 - 0.8997 = 0.1003$

29. 0.005   **31.** $1 - 0.95 = 0.05$

33. $0.975 - 0.5 = 0.475$ **35.** $0.5 - 0.0630 = 0.437$

37. $0.9750 - 0.0250 = 0.95$ **39.** $0.1003 + 0.1003 = 0.2006$

41. (a) 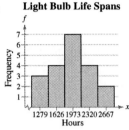   Light Bulb Life Spans

It is reasonable to assume that the life span is normally distributed since the histogram is nearly symmetric and bell-shaped.

(b) $\bar{x} = 1941.35$ $s \approx 432.385$

(c) The sample mean of 1941.35 hours is less than the claimed mean, so on the average the bulbs in the sample lasted for a shorter time. The sample standard deviation of 432 hours is greater than the claimed standard deviation, so the bulbs in the sample had a greater variation in life span than the manufacturer's claim.

43. (a) $A = 2.97$ $B = 2.98$ $C = 3.01$ $D = 3.05$

(b) $x = 3.01 \Rightarrow z = \dfrac{x - \mu}{\sigma} = \dfrac{3.01 - 3}{0.02} = 0.5$

$x = 2.97 \Rightarrow z = \dfrac{x - \mu}{\sigma} = \dfrac{2.97 - 3}{0.02} = -1.5$

$x = 2.98 \Rightarrow z = \dfrac{x - \mu}{\sigma} = \dfrac{2.98 - 3}{0.02} = -1.0$

$x = 3.05 \Rightarrow z = \dfrac{x - \mu}{\sigma} = \dfrac{3.05 - 3}{0.02} = 2.5$

(c) $x = 3.05$ is unusual due to a relatively large z-score (2, 5).

45. (a) $A = 801$ $B = 950$ $C = 1250$ $D = 1467$

(b) $x = 950 \Rightarrow z = \dfrac{x - \mu}{\sigma} = \dfrac{950 - 1026}{209} = -0.36$

$x = 1250 \Rightarrow z = \dfrac{x - \mu}{\sigma} = \dfrac{1250 - 1026}{209} = 1.07$

$x = 1467 \Rightarrow z = \dfrac{x - \mu}{\sigma} = \dfrac{1467 - 1026}{209} = 2.11$

$x = 801 \Rightarrow z = \dfrac{x - \mu}{\sigma} = \dfrac{801 - 1026}{209} = -1.08$

(c) $x = 1467$ is unusual due to a relatively large z-score (2.11).

47. 0.6915   **49.** $1 - 0.95 = 0.05$

51. $0.8413 - 0.3085 = 0.5328$ **53.** $P(z < 1.45) = 0.9265$

55. $P(z > -1.95) = 1 - P(z < -1.95) = 1 - 0.0256 = 0.9744$

57. $P(-0.89 < z < 0) = 0.5 - 0.1867 = 0.3133$

59. $P(-1.65 < z < 1.65) = 0.9505 - .0495 = 0.901$

61. $P(z < -2.58 \text{ or } z > 2.58) = 2(0.0049) = 0.0098$

63.

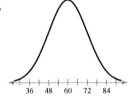

The normal distribution curve is centered at its mean (60) and has 2 points of inflection (48 and 72) representing $\mu \pm \sigma$.

65. (a) Area under curve = area of rectangle = (base)(height) = (1)(1) = 1

(b) $P(0.25 < x < 0.5) = $ (base)(height) $= (0.25)(1) = 0.25$

(c) $P(0.3 < x < 0.7) = $ (base)(height) $= (0.4)(1) = 0.4$

5.2 NORMAL DISTRIBUTIONS: FINDIING PROBABILITIES

5.2 Try It Yourself Solutions

1a.

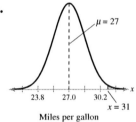

b. $z = \dfrac{x - \mu}{\sigma} = \dfrac{31 - 27}{1.6} = 2.50$

c. $P(z < 2.50) = 0.9938$

$P(z > 2.50) = 1 - 0.9938 = 0.0062$

d. The probability that a randomly selected manual transmission Focus will get more than 31 mpg in city driving is 0.0062.

2a.

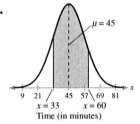

68 CHAPTER 5 | NORMAL PROBABILITY DISTRIBUTIONS

b. $z = \dfrac{x - \mu}{\sigma} = \dfrac{33 - 45}{12} = -1$

$z = \dfrac{x - \mu}{\sigma} = \dfrac{60 - 45}{12} = 1.25$

c. $P(z < -1) = 0.1587$

$P(z > 1.25) = 0.8944$

d. $P(-1 < z < 1.25) = 0.8944 - 0.1587 = 0.7357$

3a. Read user's guide for the technology tool.

b. Enter the data.

c. $P(190 < x < 225) = P(-1 < z < 0.4) = 0.4968$

The probability that a randomly selected U.S. man's cholesterol is between 190 and 225 is about 0.4968.

5.2 EXERCISE SOLUTIONS

1. $P(x < 80) = P(z < -1.2) = 0.1151$

3. $P(x > 92) = P(z > 1.2) = 1 - 0.8849 = 0.1151$

5. $P(70 < x < 80) = P(-3.2 < z < -1.2) = 0.1151 - 0.0007 = 0.1144$

7. $P(200 < x < 450) = P(-2.77 < z < -0.51) = 0.3050 - 0.0028 = 0.3022$

9. $P(200 < x < 239) = P(0.38 < z < 1.42) = 0.9222 - 0.6480 = 0.2742$

11. $P(167 < x < 174) = P(1.57 < z < 2.94) = 0.9984 - 0.9418 = 0.0566$

13. (a) $P(x < 66) = P(z < -1.10) = 0.1357$

    (b) $P(66 < x < 72) = P(-1.10 < z < 0.97) = 0.8340 - 0.1357 = 0.6983$

    (c) $P(x > 72) = P(z > 0.97) = 1 - P(z < 0.97) = 1 - 0.8340 = 0.1660$

15. (a) $P(x < 20) = P(z < -0.95) = 0.1711$

    (b) $P(20 < x < 29) = P(-0.95 < z < 1.14) = 0.8729 - 0.1711 = 0.7018$

    (c) $P(x > 29) = P(z > 1.14) = 1 - P(z < 1.14) = 1 - 0.8729 = 0.1271$

17. (a) $P(x < 4.5) = P(z < -2.5) = 0.0062$

    (b) $P(4.5 < x < 9.5) = P(-2.5 < z < 2.5) = 0.9938 - 0.0062 = 0.9876$

    (c) $P(x > 9.5) = P(z > 2.5) = 1 - P(z < 2.5) = 1 - 0.9938 = 0.0062$

19. (a) $P(x < 4) = (z < -2.44) = 0.0073$

    (b) $P(4 < x < 7) = P(-2.44 < z < 0.89) = 0.8133 - 0.0073 = 0.8060$

    (c) $P(x > 7) = P(z > 0.89) = 1 - 0.8133 = 0.1867$

21. (a) $P(x < 600) = P(z < 0.84) = 0.7995 \Rightarrow 79.95\%$

    (b) $P(x > 550) = P(z > 0.39) = 1 - P(z < 0.39) = 1 - 0.6517 = 0.3483$

    $(1000)(0.3483) = 348.3 \Rightarrow 348$

23. (a) $P(x < 200) = P(z < 0.38) = 0.6480 \Rightarrow 64.80\%$

(b) $P(x > 240) = P(z > 1.45) = 1 - P(z < 1.45) = 1 - 0.9265 = 0.0735$

$(250)(0.0735) = 18.375 \Rightarrow 18$

25. (a) $P(x > 11) = P(z > 0.5) = 1 - P(z < 0.5) = 1 - 0.6915 = 0.3085 \Rightarrow 30.85\%$

(b) $P(x < 8) = P(z < -1) = 0.1587$

$(200)(0.1587) = 31.74 \Rightarrow 31$

27. (a) $P(x > 4) = P(z > -3) = 1 - P(z < -3) = 1 - 0.0013 = 0.9987 \Rightarrow 99.87\%$

(b) $P(x < 5) = P(z < -2) = 0.0228$

$(35)(0.0228) = 0.798$

29. $P(x > 2065) = P(z > 2.17) = 1 - P(z < 2.17) = 1 - 0.9850 = 0.0150 \Rightarrow 1.5\%$

It is unusual for a battery to have a life span that is more than 2065 hours because of the relatively large z-score (2.17).

31. Out of control, since the 10th observation plotted beyond 3 standard deviations.

33. Out of control, since the first nine observations lie below the mean and since two out of three consecutive points lie more than 2 standard deviations from the mean.

5.3 NORMAL DISTRIBUTIONS: FINDING VALUES

5.3 Try It Yourself Solutions

1ab. (1) 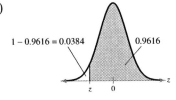 (2)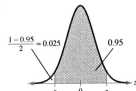

c. (1) $z = -1.77$ (2) $z = \pm 1.96$

2a. (1) 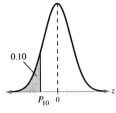 (2) 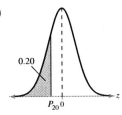 (3)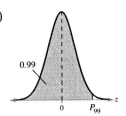

b. (1) use area = 0.1003 (2) use area = 0.2005 (3) use area = 0.9901

c. (1) $z = -1.28$ (2) $z = -0.84$ (3) $z = 2.33$

3a. $\mu = 70, \sigma = 8$

b. $z = -0.75 \Rightarrow x = \mu + z\sigma = 70 + (-0.75)(8) = 64$

$z = 4.29 \Rightarrow x = \mu + z\sigma = 70 + (4.29)(8) = 104.32$

$z = -1.82 \Rightarrow x = \mu + z\sigma = 70 + (-1.82)(8) = 55.44$

c. 64 and 55.44 are below the mean. 104.32 is above the mean.

70 CHAPTER 5 | NORMAL PROBABILITY DISTRIBUTIONS

4a.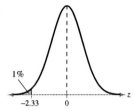

b. $z = -2.33$

c. $x = \mu + z\sigma = 158 + (-2.33)(6.51) \approx 142.83$

d. So, the longest braking distance a Ford F-150 could have and still be in the top 1% is 143 feet.

5a.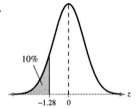

b. $z = -1.28$

c. $x = \mu + z\sigma = 11.2 + (-1.28)(2.1) = 8.512$

d. So, the maximum length of time an employee could have worked and still be laid off is 8 years.

5.3 EXERCISE SOLUTIONS

1. $z = -2.05$ **3.** $z = 0.85$ **5.** $z = -0.16$

7. $z = 2.39$ **9.** $z = -1.645$ **11.** $z = 0.99$

13. $z = -2.325$ **15.** $z = -0.25$ **17.** $z = 1.175$

19. $z = -0.675$ **21.** $z = 0.675$ **23.** $z = -0.385$

25. $z = -0.38$ **27.** $z = -0.58$ **29.** $z = \pm 1.645$

31. 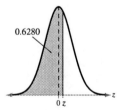 $\Rightarrow z = 0.325$ **33.** 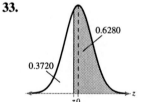 $\Rightarrow z = -0.33$

35. 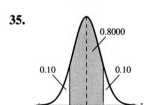 $\Rightarrow z = 1.28$ **37.** 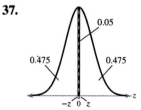 $\Rightarrow z = \pm 0.06$

39. (a) 95th percentile $\Rightarrow$ Area $= 0.95 \Rightarrow z = 1.645$

$x = \mu + z\sigma = 64 + (1.645)(2.75) = 68.52$

(b) 1st quartile $\Rightarrow$ Area $= 0.25 \Rightarrow z = -0.675$

$x = \mu + z\sigma = 64 + (-0.675)(2.75) = 62.14$

41. (a) 10th percentile $\Rightarrow$ Area $= 0.10 \Rightarrow z = -1.28$

$x = \mu + z\sigma = 17.4 + (-1.28)(4) = 12.28$

(b) 3rd quartile $\Rightarrow$ Area $= 0.75 \Rightarrow z = 0.67$

$x = \mu + z\sigma = 17.4 + (0.67)(4) = 20.08$

43. (a) Top 30% $\Rightarrow$ Area $= 0.70 \Rightarrow z = 0.52$

$x = \mu + z\sigma = 127 + (0.52)(23.5) = 139.22$

(b) Bottom 10% $\Rightarrow$ Area $= 0.10 \Rightarrow z = -1.28$

$x = \mu + z\sigma = 127 + (-1.28)(23.5) = 96.92$

45. Lower 5% $\Rightarrow$ Area $= 0.05 \Rightarrow z = -1.645$

$x = \mu + z\sigma = 20 + (-1.645)(0.07) = 19.89$

47. Bottom 10% $\Rightarrow$ Area $= 0.10 \Rightarrow z = -1.28$

$x = \mu + z\sigma = 30{,}000 + (-1.28)(2500) = 26{,}800$

Tires which wear out by 26,800 miles will be replaced free of charge.

49. Top 1% $\Rightarrow$ Area $= 0.99 \Rightarrow z = 2.33$

$x = \mu + z\sigma \Rightarrow 8 = \mu + (2.33)(0.03) \Rightarrow \mu = 7.930$

5.4 SAMPLING DISTRIBUTIONS AND THE CENTRAL LIMIT THEOREM

5.4 Try It Yourself Solutions

1a.

Sample	Mean	Sample	Mean	Sample	Mean	Sample	Mean
1, 1, 1	1	3, 1, 1	1.67	5, 1, 1	2.33	7, 1, 1	3
1, 1, 3	1.67	3, 1, 3	2.33	5, 1, 3	3	7, 1, 3	3.67
1, 1, 5	2.33	3, 1, 5	3	5, 1, 5	3.67	7, 1, 5	4.33
1, 1, 7	3	3, 1, 7	3.67	5, 1, 7	4.33	7, 1, 7	5
1, 3, 1	1.67	3, 3, 1	2.33	5, 3, 1	3	7, 3, 1	3.67
1, 3, 3	2.33	3, 3, 3	3	5, 3, 3	3.67	7, 3, 3	4.33
1, 3, 5	3	3, 3, 5	3.67	5, 3, 5	4.33	7, 3, 5	5
1, 3, 7	3.67	3, 3, 7	4.33	5, 3, 7	5	7, 3, 7	5.67
1, 5, 1	2.33	3, 5, 1	3	5, 5, 1	3.67	7, 5, 1	4.33
1, 5, 3	3	3, 5, 3	3.67	5, 5, 3	4.33	7, 5, 3	5
1, 5, 5	3.67	3, 5, 5	4.33	5, 5, 5	5	7, 5, 5	5.67
1, 5, 7	4.33	3, 5, 7	5	5, 5, 7	5.67	7, 5, 7	6.33
1, 7, 1	3	3, 7, 1	3.67	5, 7, 1	4.33	7, 7, 1	5
1, 7, 3	3.67	3, 7, 3	4.33	5, 7, 3	5	7, 7, 3	5.67
1, 7, 5	4.33	3, 7, 5	5	5, 7, 5	5.67	7, 7, 5	6.33
1, 7, 7	5	3, 7, 7	5.67	5, 7, 7	6.33	7, 7, 7	7

b.

$\bar{x}$	f	Probability
1	1	0.0156
1.67	3	0.0469
2.33	6	0.0938
3	10	0.1563
3.67	12	0.1875
4.33	12	0.1875
5	10	0.1563
5.67	6	0.0938
6.33	3	0.0469
7	1	0.0156

$\mu_{\bar{x}} = 4, \sigma_{\bar{x}}^2 \approx 1.667, \sigma_{\bar{x}} \approx 1.291$

c. $\mu_{\bar{x}} = \mu = 4, \sigma_{\bar{x}}^2 = \dfrac{\sigma^2}{n} = \dfrac{5}{3} = 1.667, \sigma_{\bar{x}} = \dfrac{\sigma}{\sqrt{n}} = \dfrac{\sqrt{5}}{\sqrt{3}} = 1.291$

2a. $\mu_{\bar{x}} = \mu = 64, \sigma_{\bar{x}} = \dfrac{\sigma}{\sqrt{n}} = \dfrac{9}{\sqrt{100}} = 0.9$

b. $n = 100$

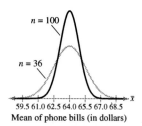

c. With a larger sample size, the mean stays the same but the standard deviation decreases.

3a. $\mu_{\bar{x}} = \mu = 3.5, \sigma_{\bar{x}} = \dfrac{\sigma}{\sqrt{n}} = \dfrac{0.2}{\sqrt{16}} = 0.05$

b.

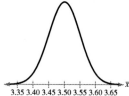

4a. $\mu_{\bar{x}} = \mu = 25, \sigma_{\bar{x}} = \dfrac{\sigma}{\sqrt{n}} = \dfrac{1.5}{\sqrt{100}} = 0.15$

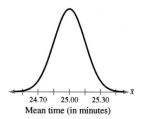

b. $\bar{x} = 24.7$: $z = \dfrac{\bar{x} - \mu}{\frac{\sigma}{\sqrt{n}}} = \dfrac{24.7 - 25}{\frac{1.5}{\sqrt{100}}} = -\dfrac{0.3}{0.15} = -2$

$\bar{x} = 25.5$: $z = \dfrac{\bar{x} - \mu}{\frac{\sigma}{\sqrt{n}}} = \dfrac{25.5 - 25}{\frac{1.5}{\sqrt{100}}} = \dfrac{0.5}{0.15} = 3.33$

c. $P(z < -2) \approx 0.0228$

$P(z < 3.33) = 0.9996$

$P(8.7 < \bar{x} < 9.5) = P(-2 < z < 3.33) = 0.9996 - 0.0228 = 0.9768$

5a. $\mu_{\bar{x}} = \mu = 243{,}756, \sigma_{\bar{x}} = \dfrac{\sigma}{\sqrt{n}} = \dfrac{44{,}000}{\sqrt{12}} \approx 12{,}701.7$

Mean sales price (in dollars)

b. $\bar{x} = 200{,}000$: $z = \dfrac{\bar{x} - \mu}{\frac{\sigma}{\sqrt{n}}} = \dfrac{200{,}000 - 243{,}756}{\frac{44{,}000}{\sqrt{12}}} = \dfrac{-43.756}{12{,}701.7} = -3.44$

c. $P(\bar{x} > 200{,}000) = P(z > -3.44) = 1 - P(z < -3.44) = 1 - 0.0003 = 0.9997$

6a. $x = 700$: $z = \dfrac{x - \mu}{\sigma} = \dfrac{700 - 625}{150} = 0.5$

$\bar{x} = 700$: $z = \dfrac{\bar{x} - \mu}{\frac{\sigma}{\sqrt{n}}} = \dfrac{700 - 625}{\frac{150}{\sqrt{10}}} = \dfrac{75}{47.43} = 1.58$

b. $P(z < 0.5) = 0.6915$

$P(z < 1.58) = 0.9429$

c. There is a 69% chance an *individual receiver* will cost less than $700. There is a 94% chance that the *mean of a sample of 10 receivers* is less than $700.

5.4 EXERCISE SOLUTIONS

1. $\mu_{\bar{x}} = \mu = 100$

$\sigma_{\bar{x}} = \dfrac{\sigma}{\sqrt{n}} = \dfrac{15}{\sqrt{50}} = 2.12$

3. $\mu_{\bar{x}} = \mu = 100$

$\sigma_{\bar{x}} = \dfrac{\sigma}{\sqrt{n}} = \dfrac{15}{\sqrt{250}} = 0.949$

5. False. As the size of the sample increases, the mean of the distribution of the sample mean does not change.

74 CHAPTER 5 | NORMAL PROBABILITY DISTRIBUTIONS

7. False. The shape of the sampling distribution of sample means is normal for large sample sizes even if the shape of the population is non-normal.

9. $\mu_{\bar{x}} = 3.5$, $\sigma_{\bar{x}} = 1.708$

   $\mu = 3.5$, $\sigma = 2.958$

Sample	Mean	Sample	Mean	Sample	Mean	Sample	Mean
0, 0, 0	0	2, 0, 0	0.67	4, 0, 0	1.33	8, 0, 0	2.67
0, 0, 2	0.67	2, 0, 2	1.33	4, 0, 2	2	8, 0, 2	3.33
0, 0, 4	1.33	2, 0, 4	2	4, 0, 4	2.67	8, 0, 4	4
0, 0, 8	2.67	2, 0, 8	3.33	4, 0, 8	4	8, 0, 8	5.33
0, 2, 0	0.67	2, 2, 0	1.33	4, 2, 0	2	8, 2, 0	3.33
0, 2, 2	1.33	2, 2, 2	2	4, 2, 2	2.67	8, 2, 2	4
0, 2, 4	2	2, 2, 4	2.67	4, 2, 4	3.33	8, 2, 4	4.67
0, 2, 8	3.33	2, 2, 8	4	4, 2, 8	4.67	8, 2, 8	6
0, 4, 0	1.33	2, 4, 0	2	4, 4, 0	2.67	8, 4, 0	4
0, 4, 2	2	2, 4, 2	2.67	4, 4, 2	3.33	8, 4, 2	4.67
0, 4, 4	2.67	2, 4, 4	3.33	4, 4, 4	4	8, 4, 4	5.33
0, 4, 8	4	2, 4, 8	4.67	4, 4, 8	5.33	8, 4, 8	6.67
0, 8, 0	2.67	2, 8, 0	3.33	4, 8, 0	4	8, 8, 0	5.33
0, 8, 2	3.33	2, 8, 2	4	4, 8, 2	4.67	8, 8, 2	6
0, 8, 4	4	2, 8, 4	4.67	4, 8, 4	5.33	8, 8, 4	6.67
0, 8, 8	5.33	2, 8, 8	6	4, 8, 8	6.67	8, 8, 8	8

11. (c) Since $\mu_{\bar{x}} = 16.5$, $\sigma_{\bar{x}} = \dfrac{\sigma}{\sqrt{n}} = \dfrac{11.9}{\sqrt{100}} = 1.19$ and the graph approximates a normal curve.

13. $\mu_{\bar{x}} = 87.5$

    $\sigma_{\bar{x}} = \dfrac{\sigma}{\sqrt{n}} = \dfrac{6.25}{\sqrt{12}} \approx 1.804$

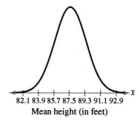
82.1 83.9 85.7 87.5 89.3 91.1 92.9
Mean height (in feet)

15. $\mu_{\bar{x}} = 349$

    $\sigma_{\bar{x}} = \dfrac{\sigma}{\sqrt{n}} = \dfrac{8}{\sqrt{40}} = 1.26$

346.5   349   351.5
Mean price (in dollars)

17. $\mu_{\bar{x}} = 113.5$

    $\sigma_{\bar{x}} = \dfrac{\sigma}{\sqrt{n}} = \dfrac{38.5}{\sqrt{20}} \approx 8.61$

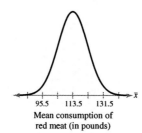
95.5   113.5   131.5
Mean consumption of
red meat (in pounds)

19. $\mu_{\bar{x}} = 87.5, \sigma_{\bar{x}} = \dfrac{\sigma}{\sqrt{n}} = \dfrac{6.25}{\sqrt{24}} \approx 1.276$

$\mu_{\bar{x}} = 87.5, \sigma_{\bar{x}} = \dfrac{\sigma}{\sqrt{n}} = \dfrac{6.25}{\sqrt{36}} \approx 1.042$

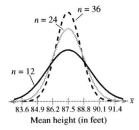
Mean height (in feet)

As the sample size increases, the standard error decreases, while the mean of the sample means remains constant.

21. $z = \dfrac{\bar{x} - \mu}{\dfrac{\sigma}{\sqrt{n}}} = \dfrac{38{,}000 - 40{,}500}{\dfrac{5600}{\sqrt{42}}} = \dfrac{-2500}{864.10} \approx -2.89$

$P(\bar{x} < 38{,}000) = P(z < -2.89) = 0.0019$

23. $z = \dfrac{\bar{x} - \mu}{\dfrac{\sigma}{\sqrt{n}}} = \dfrac{1.684 - 1.689}{\dfrac{0.045}{\sqrt{32}}} = \dfrac{-0.005}{0.00795} \approx -0.63$

$z = \dfrac{\bar{x} - \mu}{\dfrac{\sigma}{\sqrt{n}}} = \dfrac{1.699 - 1.689}{\dfrac{0.045}{\sqrt{32}}} = \dfrac{0.01}{0.00795} \approx 1.26$

$P(1.684 < \bar{x} < 1.699) = P(-0.63 < z < 1.26) = 0.8962 - 0.2643 = 0.6319$

25. $z = \dfrac{\bar{x} - \mu}{\dfrac{\sigma}{\sqrt{n}}} = \dfrac{66 - 64}{\dfrac{2.75}{\sqrt{60}}} = \dfrac{2}{0.355} \approx 5.63$

$P(\bar{x} > 68) = P(z > 5.63) \approx 0$

27. $z = \dfrac{\bar{x} - \mu}{\sigma} = \dfrac{70 - 64}{2.75} \approx 2.18$

$P(x < 70) = P(z < 2.18) = 0.9854$

$z = \dfrac{\bar{x} - \mu}{\dfrac{\sigma}{\sqrt{n}}} = \dfrac{70 - 64}{\dfrac{2.75}{\sqrt{20}}} = \dfrac{6}{0.615} \approx 9.76$

$P(\bar{x} < 70) = P(z < 9.76) \approx 1$

It is more likely to select a sample of 20 women with a mean height less than 70 inches, because the sample of 20 has a higher probability.

29. $z = \dfrac{\bar{x} - \mu}{\dfrac{\sigma}{\sqrt{n}}} = \dfrac{127.9 - 128}{\dfrac{0.20}{\sqrt{40}}} = \dfrac{-0.1}{0.032} \approx 3.16$

$P(\bar{x} < 127.9) = P(z < -3.16) = 0.0008$

Yes, it is very unlikely that we would have randomly sampled 40 cans with a mean equal to 127.9 ounces.

76 CHAPTER 5 | NORMAL PROBABILITY DISTRIBUTIONS

31. (a) $\mu = 96$

$\sigma = 0.5$

$z = \dfrac{\bar{x} - \mu}{\dfrac{\sigma}{\sqrt{\mu}}} = \dfrac{96.25 - 96}{\dfrac{0.5}{\sqrt{90}}} = \dfrac{0.25}{0.079} \approx 3.16$

$P(\bar{x} \geq 96.25) = P(z > 3.16) = 1 - P(z < 3.16) = 1 - 0.9992 = 0.0008$

(b) Claim is inaccurate

(c) Assuming the distribution is normally distributed:

$z = \dfrac{\bar{x} - \mu}{\sigma} = \dfrac{96.25 - 96}{0.5} = 0.5$

$P(x > 96.25) = P(z > 0.5) = 1 - P(z < 0.5) = 1 - 0.6915 = 0.3085$

Assuming the manufacturer's claim is true, an individual board with a length of 96.25 would not be unusual. It is within 1 standard deviation of the mean for an individual board.

33. (a) $\mu = 50{,}000$

$\sigma = 800$

$z = \dfrac{\bar{x} - \mu}{\dfrac{\sigma}{\sqrt{n}}} = \dfrac{49{,}721 - 50{,}000}{\dfrac{800}{\sqrt{100}}} = \dfrac{-279}{80} = -3.49$

$P(\bar{x} \leq 49{,}721) = P(z \leq -3.49) = 0.0002$

(b) The manufacturer's claim is inaccurate.

(c) Assuming the distribution is normally distributed:

$z = \dfrac{x - \mu}{\sigma} = \dfrac{49{,}721 - 50{,}000}{800} = -0.35$

$P(x < 49{,}721) = P(z < -0.35) = 0.3669$

Assuming the manufacturer's claim is true, an individual tire with a life span of 49,721 miles is not unusual. It is within 1 standard deviation of the mean for an individual tire.

35. $\mu = 500$

$\sigma = 100$

$z = \dfrac{\bar{x} - \mu}{\sigma/\sqrt{n}} = \dfrac{530 - 500}{100/\sqrt{50}} = \dfrac{30}{14.14} = 2.12$

$P(\bar{x} \geq 530) = P(z \geq 2.12) = 1 - P(z \leq 2.12) = 1 - 0.9830 = 0.017$

The high school's claim is justified since it is rare to find a sample mean as large as 530.

37. Use the finite correction factor since $n = 55 > 40 = 0.05N$.

$z = \dfrac{\bar{x} - \mu}{\dfrac{\sigma}{\sqrt{n}}\sqrt{\dfrac{N-n}{N-1}}} = \dfrac{1.683 - 1.688}{\dfrac{0.009}{\sqrt{55}}\sqrt{\dfrac{800-55}{800-1}}} = \dfrac{-0.005}{(0.00121)\sqrt{0.9324}} \approx -4.27$

$P(\bar{x} < 1.683) = P(z < -4.27) \approx 0$

5.5 NORMAL APPROXIMATIONS TO BINOMIAL DISTRIBUTIONS

5.5 Try It Yourself Solutions

1a. $n = 70, p = 0.61, q = 0.39$

b. $np = 42.7, nq = 27.3$

c. Since $np \geq 5$ and $nq \geq 5$, the normal distribution can be used.

d. $\mu = np = (0.70)(0.61) = 42.7$

$\sigma = \sqrt{npq} = \sqrt{(70)(0.61)(0.39)} \approx 4.08$

2a. (1) $57, 58, \ldots, 83$ (2) $\ldots, 52, 53, 54$

b. (1) $56.5 < x < 83.5$ (2) $x < 54.5$

3a. $n = 70, p = 0.61$

$np = 42.7 \geq 5$ and $nq = 27.3 \geq 5$

The normal distribution can be used.

b. $\mu = np = 42.7$

$\sigma = \sqrt{npq} \approx 4.08$

c. $x > 50.5$

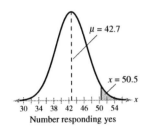

d. $z = \dfrac{x - \mu}{\sigma} = \dfrac{50.5 - 42.7}{4.08} \approx 1.91$

e. $P(z < 1.91) = 0.9719$

$P(x > 50.5) = P(z > 1.91) = 1 - P(z < 1.91) = 0.0281$

The probability that more than 50 respond yes is 0.0281.

4a. $n = 200, p = 0.38$

$np = 76 \geq 5$ and $nq = 124 \geq 5$

The normal distribution can be used.

b. $\mu = np = 76$

$\sigma = \sqrt{npq} \approx 6.86$

c. $P(x < 85.5)$

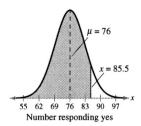

d. $z = \dfrac{x - \mu}{\sigma} = \dfrac{85.5 - 76}{6.86} \approx 1.38$

e. $P(x < 65.5) = P(z < 1.38) = 0.9162$

The probability that at most 65 people will say yes is 0.9162.

5a. $n = 200, p = 0.95$

$np = 190 \geq 5$ and $nq = 10 \geq 5$

The normal distribution can be used.

b. $\mu = np = 190$

$\sigma = \sqrt{npq} \approx 3.08$

c. $P(190.5 < x < 191.5)$

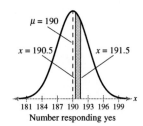

d. $z = \dfrac{x - \mu}{\sigma} = \dfrac{190.5 - 190}{3.08} \approx 0.16$

$z = \dfrac{x - \mu}{\sigma} = \dfrac{191.5 - 190}{308.6} \approx 0.49$

e. $P(z < 0.16) = 0.5636$

$P(z < 0.49) = 0.6879$

$P(0.16 < z < 0.49) = 0.6879 - 0.5636 = 0.1243$

The probability that exactly 191 people will respond yes is 0.1243.

5.5 EXERCISE SOLUTIONS

1. $np = (20)(0.80) = 16 \geq 5$

$nq = (20)(0.20) = 4 < 5$

Cannot use normal distribution.

3. $np = (15)(0.65) = 9.75 \geq 5$

$nq = (15)(0.35) = 5.25 \geq 5$

Use normal distribution.

5. $n = 10, p = 0.44, q = 0.56$

$np = 4.4 < 5, nq = 5.6 \geq 5$

Cannot use normal distribution.

7. $n = 10, p = 0.97, q = 0.03$

$np = 9.7 \geq 5, nq = 0.3 < 5$

Cannot use normal distribution.

9. d   **11.** a   **13.** a   **15.** c

17. Binomial: $P(5 \leq x \leq 7) = 0.162 + 0.198 + 0.189 = 0.549$

Normal: $P(4.5 \leq x \leq 7.5) = P(-0.97 < z < 0.56) = 0.7123 - 0.1660 = 0.5463$

19. $n = 30, p = 0.07 \rightarrow np = 2.1$ and $nq = 27.9$

Cannot use normal distribution.

(a) $P(x = 10) = {}_{30}C_{10}(0.07)^{10}(0.93)^{20} \approx 0.0000199$

(b) $P(x \geq 10) = 1 - P(x < 10)$

$= 1 - [{}_{30}C_0(0.07)^0(0.93)^{30} + {}_{30}C_1(0.07)^1(0.93)^{29} + \cdots + {}_{30}C_9(0.07)^9(0.93)^{21}]$

$= 1 - .999977 \approx 0.000023$

(c) $P(x < 10) \approx 0.999977$ (see part b)

(d) $n = 100, p = 0.07 \rightarrow np = 7$ and $nq = 93$

Use normal distribution.

$z = \dfrac{x - \mu}{\sigma} = \dfrac{4.5 - 7}{2.55} \approx -0.98$

$P(x < 5) = P(x < 4.5) = P(z < -0.98) = 0.1635$

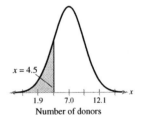
Number of donors

21. $n = 120, p = 0.05, q = 0.95$

$np = 6 \geq 5, nq = 114 \geq 5$

Use the normal distribution.

(a) $z = \dfrac{x - \mu}{\sigma} = \dfrac{7.5 - 6}{2.39} = 0.63$

$z = \dfrac{x - \mu}{\sigma} = \dfrac{8.5 - 6}{2.30} = 1.05$

$P(x = 8) \approx P(7.5 \leq x \leq 8.5) = P(0.63 \leq z \leq 1.05)$

$= 0.8531 - 0.7357 = 0.1174$

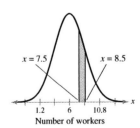
Number of workers

(b) $P(x \geq 8) \approx P(x \geq 7.5) = P(z \geq 0.63) = 1 - P(z \leq 0.63) = 1 - 0.7357 = 0.2643$

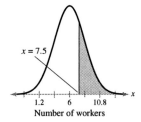

(c) $P(x < 8) \approx P(x \leq 7.5) = P(z \leq 0.63) = 0.7357$

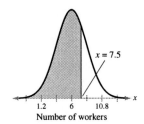

(d) $n = 250, p = 0.05, q = 0.95$

$np = 12.5 \geq 5, nq = 237.5 \geq 5$

Use normal distribution.

$$z = \frac{x - \mu}{\sigma} = \frac{14.5 - 12.5}{3.45} = 0.58$$

$P(x < 15) \approx P(x \leq 14.5) = P(z \leq 0.58) = 0.7190$

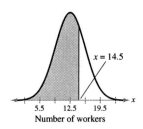

23. $n = 40, p = 0.52 \rightarrow np = 20.8$ and $nq = 19.2$

Use the normal distribution.

(a) $z = \dfrac{x - \mu}{\sigma} = \dfrac{15.5 - 20.8}{3.160} \approx -1.68$

$P(x \leq 15) = P(x < 15.5) = P(z < -1.68) = 0.0465$

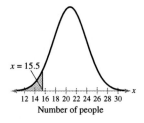

(b) $z = \dfrac{x - \mu}{\sigma} = \dfrac{14.5 - 20.8}{3.160} \approx 1.99$

$P(x \geq 15) = P(x > 14.5) = P(z > -1.99) = 1 - P(z < 1.99) = 1 - 0.0233 = 0.9767$

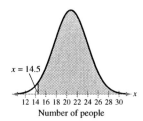

(c) $P(x > 15) = P(x > 15.5) = P(z > -1.68) = 1 - P(z < -1.68) = 1 - 0.0465 = 0.9535$

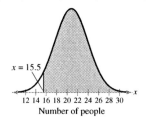

(d) $n = 650, p = 0.52 \rightarrow np = 338$ and $nq = 312$

Use normal distribution.

$z = \dfrac{x - \mu}{\sigma} = \dfrac{350.5 - 338}{12.74} \approx 0.98$

$P(x > 350) = P1(x > 350.5) = P(z > 0.98) = 1 - P(z < 0.98) = 1 - 0.8365 = 0.1635$

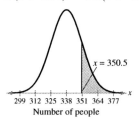

25. (a) $n = 25, p = 0.24, q = 0.76$

$np = 6 \geq 5, nq = 19 \geq 5$

Use normal distribution.

(b) $z = \dfrac{x - \mu}{\sigma} = \dfrac{8.5 - 6}{2.135} = 1.17$

$P(x > 8) \approx P(x \geq 8.5) = P(z \geq 1.17) = 1 - P(z \leq 1.17) = 1 - 0.8790 = 0.121$

(c) $z = \dfrac{x - \mu}{\sigma} = \dfrac{7.5 - 6}{2.135} = 0.70$

$z = \dfrac{x - \mu}{\sigma} = \dfrac{8.5 - 6}{2.135} = 1.17$

$P(x = 8) \approx P(7.5 \leq x \leq 8.5) = P(0.07 \leq z \leq 1.17) = 0.8790 - 0.7580 = 0.121$

It is not unusual for 8 out of 25 homeowners to say their home is too small.

82 CHAPTER 5 | NORMAL PROBABILITY DISTRIBUTIONS

27. $n = 250, p = 0.70$

60% say no $\rightarrow 250(0.6) = 150$ say no while 100 say yes.

$$z = \frac{x - \mu}{\sigma} = \frac{99.5 - 175}{7.25} = -10.41$$

$P(\text{less than 100 yes}) = P(x < 100) = P(x < 99.5) = P(z < -10.41) \approx 0$

It is highly unlikely that 60% responded no. Answers will vary.

29. $n = 100, p = 0.75$

$$z = \frac{x - \mu}{\sigma} = \frac{69.5 - 75}{4.33} \approx 1.27$$

$P(\text{reject claim}) = P(x < 70) = P(x < 69.5) = P(z < -1.27) = 0.1020$

CHAPTER 5 REVIEW EXERCISE SOLUTIONS

1. $\mu = 15, \sigma = 3$

3. $x = 1.32: z = \frac{x - \mu}{\sigma} = \frac{1.32 - 1.5}{0.08} = -2.25$

$x = 1.54: z = \frac{x - \mu}{\sigma} = \frac{1.54 - 1.5}{0.08} = 0.5$

$x = 1.66: z = \frac{x - \mu}{\sigma} = \frac{1.66 - 1.5}{0.08} = 2$

$x = 1.78: z = \frac{x - \mu}{\sigma} = \frac{1.78 - 1.5}{0.08} = 3.5$

5. 0.2005

7. 0.3936

9. $1 - 0.9535 = 0.0465$

11. $0.5 - 0.0505 = 0.4495$

13. $0.9115 - 0.5596 = 0.3519$

15. $0.0668 + 0.0668 = 0.1336$

17. $P(z < 1.28) = 0.8997$

19. $P(-2.15 < x < 1.55) = 0.9394 - 0.0158 = 0.9236$

21. $P(z < -2.50 \text{ or } z > 2.50) = 2(0.0062) = 0.0124$

23. (a) $z = \frac{x - \mu}{\sigma} = \frac{1900 - 2200}{625} \approx -0.48$

$P(x < 1900) = P(z < -0.48) = 0.3156$

(b) $z = \frac{x - \mu}{\sigma} = \frac{2000 - 2200}{625} \approx -0.32$

$z = \frac{x - \mu}{\sigma} = \frac{25900 - 2200}{625} \approx 0.48$

$P(2000 < x < 2500) = P1(-0.32 < z < 0.48) = 0.6844 - 0.3745 = 0.3099$

(c) $z = \dfrac{x - \mu}{\sigma} = \dfrac{2450 - 2200}{625} = -0.4$

$P(x > 2450) = P1(z > 0.4) = 0.3446$

25. $z = -0.07$

27. $z = 1.13$

29. $z = 1.04$

31. $x = \mu + z\sigma = 45.1 + (-2.4)(0.5) = 43.9$ meters

33. 95th percentile $\Rightarrow$ Area $= 0.95 \Rightarrow z = 1.645$

   $x = \mu + z\sigma = 45.1 + (1.645)(0.5) = 45.9$ meters

35. Top 10% $\Rightarrow$ Area $= 0.90 \Rightarrow z = 1.28$

   $x = \mu + z\sigma = 45.1 + (1.28)(0.5) = 45.74$ meters

37. {0 0 0, 0 0 200, 0 0 40, 0 0 600, 0 0 80, 0 200 0, 0 200 200, 0 200 40, 0 200 600, 0 200 80, 0 40 0, 0 40 200, 0 40 40, 0 40 600, 0 40 80, 0 600 0, 0 600 200, 0 600 40, 0 600 600, 0 600 80, 0 80 0, 0 80 200, 0 80 40, 0 80 600, 0 80 80, 200 0 0, 200 0 200, 200 0 40, 200 0 600, 200 0 80, 200 200 0, 200 200 200, 200 200 40, 200 200 600, 200 200 80, 200 40 0, 200 40 200, 200 40 40, 200 40 600, 200 40 80, 200 600 0, 200 600 200, 200 600 40, 200 600 600, 200 600 80, 200 80 0, 200 80 200, 200 80 40, 200 80 600, 200 80 80, 40 0 0, 40 0 200, 40 0 40, 40 0 600, 40 0 80, 40 200 0, 40 200 200, 40 200 40, 40 200 600, 40 200 80, 40 40 0, 40 40 200, 40 40 40, 40 40, 600, 40 40 80, 40 600 0, 40 600 200, 40 600 40, 40 600 600, 40 600 80, 40 80 0, 40 80 200, 40 80 40, 40 80 600, 40 80 80, 600 0 0, 600 0 200, 600 0 40, 600 0 600, 600 0 80, 600 200 0, 600 200 200, 600 200 40, 600 200 600, 600 200 80, 600 40 0, 600 40 200, 600 40 40, 600 40 600, 600 40 80, 600 600 0, 600 600 200, 600 600 40, 600 600 600, 600 600 80, 600 80 0, 600 80 200, 600 80 40, 600 80 600, 600 80 80, 80 0 0, 80 0 200, 80 0 40, 80 0 600, 80 0 80, 80 200 0, 80 200 200, 80 200 40, 80 200 600, 80 200 80, 80 40 0, 80 40 200, 80 40 40, 80 40 600, 80 40 80, 80 600 0, 80 600 200, 80 600 40, 80 600 600, 80 600 80, 80 80 0, 80 80 200, 80 80 40, 80 80 600, 80 80 80}

$\mu = 184, \sigma \approx 218.504$

$\mu_{\bar{x}} = 184, \sigma_{\bar{x}} \approx 126.153$

39. $\mu_{\bar{x}} = 152.7$, $\sigma_{\bar{x}} \dfrac{\sigma}{\sqrt{n}} = \dfrac{51.6}{\sqrt{35}} \approx 8.72$

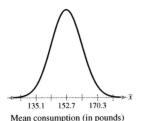

135.1 152.7 170.3
Mean consumption (in pounds)

41. (a) $z = \dfrac{\bar{x} - \mu}{\dfrac{\sigma}{\sqrt{n}}} = \dfrac{1900 - 2200}{\dfrac{625}{\sqrt{12}}} = \dfrac{-300}{180.42} \approx -1.66$

   $P(\bar{x} < 1900) = P(z < -1.66) = 0.0485$

(b) $z = \dfrac{\bar{x}-\mu}{\frac{\sigma}{\sqrt{n}}} = \dfrac{2000-2200}{\frac{625}{\sqrt{12}}} = \dfrac{-200}{180.42} \approx -1.11$

$z = \dfrac{\bar{x}-\mu}{\frac{\sigma}{\sqrt{n}}} = \dfrac{2500-2200}{\frac{625}{\sqrt{12}}} = \dfrac{300}{180.42} \approx 1.66$

$P(2000 < \bar{x} < 2500) = P(-1.11 < z < 1.66) = 0.9515 - 0.1335 = 0.8180$

(c) $z = \dfrac{\bar{x}-\mu}{\frac{\sigma}{\sqrt{n}}} = \dfrac{2450-2200}{\frac{625}{\sqrt{12}}} = \dfrac{250}{180.42} \approx 1.39$

$P(\bar{x} > 2450) = P(z > 1.39) = 0.0823$

(a) and (c) are smaller, (b) is larger. This is to be expected because the standard error of the sample mean is smaller.

43. (a) $z = \dfrac{\bar{x}-\mu}{\frac{\sigma}{\sqrt{n}}} = \dfrac{23{,}700-24{,}700}{\frac{1500}{\sqrt{45}}} = \dfrac{-1000}{223.61} \approx -4.47$

$P(\bar{x} < 23{,}700) = P(z < -4.47) \approx 0$

(b) $z = \dfrac{\bar{x}-\mu}{\frac{\sigma}{\sqrt{n}}} = \dfrac{26{,}200-24{,}700}{\frac{1500}{\sqrt{45}}} = \dfrac{1500}{223.61} \approx 6.71$

$P(\bar{x} > 26{,}200) = P(z > 6.71) \approx 0$

45. Assuming the distribution is normally distributed:

$z = \dfrac{\bar{x}-\mu}{\frac{\sigma}{\sqrt{n}}} = \dfrac{1.125-1.5}{\frac{.5}{\sqrt{15}}} = \dfrac{-0.375}{0.129} \approx -2.90$

$P(\bar{x} < 1.125) = P(z < -2.90) = 0.0019$

47. $n = 12, p = 0.90, q = 0.10$

$np = 10.8 > 5$, but $nq = 1.2 < 5$

Cannot use the normal distribution.

49. $P(x \geq 25) = P(x > 24.5)$

51. $P(x = 45) = P(44.5 < x < 45.5)$

53. $n = 45, p = 0.65 \rightarrow np = 29.25, nq = 15.75$

Use normal distribution.

$\mu = np = 29.25, \sigma = \sqrt{npq} = \sqrt{45(0.65)(0.35)} \approx 3.200$

$z = \dfrac{x-\mu}{\sigma} = \dfrac{20.5-29.25}{3.200} \approx -2.73$

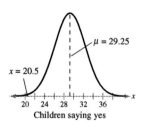

$P(x \leq 20) = P(x < 20.5) = P(z < -2.73) = 0.0032$

CHAPTER 5 QUIZ SOLUTIONS

1. (a) $P(z > -2.10) = 0.9821$

   (b) $P(z < 3.22) = 0.9994$

   (c) $P(-2.33 < z < 2.33) = 0.9901 - 0.0099 = 0.9802$

   (d) $P(z < -1.75 \text{ or } z > -0.75) = 0.0401 + 0.7734 = 0.8135$

2. (a) $z = \dfrac{x - \mu}{\sigma} = \dfrac{5.36 - 5.5}{0.08} \approx -1.75$

   $z = \dfrac{x - \mu}{\sigma} = \dfrac{5.64 - 5.5}{0.08} \approx 1.75$

   $P(5.36 < x < 5.64) = P(-1.75 < z < 1.75) = 0.9599 - 0.0401 = 0.9198$

   (b) $z = \dfrac{x - \mu}{\sigma} = \dfrac{-5.00 - (-8.2)}{7.84} \approx 0.41$

   $z = \dfrac{x - \mu}{\sigma} = \dfrac{0 - (-8.2)}{7.84} \approx 1.05$

   $P(-5.00 < x < 0) = P(0.41 < z < 1.05) = 0.8531 - 0.6591 = 0.1940$

   (c) $z = \dfrac{x - \mu}{\sigma} = \dfrac{0 - 18.5}{9.25} = -2$

   $z = \dfrac{x - \mu}{\sigma} = \dfrac{37 - 18.5}{9.25} = 2$

   $P(x < 0 \text{ or } x > 37) = P(z < -2 \text{ or } z > 2) = 2(0.0228) = 0.0456$

3. $z = \dfrac{x - \mu}{\sigma} = \dfrac{315 - 281}{34.4} \approx 0.99$

   $P(x > 315) = P(z > 0.99) = 0.1611$

4. $z = \dfrac{x - \mu}{\sigma} = \dfrac{250 - 281}{34.4} \approx -0.90$

   $z = \dfrac{x - \mu}{\sigma} = \dfrac{305 - 281}{34.4} \approx 0.70$

   $P(250 < x < 305) = P(-0.90 < z < 0.70) = 0.7580 - 0.1841 = 0.5739$

5. $P(x > 250) = P(z > -0.90) = 0.8159 \rightarrow 81.59\%$

6. $z = \dfrac{x - \mu}{\sigma} = \dfrac{300 - 281}{34.4} \approx 0.55$

   $P(x < 300) = P(z < 0.55) = 0.7088$

   $(2000)(0.7088) = 1417.6$

7. top 5% $\rightarrow z \approx 1.645$

   $\mu + z\sigma = 281 + (1.645)(34.4) = 337.588$

8. bottom 25% $\rightarrow z \approx -0.67$

   $\mu + z\sigma = 281 + (-0.67)(34.4) = 257.952$

86 CHAPTER 5 | NORMAL PROBABILITY DISTRIBUTIONS

9. $z = \dfrac{\bar{x} - \mu}{\dfrac{\sigma}{\sqrt{n}}} = \dfrac{300 - 281}{\dfrac{34.4}{\sqrt{60}}} = \dfrac{19}{4.44} \approx 4.28$

   $P(\bar{x} > 300) = P(z > 4.28) \approx 0$

10. $z = \dfrac{x - \mu}{\sigma} = \dfrac{300 - 281}{34.4} \approx 0.55$

    $P(\bar{x} > 300) = P(z > 0.55) = 0.2912$

    $z = \dfrac{\bar{x} - \mu}{\dfrac{\sigma}{\sqrt{n}}} = \dfrac{305 - 281}{\dfrac{34.4}{\sqrt{15}}} = \dfrac{24}{8.88} \approx 2.70$

    $P(\bar{x} > 300) = P(z > 2.70) = 0.0035$

    You are more likely to select one student with a test score greater than 300 because the standard error of the mean is less than the standard deviation.

11. $n = 24, p = 0.68 \rightarrow np = 16.32, nq = 7.68$

    Use normal distribution.

    $\mu = np = 16.32 \qquad \sigma = \sqrt{npq} \approx 2.285$

12. $z = \dfrac{x - \mu}{\sigma} = \dfrac{15.5 - 16.32}{2.285} \approx -0.36$

    $P(x \leq 15) = P(x < 15.5) = P(z < -0.36) = 0.3594$

Confidence Intervals

CHAPTER 6

6.1 CONFIDENCE INTERVALS FOR THE MEAN (LARGE SAMPLES)

6.1 Try It Yourself Solutions

1a. $\bar{x} \approx 14.767$

  b. The mean number of sentences per magazine advertisement is 14.767.

2a. $Z_c = 1.96, n = 30, s = 16.536$

  b. $E = Z_c \dfrac{s}{\sqrt{n}} = 1.96 \dfrac{16.536}{\sqrt{30}} \approx 5.917$

  c. You are 95% confident that the maximum error of the estimate is about 5.917 sentences per magazine advertisement.

3a. $\bar{x} \approx 14.767, E \approx 5.917$

  b. $\bar{x} - E = 14.767 - 5.917 = 8.850$
  $\bar{x} + E = 14.767 + 5.917 = 20.684$

  c. You are 95% confident that the mean number of sentences per magazine advertisements is between 8.850 and 20.684.

4b. 75% CI: (11.6, 13.2)
  85% CI: (11.4, 13.4)

  c. The width of the interval increases as the level of confidence increases.

5a. $n = 30, \bar{x} = 22.9, \sigma = 1.5, Z_c = 1.645$

  b. $E = Z_c \dfrac{\sigma}{\sqrt{n}} = 1.645 \dfrac{1.5}{\sqrt{30}} \approx 0.451$

  $\bar{x} - E = 22.9 - 0.451 \approx 22.4$
  $\bar{x} + E = 22.9 + 0.451 \approx 23.4$

  c. You are 90% confident that the mean age of the students is between 22.4 and 23.4 years.

6a. $Z_c = 1.96, E = 2, s \approx 5.0$

  b. $n = \left(\dfrac{Z_c s}{E}\right)^2 = \left(\dfrac{1.96 \cdot 5.0}{2}\right)^2 = 24.01 \rightarrow 25$

  c. You should have at least 25 magazine advertisements in your sample.

6.1 EXERCISE SOLUTIONS

1. You are more likely to be correct using an interval estimate since it is unlikely that a point estimate will equal the population mean exactly.

3. d; As the level of confidence increases, z_c increases therefore creating wider intervals.

5. 1.28   7. 1.15

9. $|\bar{x} - \mu| = |3.8 - 4.27| = 0.47$

11. $|\bar{x} - \mu| = |26.43 - 24.67| = 1.76$

13. $|\bar{x} - \mu| = |0.7 - 1.3| = 0.60$

15. $E = z_c \dfrac{s}{\sqrt{n}} = 1.645 \dfrac{2.5}{\sqrt{36}} \approx 0.685$

17. $E = z_c \dfrac{s}{\sqrt{n}} = 0.93 \dfrac{1.5}{\sqrt{50}} = 0.197$

19. $c = 0.88 \Rightarrow z_c = 1.55$

$\bar{x} = 57.2, s = 7.1, n = 50$

$\bar{x} \pm z_c \dfrac{s}{\sqrt{n}} = 57.2 \pm 1.55 \dfrac{7.1}{\sqrt{50}} = 57.2 \pm 1.556 \approx (55.6, 58.8)$

Answer: (c)

21. $c = 0.95 \Rightarrow z_c = 1.96$

$\bar{x} = 57.2, s = 7.1, n = 50$

$\bar{x} \pm z_c \dfrac{s}{\sqrt{n}} = 57.2 \pm 1.96 \dfrac{7.1}{\sqrt{50}} = 57.2 \pm 1.968 \approx (55.2, 59.2)$

Answer: (b)

23. $\bar{x} \pm z_c \dfrac{s}{\sqrt{n}} = 15.2 \pm 1.645 \dfrac{2.0}{\sqrt{60}} = 15.2 \pm 0.425 \approx (14.8, 15.6)$

25. $\bar{x} \pm z_c \dfrac{s}{\sqrt{n}} = 4.27 \pm 1.96 \dfrac{0.3}{\sqrt{42}} = 4.27 \pm 0.091 \approx (4.18, 4.36)$

27. $c = 0.90 \Rightarrow z_c = 1.645$

$n = \left(\dfrac{z_c \sigma}{E}\right)^2 = \left(\dfrac{(1.645)(6.8)}{1}\right)^2 = 125.13 \Rightarrow 126$

29. $c = 0.80 \Rightarrow z_c = 1.28$

$n = \left(\dfrac{z_c \sigma}{E}\right)^2 = \left(\dfrac{(1.28)(4.1)}{2}\right)^2 = 6.89 \Rightarrow 7$

31. $(2.1, 3.5) \Rightarrow 2E = 3.5 - 2.1 = 1.4 \Rightarrow E = 0.7$ and $\bar{x} = 2.1 + E$
$= 2.1 + 0.7 = 2.8$

33. 90% CI: $\bar{x} \pm z_c \dfrac{s}{\sqrt{n}} = 630.90 \pm 1.645 \dfrac{56.70}{\sqrt{32}} = 630.9 \pm 16.49 \approx (614.41, 647.39)$

95% CI: $\bar{x} \pm z_c \dfrac{s}{\sqrt{n}} = 630.90 \pm 1.96 \dfrac{56.70}{\sqrt{32}} = 630.9 \pm 19.65 \approx (611.25, 650.55)$

The 95% confidence interval is wider.

35. 90% CI: $\bar{x} \pm z_c \dfrac{s}{\sqrt{n}} = 26.8 \pm 1.645 \dfrac{8.0}{\sqrt{156}} = 26.8 \pm 1.054 \approx (25.7, 27.9)$

95% CI: $\bar{x} \pm z_c \dfrac{s}{\sqrt{n}} = 26.8 \pm 1.96 \dfrac{8.0}{\sqrt{156}} = 26.8 \pm 1.255 \approx (25.5, 28.1)$

The 95% confidence interval is wider.

37. $\bar{x} \pm z_c \dfrac{s}{\sqrt{n}} = 100 \pm 1.96 \dfrac{17.50}{\sqrt{40}} = 100 \pm 5.423 \approx (95.58, 105.42)$

39. $\bar{x} \pm z_c \dfrac{s}{\sqrt{n}} = 100 \pm 1.96 \dfrac{17.50}{\sqrt{80}} = 100 \pm 3.835 \approx (96.17, 104.84)$

$n = 40$ CI is wider because we have taken a smaller sample giving us less information about the population.

41. $\bar{x} \pm z_c \dfrac{s}{\sqrt{n}} = 10.452 \pm 2.575 \dfrac{2.130}{\sqrt{56}} = 10.452 \pm 0.733 \approx (9.719, 11.185)$

43. $\bar{x} \pm z_c \dfrac{s}{\sqrt{n}} = 10.452 \pm 2.575 \dfrac{5.130}{\sqrt{56}} = 10.452 \pm 1.765 \approx (8.687, 12.217)$

$s = 5.130$ CI is wider because of the increased variability within the population.

45. (a) An increase in the level of confidence will widen the confidence interval.

    (b) An increase in the sample size will narrow the confidence interval.

    (c) An increase in the standard deviation will widen the confidence interval.

47. $\bar{x} = \dfrac{\Sigma x}{n} = \dfrac{136}{15} \approx 9.1$

    90% CI: $\bar{x} \pm z_c \dfrac{\sigma}{\sqrt{n}} = 9.1 \pm 1.645 \dfrac{1.5}{\sqrt{15}} = 9.1 \pm 0.637 \approx (8.4, 9.7)$

    99% CI: $\bar{x} \pm z_c \dfrac{\sigma}{\sqrt{n}} = 9.1 \pm 2.575 \dfrac{1.5}{\sqrt{15}} = 9.1 \pm 0.997 \approx (8.1, 10.1)$

    99% CI is wider.

49. $n = \left(\dfrac{z_c \sigma}{E}\right)^2 = \left(\dfrac{1.96 \cdot 4.8}{1}\right)^2 \approx 88.510 \rightarrow 89$

51. (a) $n = \left(\dfrac{z_c \sigma}{E}\right)^2 = \left(\dfrac{1.96 \cdot 2.8}{0.5}\right)^2 \approx 120.473 \rightarrow 121$

    (b) $n = \left(\dfrac{z_c \sigma}{E}\right)^2 = \left(\dfrac{2.575 \cdot 2.8}{0.5}\right)^2 \approx 207.936 \rightarrow 208$

    99% CI requires larger sample because more information is needed from the population to be 99% confident.

53. (a) $n = \left(\dfrac{z_c \sigma}{E}\right)^2 = \left(\dfrac{1.645 \cdot 0.85}{0.25}\right)^2 \approx 31.282 \rightarrow 32$

    (b) $n = \left(\dfrac{z_c \sigma}{E}\right)^2 = \left(\dfrac{1.645 \cdot 0.85}{0.15}\right)^2 \approx 86.893 \rightarrow 87$

    $E = 0.15$ requires a larger sample size. As the error size decreases, a larger sample must be taken to obtain enough information from the population to ensure desired accuracy.

55. $n = \left(\dfrac{z_c \sigma}{E}\right)^2 = \left(\dfrac{1.960 \cdot 0.25}{0.125}\right)^2 = 15.3664 \rightarrow 16$

    $n = \left(\dfrac{z_c \sigma}{E}\right)^2 = \left(\dfrac{1.960 \cdot 0.25}{0.0625}\right)^2 = 61.4656 \rightarrow 62$

    $E = 0.0625$ requires a larger sample size. As the error size decreases, a larger sample must be taken to obtain enough information from the population to ensure desired accuracy.

57. (a) $n = \left(\dfrac{z_c \sigma}{E}\right)^2 = \left(\dfrac{2.575 \cdot 0.25}{0.1}\right)^2 \approx 41.441 \to 42$

(b) $n = \left(\dfrac{z_c \sigma}{E}\right)^2 = \left(\dfrac{2.575 \cdot 0.30}{0.1}\right)^2 \approx 59.676 \to 60$

$\sigma = 0.30$ requires a larger sample size. Due to the increased variability in the population, a larger sample size is needed to ensure the desired accuracy.

59. (a) An increase in the level of confidence will increase the minimum sample size required.

(b) An increase (larger E) in the error tolerance will decrease the minimum sample size required.

(c) An increase in the population standard deviation will increase the minimum sample size required.

61. $\bar{x} = 307.4, s \approx 11.190, n = 32$

$\bar{x} \pm z_c \dfrac{s}{\sqrt{n}} = 307.4 \pm 1.96 \dfrac{11.190}{\sqrt{32}} = 307.4 \pm 3.877 \approx (303.5, 311.3)$

63. $\bar{x} = 14.440, s \approx 2.124, n = 30$

$\bar{x} \pm z_c \dfrac{s}{\sqrt{n}} = 14.440 \pm 1.96 \dfrac{2.124}{\sqrt{30}} = 14.440 \pm 0.760 \approx (13.680, 15.200)$

65. (a) $\sqrt{\dfrac{N-n}{N-1}} = \sqrt{\dfrac{1000-500}{1000-1}} \approx 0.707$

(b) $\sqrt{\dfrac{N-n}{N-1}} = \sqrt{\dfrac{1000-100}{1000-1}} \approx 0.949$

(c) $\sqrt{\dfrac{N-n}{N-1}} = \sqrt{\dfrac{1000-75}{1000-1}} \approx 0.962$

(d) $\sqrt{\dfrac{N-n}{N-1}} = \sqrt{\dfrac{1000-50}{1000-1}} \approx 0.975$

(e) The finite population correction factor approaches 1 as the sample size decreases while the population size remains the same.

67. $n = \left(\dfrac{z_c \sigma}{E}\right)^2 \to \sqrt{n} = \dfrac{z_c \sigma}{E} \to E = \dfrac{z_c \sigma}{\sqrt{n}}$

6.2 CONFIDENCE INTERVALS FOR THE MEAN (SMALL SAMPLES)

6.2 Try It Yourself Solutions

1a. d.f. $= n - 1 = 22 - 1 = 21$ **b.** $c = 0.90$ **c.** 1.721

2a. 90% CI: $t_c = 1.753$

$E = t_c \dfrac{s}{\sqrt{n}} = 1.753 \dfrac{10}{\sqrt{16}} \approx 4.383$

99% CI: $t_c = 2.947$

$E = t_c \dfrac{s}{\sqrt{n}} = 2.947 \dfrac{10}{\sqrt{16}} \approx 7.368$

b. 90% CI: $\bar{x} \pm E = 162 \pm 4.383 \approx (157.6, 166.4)$

99% CI: $\bar{x} \pm E = 162 \pm 7.368 \approx (154.6, 169.4)$

c. You are 90% confident that the mean temperature of coffee sold is between 157.6° and 166.4°.

You are 99% confident that the mean temperature of coffee sold is between 154.6° and 169.4°.

3a. 90% CI: $t_c = 1.729$

$$E = t_c \frac{s}{\sqrt{n}} = 1.729 \frac{0.42}{\sqrt{20}} \approx 0.162$$

95% CI: $t_c = 2.093$

$$E = t_c \frac{s}{\sqrt{n}} = 2.093 \frac{0.42}{\sqrt{20}} \approx 0.197$$

b. 90% CI: $\bar{x} \pm E = 6.93 \pm 0.162 \approx (6.77, 7.09)$

95% CI: $\bar{x} \pm E = 6.93 \pm 0.197 \approx (6.73, 7.13)$

c. You are 90% confident that the mean mortgage interest rate is contained between 6.77% and 7.10%.

You are 95% confident that the mean mortgage interest rate is contained between 6.73% and 7.13%.

4a. Is $n \geq 30$? No

Is the population normally distributed? Yes

Is σ known? No

Use the t-distribution to construct the 90% CI.

6.2 EXERCISE SOLUTIONS

1. 1.833   **3.** 2.947

5. (a) $E = z_c \frac{s}{\sqrt{n}} = 1.96 \frac{5}{\sqrt{16}} \approx 2.450$ (b) $E = t_c \frac{s}{\sqrt{n}} = 2.131 \frac{5}{\sqrt{16}} \approx 2.664$

7. (a) $\bar{x} \pm t_c \frac{s}{\sqrt{n}} = 12.5 \pm 2.015 \frac{2.0}{\sqrt{6}} = 12.5 \pm 1.645 \approx (10.9, 14.1)$

(b) $\bar{x} \pm t_c \frac{s}{\sqrt{n}} = 12.5 \pm 1.645 \frac{2.0}{\sqrt{6}} = 12.5 \pm 1.343 \approx (11.2, 13.8)$

t-CI is wider.

9. (a) $\bar{x} \pm t_c \frac{s}{\sqrt{n}} = 4.3 \pm 2.650 \frac{0.34}{\sqrt{14}} = 4.3 \pm 0.241 \approx (4.1, 4.5)$

(b) $\bar{x} \pm z_c \frac{s}{\sqrt{n}} = 4.3 \pm 2.326 \frac{0.34}{\sqrt{14}} = 4.3 \pm 0.211 \approx (4.1, 4.5)$

t-CI is wider.

11. $\bar{x} \pm t_c \frac{s}{\sqrt{n}} = 75 \pm 2.776 \frac{12.50}{\sqrt{5}} = 75 \pm 15.518 \approx (59.48, 90.52)$

$$E = t_c \frac{s}{\sqrt{n}} = 2.776 \frac{12.50}{\sqrt{5}} \approx 15.518$$

92 CHAPTER 6 | CONFIDENCE INTERVALS

13. $\bar{x} \pm z_c \dfrac{\sigma}{\sqrt{n}} = 75 \pm 1.96 \dfrac{15}{\sqrt{5}} = 75 \pm 13.148 \approx (61.85, 88.15)$

    $E = z_c \dfrac{\sigma}{\sqrt{n}} = 1.96 \dfrac{15}{\sqrt{5}} \approx 13.148$

    t-CI is wider.

15. (a) $\bar{x} \pm t_c \dfrac{s}{\sqrt{n}} = 4.5 \pm 1.833 \dfrac{1.2}{\sqrt{10}} = 4.5 \pm 0.696 \approx (3.8, 5.2)$

    (b) $\bar{x} \pm z_c \dfrac{s}{\sqrt{n}} = 4.5 \pm 1.645 \dfrac{1.2}{\sqrt{500}} = 4.5 \pm 0.088 \approx (4.4, 4.6)$

    t-CI is wider.

17. (a) $\bar{x} = 4174.75$ (b) $s \approx 100.341$

    (c) $\bar{x} \pm t_c \dfrac{s}{\sqrt{n}} = 4174.75 \pm 3.250 \dfrac{100.341}{\sqrt{10}} = 4174.75 \pm 103.124 \approx (4071.63, 4277.87)$

19. (a) $\bar{x} \approx 909.1$ (b) $s \approx 305.266$

    (c) $\bar{x} \pm t_c \dfrac{s}{\sqrt{n}} = 909.1 \pm 3.106 \dfrac{305.266}{\sqrt{12}} = 909.1 \pm 273.709 \approx (635.4, 1182.8)$

21. $n \geq 30 \to$ use normal distribution

    $\bar{x} \pm z_c \dfrac{s}{\sqrt{n}} = 1.25 \pm 1.96 \dfrac{0.01}{\sqrt{70}} = 1.25 \pm 0.002 \approx (1.248, 1.252)$

23. $\bar{x} = 24, s = 3, n < 30, \sigma$ known, and pop normally distributed $\to$ use t-distribution

    $\bar{x} \pm t_c \dfrac{s}{\sqrt{n}} = 24 \pm 2.064 \dfrac{3}{\sqrt{25}} = 24 \pm 1.238 \approx (23, 25)$

25. $n < 30, \sigma$ unknown, and pop *not* normally distributed $\to$ cannot use either the normal or t-distributions.

27. $n = 25, \bar{x} = 56.0, s = 0.25$

    $\pm t_{0.99} \to 99\%$ t-CI

    $\bar{x} \pm t_c \dfrac{s}{\sqrt{n}} = 56.0 \pm 2.797 \dfrac{0.25}{\sqrt{25}} = 56.0 \pm 0.140 \approx (55.9, 56.1)$

    They are not making good tennis balls since desired bounce height of 55.5 inches is not contained between 55.9 and 56.1 inches.

6.3 CONFIDENCE INTERVALS FOR POPULATION PROPORTIONS

6.3 Try It Yourself Solutions

1a. $x = 1177, n = 3859$

b. $\hat{p} = \dfrac{1177}{3859} \approx 0.305$

2a. $\hat{p} \approx 0.305, \hat{q} \approx 0.695$

b. $n\hat{p} = (3859)(0.305) = 1175.995 > 5$

$n\hat{q} = (3859)(0.695) = 2682.005 > 5$

c. $z_c = 1.645$

$$E = z_c\sqrt{\frac{\hat{p}\hat{q}}{n}} = 1.645\sqrt{\frac{0.305 \cdot 0.695}{3859}} \approx 0.011$$

d. $\hat{p} \pm E = 0.305 \pm 0.011 \approx (0.293, 0.317)$

e. You are 90% confident that the proportion of adults that admired Franklin D. Roosevelt the most is contained between 29.3% and 31.7%.

3a. $n = 900, \hat{p} \approx 0.33$

b. $\hat{q} = 1 - \hat{p} = 1 - 0.33 \approx 0.67$

c. $n\hat{p} = 900 \cdot 0.33 \approx 297 > 5$

$n\hat{q} = 900 \cdot 0.67 \approx 603 > 5$

Distribution of $\hat{p}$ is approximately normal.

d. $z_c = 2.575$

e. $\hat{p} \pm z_c\sqrt{\frac{\hat{p}\hat{q}}{n}} = 0.33 \pm 2.575\sqrt{\frac{0.33 \cdot 0.67}{900}} = 0.33 \pm 0.04 \approx (0.290, 0.370)$

f. You are 99% confident that the proportion of adults who think that people over 75 are more dangerous drivers is contained between 29.0% and 37.0%.

4. (1) **a.** $\hat{p} = 0.50, \hat{q} = 0.50$

$z_c = 1.645, E = 0.02$

b. $n = \hat{p}\hat{q}\left(\frac{z_c}{E}\right)^2 = (0.50)(0.50)\left(\frac{1.645}{0.02}\right)^2 = 1691.266 \rightarrow 1692$

c. At least 1692 adults should be included in the sample.

(2) **a.** $\hat{p} = 0.04, \hat{q} = 0.96$

$z_c = 1.645, E = 0.02$

b. $n = \hat{p}\hat{q}\left(\frac{z_c}{E}\right)^2 = 0.04 \cdot 0.96\left(\frac{1.645}{0.02}\right)^2 \approx 259.778 \rightarrow 260$

c. At least 260 adults should be included in the sample.

6.3 EXERCISE SOLUTIONS

1. False. To estimate the value of p, the population proportion of successes, use the point estimate $\hat{p} = \frac{x}{n}$.

3. $\hat{p} = \frac{x}{n} = \frac{122}{1024} \approx 0.119$

$\hat{q} = 1 - \hat{p} \approx 0.881$

5. $\hat{p} = \frac{x}{n} = \frac{2634}{4115} \approx 0.640$

$\hat{q} = 1 - \hat{p} \approx 0.360$

7. $\hat{p} = \frac{x}{n} = \frac{102}{848} \approx 0.120$

$\hat{q} = 1 - \hat{p} \approx 0.880$

9. $\hat{p} = \frac{x}{n} = \frac{197}{284} \approx 0.694$

$\hat{q} = 1 - \hat{p} \approx 0.306$

11. $\hat{p} = \frac{x}{n} = \frac{264}{586} = 0.451$

$q = 1 - \hat{p} = 1 - 0.451 = 0.549$

94 CHAPTER 6 | CONFIDENCE INTERVALS

13. $\hat{p} = 0.48, E = 0.03$

$\hat{p} \pm E = 0.48 \pm 0.03 = (0.45, 0.51)$

15. 95% CI: $\hat{p} \pm z_c\sqrt{\dfrac{\hat{p}\hat{q}}{n}} = 0.119 \pm 1.96\sqrt{\dfrac{0.119 \cdot 0.881}{1024}} = 0.119 \pm 0.020 \approx (0.099, 0.139)$

99% CI: $\hat{p} \pm z_c\sqrt{\dfrac{\hat{p}\hat{q}}{n}} = 0.119 \pm 2.575\sqrt{\dfrac{0.119 \cdot 0.881}{1024}} = 0.119 \pm 0.026 \approx (0.093, 0.145)$

99% CI is wider.

17. 95% CI: $\hat{p} \pm z_c\sqrt{\dfrac{\hat{p}\hat{q}}{n}} = 0.640 \pm 1.96\sqrt{\dfrac{0.64 \cdot 0.36}{4115}} = 0.640 \pm 0.015 \approx (0.625, 0.655)$

99% CI: $\hat{p} \pm z_c\sqrt{\dfrac{\hat{p}\hat{q}}{n}} = 0.640 \pm 2.575\sqrt{\dfrac{0.64 \cdot 0.36}{4115}} = 0.640 \pm 0.019 \approx (0.621, 0.659)$

99% CI is wider.

19. 95% CI: $\hat{p} \pm z_c\sqrt{\dfrac{\hat{p}\hat{q}}{n}} = 0.120 \pm 1.96\sqrt{\dfrac{0.120 \cdot 0.880}{848}} = 0.120 \pm 0.022 \approx (0.098, 0.142)$

99% CI: $\hat{p} \pm z_c\sqrt{\dfrac{\hat{p}\hat{q}}{n}} = 0.120 \pm 2.575\sqrt{\dfrac{0.120 \cdot 0.880}{848}} = 0.120 \pm 0.029 \approx (0.091, 0.149)$

99% CI is wider.

21. (a) $n = \hat{p}\hat{q}\left(\dfrac{z_c}{E}\right)^2 = 0.5 \cdot 0.5\left(\dfrac{1.96}{0.03}\right)^2 \approx 1067.111 \rightarrow 1068$

(b) $n = \hat{p}\hat{q}\left(\dfrac{z_c}{E}\right)^2 = 0.26 \cdot 0.74\left(\dfrac{1.96}{0.03}\right)^2 \approx 821.249 \rightarrow 822$

(c) Having an estimate of the proportion reduces the minimum sample size needed.

23. (a) $n = \hat{p}\hat{q}\left(\dfrac{z_c}{E}\right)^2 = 0.5 \cdot 0.5\left(\dfrac{2.055}{0.025}\right)^2 = 1689.21 \rightarrow 1690$

(b) $n = \hat{p}\hat{q}\left(\dfrac{z_c}{E}\right)^2 = 0.25 \cdot 0.75\left(\dfrac{2.055}{0.025}\right)^2 \approx 1266.91 \rightarrow 1267$

(c) Having an estimate of the proportion reduces the minimum sample size needed.

25. (a) $\hat{p} = 0.61, n = 500$

$\hat{p} \pm z_c\sqrt{\dfrac{\hat{p}\hat{q}}{n}} = 0.61 \pm 2.575\sqrt{\dfrac{0.61 \cdot 0.39}{500}} = 0.61 \pm 0.056 \approx (0.554, 0.666)$

(b) $\hat{p} = 0.44, n = 500$

$\hat{p} \pm z_c\sqrt{\dfrac{\hat{p}\hat{q}}{n}} = 0.44 \pm 2.575\sqrt{\dfrac{0.44 \cdot 0.56}{500}} = 0.44 \pm 0.057 \approx (0.383, 0.497)$

It is unlikely that the two proportions are equal because the confidence intervals estimating the proportions do not overlap.

27. (a) $\hat{p} = 0.24, \hat{q} = 0.76, n = 600$

$\hat{p} \pm z_c\sqrt{\dfrac{\hat{p}\hat{q}}{n}} = 0.24 \pm 2.575\sqrt{\dfrac{(0.24)(0.76)}{600}} = 0.24 \pm 0.45 = (0.195, 0.285)$

(b) $\hat{p} = 0.30$, $\hat{q} = 0.70$, $n = 600$

$$\hat{p} \pm z_c \sqrt{\frac{\hat{p}\hat{q}}{n}} = 0.30 \pm 2.575 \sqrt{\frac{(0.30)(0.70)}{600}} = 0.30 \pm 0.048 = (0.252, 0.348)$$

(c) $\hat{p} = 0.37$, $\hat{q} = 0.63$, $n = 429$

$$\hat{p} \pm z_c \sqrt{\frac{\hat{p}\hat{q}}{n}} = 0.37 \pm 2.575 \sqrt{\frac{(0.37)(0.63)}{429}} = 0.37 \pm 0.060 = (0.310, 0.430)$$

29. $31.4\% \pm 1\% \to (30.4\%, 32.4\%)$

$$E = z_c \sqrt{\frac{\hat{p}\hat{q}}{n}} \to z_c = E \sqrt{\frac{n}{\hat{p}\hat{q}}} = 0.01 \sqrt{\frac{8451}{0.314 \cdot 0.686}} \approx 1.981 \to z_c = 1.98 \to c = 0.952$$

(30.4%, 32.4%) is approximately a 95.2% CI.

31. If $n\hat{p} < 5$ or $n\hat{q} < 5$, the sampling distribution of $\hat{p}$ may not be normally distributed; therefore preventing the use of z_c when calculating the confidence interval.

33.

p	q = 1 − p	pq	p	q = 1 − p	pq
0.0	1.0	0.00	0.45	0.55	0.2475
0.1	0.9	0.09	0.46	0.54	0.2484
0.2	0.8	0.16	0.47	0.53	0.2491
0.3	0.7	0.21	0.48	0.52	0.2496
0.4	0.6	0.24	0.49	0.51	0.2499
0.5	0.5	0.25	0.50	0.50	0.2500
0.6	0.4	0.24	0.51	0.49	0.2499
0.7	0.3	0.21	0.52	0.48	0.2496
0.8	0.2	0.16	0.53	0.47	0.2491
0.9	0.1	0.09	0.54	0.46	0.2484
1.0	0.0	0.00	0.55	0.45	0.2475

$\hat{p} = 0.5$ give the maximum value of $\hat{p}\hat{q}$.

6.4 CONFIDENCE INTERVALS FOR VARIANCE AND STANDARD DEVIATION

6.4 Try It Yourself Solutions

1a. d.f. $= n - 1 = 24$

level of confidence $= 0.95$

b. Area to the right of χ_R^2 is 0.025.

Area to the left of χ_L^2 is 0.975.

c. $\chi_R^2 = 39.364$, $\chi_L^2 = 12.401$

2a. 90% CI: $\chi_R^2 = 42.557$, $\chi_L^2 = 17.708$

95% CI: $\chi_R^2 = 45.722$, $\chi_L^2 = 16.047$

b. 90% CI for σ^2: $\left(\frac{(n-1)s^2}{\chi_R^2}, \frac{(n-1)s^2}{\chi_L^2}\right) = \left(\frac{29 \cdot (1.2)^2}{42.557}, \frac{29 \cdot (1.2)^2}{17.708}\right) \approx (0.98, 2.36)$

95% CI for σ^2: $\left(\frac{(n-1)s^2}{\chi_R^2}, \frac{(n-1)s^2}{\chi_L^2}\right) = \left(\frac{29 \cdot (1.2)^2}{45.722}, \frac{29 \cdot (1.2)^2}{16.047}\right) \approx (0.91, 2.60)$

96 CHAPTER 6 | CONFIDENCE INTERVALS

c. 90% CI for σ: $\left(\sqrt{0.981}, \sqrt{2.358}\right) = (0.99, 1.54)$

95% CI for σ: $\left(\sqrt{0.913}, \sqrt{2.602}\right) = (0.96, 1.61)$

d. You are 90% confident that the population variance is between 0.98 and 2.36, and that the population standard deviation is between 0.99 and 1.54. You are 95% confident that the population variance is between 0.91 and 2.60, and that the population standard deviation is between 0.96 and 1.61.

6.4 EXERCISE SOLUTIONS

1. $\chi_R^2 = 16.919$, $\chi_L^2 = 3.325$

3. $\chi_R^2 = 35.479$, $\chi_L^2 = 10.283$

5. $\chi_R^2 = 52.336$, $\chi_L^2 = 13.121$

7. (a) $s = 0.00843$

$$\left(\frac{(n-1)s^2}{\chi_R^2}, \frac{(n-1)s^2}{\chi_L^2}\right) = \left(\frac{13 \cdot (0.00843)^2}{22.362}, \frac{13 \cdot (0.00843)^2}{5.892}\right) \approx (0.0000413, 0.000157)$$

(b) $\left(\sqrt{0.0000413}, \sqrt{0.000157}\right) \approx (0.00643, 0.0125)$

9. (a) $s = 0.253$

$$\left(\frac{(n-1)s^2}{\chi_R^2}, \frac{(n-1)s^2}{\chi_L^2}\right) = \left(\frac{17 \cdot (0.253)^2}{35.718}, \frac{17 \cdot (0.253)^2}{5.697}\right) \approx (0.031, 0.191)$$

(b) $\left(\sqrt{0.0305}, \sqrt{0.191}\right) \approx (0.175, 0.437)$

11. (a) $\left(\frac{(n-1)s^2}{\chi_R^2}, \frac{(n-1)s^2}{\chi_L^2}\right) = \left(\frac{11 \cdot (3.25)^2}{26.757}, \frac{11 \cdot (3.25)^2}{2.603}\right) \approx (4.34, 44.64)$

(b) $\left(\sqrt{4.342}, \sqrt{44.636}\right) \approx (2.08, 6.68)$

13. (a) $\left(\frac{(n-1)s^2}{\chi_R^2}, \frac{(n-1)s^2}{\chi_L^2}\right) = \left(\frac{9 \cdot (26)^2}{16.919}, \frac{9 \cdot (26)^2}{3.325}\right) \approx (360, 1830)$

(b) $\left(\sqrt{359.596}, \sqrt{1829.774}\right) \approx (19, 43)$

15. (a) $\left(\frac{(n-1)s^2}{\chi_R^2}, \frac{(n-1)s^2}{\chi_L^2}\right) = \left(\frac{18 \cdot (15)^2}{31.526}, \frac{18 \cdot (15)^2}{8.231}\right) \approx (128, 492)$

(b) $\left(\sqrt{128.465}, \sqrt{492.042}\right) \approx (11, 22)$

17. (a) $\left(\frac{(n-1)s^2}{\chi_R^2}, \frac{(n-1)s^2}{\chi_L^2}\right) = \left(\frac{19 \cdot (107)^2}{32.852}, \frac{19 \cdot (107)^2}{8.907}\right) \approx (6622, 24{,}422)$

(b) $\left(\sqrt{6621.545}, \sqrt{24{,}422.477}\right) \approx (81, 156)$

19. (a) $\left(\frac{(n-1)s^2}{\chi_R^2}, \frac{(n-1)s^2}{\chi_L^2}\right) = \left(\frac{(14)(0.42)^2}{31.319}, \frac{(14)(0.42)^2}{4.075}\right) = (0.08, 0.61)$

(b) $\left(\sqrt{0.08}, \sqrt{0.61}\right) = (0.28, 0.78)$

21. 90% CI for σ: $(0.00643, 0.0125)$

Yes, because the confidence interval is below 0.015.

CHAPTER 6 REVIEW EXERCISE SOLUTIONS

1. (a) $\bar{x} \approx 103.5$

   (b) $s \approx 34.663$

   $$E = z_c \frac{s}{\sqrt{n}} = 1.645 \frac{34.663}{\sqrt{40}} \approx 9.016$$

3. $\bar{x} \pm z_c \frac{s}{\sqrt{n}} = 10.3 \pm 1.96 \frac{0.277}{\sqrt{100}} = 10.3 \pm 0.054 \approx (10.2, 10.4)$

5. $s = 34.663$

   $$n = \left(\frac{z_c \sigma}{E}\right)^2 = \left(\frac{1.96 \cdot 34.663}{10}\right)^2 \approx 46.158 \Rightarrow 47$$

7. $n = \left(\frac{z_c \sigma}{E}\right)^2 = \left(\frac{(1.96)(7.098)}{2}\right)^2 = 48.39 \Rightarrow 49$

9. $t_c = 1.415$

11. $t_c = 2.624$

13. $E = t_c \frac{s}{\sqrt{n}} = 1.753 \frac{23.4}{\sqrt{16}} \approx 10.255$

15. $E = t_c \frac{s}{\sqrt{n}} = 2.718 \left(\frac{0.9}{\sqrt{12}}\right) = 0.7$

17. $\bar{x} \pm z_c \frac{s}{\sqrt{n}} = 52.8 \pm 1.753 \frac{23.4}{\sqrt{16}} = 52.8 \pm 10.255 \approx (42.5, 63.1)$

19. $\bar{x} \pm t_c \frac{s}{\sqrt{n}} = 6.8 \pm 2.718 \left(\frac{0.9}{\sqrt{12}}\right) = 6.8 \pm 0.706 = (6.1, 7.5)$

21. $\bar{x} \pm t_c \frac{s}{\sqrt{n}} = 80 \pm 1.761 \frac{14}{\sqrt{15}} = 80 \pm 6.366 \approx (74, 86)$

23. $\hat{p} = \frac{x}{n} = \frac{1378}{2120} = 0.65, \hat{q} = 0.35$

25. $\hat{p} = \frac{x}{n} = \frac{597}{1011} = 0.591, \hat{q} = 0.409$

27. $\hat{p} = \frac{x}{n} = \frac{116}{644} = 0.180, \hat{q} = 0.820$

29. $\hat{p} = \frac{x}{n} = \frac{548}{1371} = 0.400, \hat{q} = 0.600$

31. $\hat{p} \pm z_c \sqrt{\frac{\hat{p}\hat{q}}{n}} = 0.650 \pm 1.96 \sqrt{\frac{0.650 \cdot 0.350}{2120}} = 0.650 \pm 0.020 \approx (0.630, 0.670)$

33. $\hat{p} \pm z_c \sqrt{\frac{\hat{p}\hat{q}}{n}} = 0.591 \pm 1.645 \sqrt{\frac{0.591 \cdot 0.409}{1011}} = 0.591 \pm 0.025 \approx (0.566, 0.616)$

35. $\hat{p} \pm z_c \sqrt{\frac{\hat{p}\hat{q}}{n}} = 0.180 \pm 2.575 \sqrt{\frac{0.180 \cdot 0.820}{644}} = 0.180 \pm 0.039 = (0.141, 0.219)$

37. $\hat{p} \pm z_c \sqrt{\frac{\hat{p}\hat{q}}{n}} = 0.400 \pm 1.96 \sqrt{\frac{(0.400)(0.600)}{1371}} = 0.400 \pm 0.026 = (0.374, 0.426)$

39. $n = \hat{p}\hat{q} \left(\frac{z_c}{E}\right)^2 = 0.23 \cdot 0.77 \left(\frac{1.96}{0.05}\right)^2 \approx 272.139 \to 273$

41. $\chi_R^2 = 23.337, \chi_L^2 = 4.404$

43. $\chi_R^2 = 14.067, \chi_L^2 = 2.167$

45. $s = 0.0727$

   95% CI for σ^2: $\left(\frac{(n-1)s^2}{\chi_R^2}, \frac{(n-1)s^2}{\chi_L^2}\right) = \left(\frac{15 \cdot (0.0727)^2}{27.488}, \frac{15 \cdot (0.0727)^2}{6.262}\right) \approx (0.0029, 0.0127)$

   95% CI for σ: $\left(\sqrt{0.0029}, \sqrt{0.0127}\right) \approx (0.0539, 0.1127)$

47. $s = 1.12$

90% CI for σ^2: $\left(\dfrac{(n-1)s^2}{\chi_R^2}, \dfrac{(n-1)s^2}{\chi_L^2}\right) = \left(\dfrac{(23)(1.12)^2}{35.172}, \dfrac{(23)(1.12)^2}{13.091}\right) = (0.82, 2.20)$

90% CI for σ: $\left(\sqrt{0.82}, \sqrt{2.20}\right) \approx (0.91, 1.48)$

CHAPTER 6 QUIZ SOLUTIONS

1. (a) $\bar{x} \approx 98.110$

(b) $s \approx 24.722$

$E = t_c \dfrac{s}{\sqrt{n}} = 1.960 \dfrac{24.722}{\sqrt{30}} \approx 8.847$

(c) $\bar{x} \pm t_c \dfrac{s}{\sqrt{n}} = 98.110 \pm 1.960 \dfrac{24.722}{\sqrt{30}} = 98.110 \pm 8.847 \approx (89.263, 106.957)$

You are 95% confident that the population mean repair costs is contained between \$89.26 and \$106.96.

2. $n = \left(\dfrac{z_c \sigma}{E}\right)^2 = \left(\dfrac{2.575 \cdot 22.50}{10}\right)^2 \approx 33.568 \rightarrow$ sample size must be 34

3. (a) $\bar{x} = 6.61$ (b) $s \approx 3.38$

(c) $\bar{x} \pm t_c \dfrac{s}{\sqrt{n}} = 6.61 \pm 1.833 \dfrac{3.38}{\sqrt{10}} = 6.61 \pm 1.957 \approx (4.65, 8.57)$

(d) $\bar{x} \pm z_c \dfrac{\sigma}{\sqrt{n}} = 6.61 \pm 1.645 \dfrac{3.5}{\sqrt{10}} = 6.61 \pm 1.821 \approx (4.79, 8.43)$

The t-CI is wider since less information is available.

4. $\bar{x} \pm t_c \dfrac{s}{\sqrt{n}} = 4744 \pm 2.365 \dfrac{580}{\sqrt{8}} = 4744 \pm 484.97 \approx (4259, 5229)$

5. (a) $\hat{p} = \dfrac{x}{n} = \dfrac{643}{1037} = 0.620$

(b) $\hat{p} \pm z_c \sqrt{\dfrac{\hat{p}\hat{q}}{n}} = 0.620 \pm 1.645 \sqrt{\dfrac{0.620 \cdot 0.38}{1037}} = 0.620 \pm 0.025 \approx (0.595, 0.645)$

(c) $n = \hat{p}\hat{q}\left(\dfrac{z_c}{E}\right)^2 = 0.620 \cdot 0.38 \left(\dfrac{2.575}{0.04}\right)^2 \approx 976.36 \rightarrow 977$

6. (a) $\left(\dfrac{(n-1)s^2}{\chi_R^2}, \dfrac{(n-1)s^2}{\chi_L^2}\right) = \left(\dfrac{29 \cdot (24.722)^2}{45.722}, \dfrac{29 \cdot (24.722)^2}{16.047}\right) \approx (387.650, 1104.514)$

(b) $\left(\sqrt{387.650}, \sqrt{1104.514}\right) \approx (19.689, 33.234)$

Hypothesis Testing with One Sample

CHAPTER 7

7.1 INTRODUCTION TO HYPOTHESIS TESTING

7.1 Try It Yourself Solutions

1a. (1) The mean ... is not 74 months.
$\mu \neq 74$

(2) The variance ... is less than or equal to 3.5.
$\sigma^2 \leq 3.5$

(3) The proportion ... is greater than 39%.
$p > 0.39$

b. (1) $\mu = 74$ (2) $\sigma^2 > 3.5$ (3) $p \leq 0.39$

c. (1) $H_0: \mu = 74$; $H_a: \mu \neq 74$; (claim)

(2) $H_0: \sigma^2 \leq 3.5$ (claim); $H_a: \sigma^2 > 3.5$

(3) $H_0: p \leq 0.39$; $H_a: p > 0.39$ (claim)

2a. $H_0: p \leq 0.01$; $H_a: p > 0.01$

b. Type I error will occur if the actual proportion is less than or equal to 0.01, but you decided to reject H_0.

Type II error will occur if the actual proportion is greater than 0.01, but you do not reject H_0.

c. Type II error is more serious since you would be misleading the consumer, possibly causing serious injury or death.

3a. (1) $H_0: \mu = 74$; $H_a: \mu \neq 74$
(2) $H_0: p \leq 0.39$; $H_a: p > 0.39$

b. (1) Two-tailed (2) Right-tailed

c. (1) 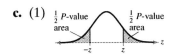 (2)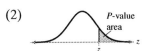

4a. There is enough evidence to support the radio station's claim.

b. There is not enough evidence to support the radio station's claim.

5a. (No answer required)

b. (1) Support claim. (2) Reject claim.

c. (1) $H_0: \mu \geq 650$; $H_a: \mu < 650$ (claim)

(2) $H_0: \mu = 98.6$ (claim); $H_a: \mu \neq 98.6$

7.1 EXERCISE SOLUTIONS

1. Null hypothesis (H_0) and alternative hypothesis (H_a). One represents the claim, the other, its complement.

3. False. In a hypothesis test, you assume the null hypothesis is true.

5. True

7. False. A small P-value in a test will favor a rejection of the null hypothesis.

9. $H_0: \mu \le 645$; $H_a: \mu > 645$

11. $H_0: \sigma = 5$; $H_a: \sigma \ne 5$

13. $H_0: p \ge 0.45$: $H_a: p < 0.45$

15. c, $H_0: \mu \le 3$

17. b, $H_0: \mu = 3$

19. Right-tailed

21. Two-tailed

23. $\mu > 750$
    $H_0: \mu \le 750$; $H_a: \mu > 750$ (claim)

25. $\sigma \le 1220$
    $H_0: \sigma \le 1220$ (claim); $H_a: \sigma > 1220$

27. $p = 0.84$
    $H_0: p = 0.84$ (claim); $H_a: p \ne 0.84$

29. Type I: Rejecting $H_0: p \ge 0.24$ when actually $p \ge 0.24$.
    Type II: Not rejecting $H_0: p \ge 0.24$ when actually $p < 0.24$.

31. Type I: Rejecting $H_0: \sigma \le 23$ when actually $\sigma \le 23$.
    Type II: Not rejecting $H_0: \sigma \le 23$ when actually $\sigma > 23$.

33. Type I: Rejecting $H_0: p = 0.88$ when actually $p = 0.88$.
    Type II: Not rejecting $H_0: p = 0.88$ when actually $p \ne 0.88$.

35. The null hypothesis is $H_0: p \ge 0.14$, the alternative hypothesis is $H_a: p < 0.14$. Therefore, because the alternative hypothesis contains <, the test is a left-tailed test.

37. The null hypothesis is $H_0: p = 0.90$, the alternative hypothesis is $H_a: p \ne 0.90$. Therefore, because the alternative hypothesis contains $\ne$, the test is a two-tailed test.

39. The null hypothesis is $H_0: p = 0.053$, the alternative hypothesis is $H_a: p \ne 0.053$. Therefore, because the alternative hypothesis contains $\ne$, the test is a two-tailed test.

41. (a) There is enough evidence to support the company's claim.
    (b) There is not enough evidence to support the company's claim.

43. (a) There is enough evidence to support the Dept of Labor's claim.
    (b) There is not enough evidence to decide that the Dept of Labor's claim is true.

45. (a) There is enough evidence to reject the manufacturer's claim.
    (b) There is not enough evidence to reject the manufacturer's claim.

47. $H_0: \mu = 10$; and $H_a: \mu \ne 10$

49. (a) $H_0: \mu \le 15$ $H_a: \mu > 15$ (b) $H_0: \mu \ge 15$; $H_a: \mu < 15$

51. If you decrease, you are decreasing the probability that you reject H_0. Therefore, you are increasing the probability of failing to reject H_0. This could increase β, the probability of failing to reject H_0 when H_0 is false.

53. (a) Reject H_0 since the CI is located below 70.

    (b) Do not reject H_0 since the CI includes values larger than 70.

    (c) Do not reject H_0 since the CI includes values larger than 70.

55. (a) Reject H_0 since the CI is located above 0.20.

    (b) Do not reject H_0 since the CI includes values less than 0.20.

    (c) Do not reject H_0 since the CI includes values less than 0.20.

7.2 HYPOTHESIS TESTING FOR THE MEAN (LARGE SAMPLES)

7.2 Try It Yourself Solutions

1a. (1) $P = 0.0347 > 0.01 = \alpha$ (2) $P = 0.0347 < 0.05 = \alpha$

  b. (1) Fail to reject H_0 because $0.0347 > 0.01$.

      (2) Reject H_0 because $0.0347 < 0.05$.

2a. 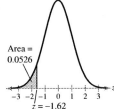   **b.** $P = 0.0526$

  c. Fail to reject H_0 since $P = 0.0526 > 0.05 = \alpha$.

3a. Area that corresponds to $z = 2.31$ is 0.9896.

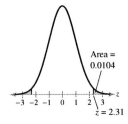

  b. $P = 2\,(\text{area}) = 2(0.0104) = 0.0208$

  c. Fail to reject H_0 since $P = 0.0208 > 0.01 = \alpha$.

4a. The claim is "the mean speed is greater than 35 miles per hour."

    $H_0: \mu \leq 35$; $H_a: \mu > 35$ (claim)

  b. $\alpha = 0.05$

  c. $z = \dfrac{\bar{x} - \mu}{\dfrac{s}{\sqrt{n}}} = \dfrac{36 - 35}{\dfrac{4}{\sqrt{100}}} = \dfrac{1}{0.4} = 2.500$

  d. P-value = Area right of $z = 2.50 = 0.0062$

  e. Reject H_0 since P-value $= 0.0062 < 0.05 = \alpha$.

  f. Because you reject H_0, there is enough evidence to claim the average speed limit is greater than 35 miles per hour.

5a. $H_0: \mu = 150$ (claim); $H_a: \mu \neq 150$

b. $\alpha = 0.01$

c. $z = \dfrac{\bar{x} - \mu}{\frac{\sigma}{\sqrt{n}}} = \dfrac{143 - 150}{\frac{15}{\sqrt{35}}} = \dfrac{-7}{2.535} \approx -2.76$

d. P-value $= 0.0058$

e. Reject H_0 since P-value $= 0.0058 < 0.01 = \alpha$.

f. There is enough evidence to state the claim is false.

6a. $P = 0.0391 > 0.01 = \alpha$

b. Fail to reject H_0.

7a.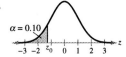

b. Area $= 0.1003$

c. $z_0 = -1.28$

d. $z < -1.28$

8a.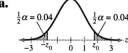

b. 0.0401 and 0.9599

c. $z_0 = -1.75$ and 1.75

d. $z < -1.75$, $z > 1.75$

e. Reject H_0.

9a. $H_0: \mu \geq 8.5$; $H_a: \mu < 8.5$ (claim)

b. $\alpha = 0.01$

c. $z_0 = -2.33$; Rejection region: $z < -2.33$

d. $z = \dfrac{\bar{x} - \mu}{\frac{s}{\sqrt{n}}} = \dfrac{8.2 - 8.5}{\frac{0.5}{\sqrt{35}}} = \dfrac{-0.300}{0.0845} \approx -3.550$

e.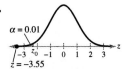

Reject H_0.

f. There is enough evidence to support the claim.

10a. $\alpha = 0.01$

b. $\pm z_0 = \pm 2.575$; Rejection regions: $z < -2.575$, $z > 2.575$

c. Fail to reject H_0.

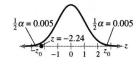

d. There is not enough evidence to support the claim that the mean cost is significantly different from $9405 at the 1% level of significance.

7.2 EXERCISE SOLUTIONS

1.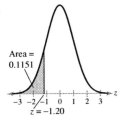

$P = 0.1151$; Fail to reject H_0 since $P = 0.1151 > 0.10 = \alpha$.

3.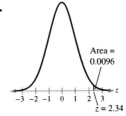

$P = 0.0096$; Reject H_0 since $P = 0.0096 < 0.01 = \alpha$.

5.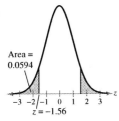

$P = 2(\text{Area}) = 2(0.0594) = 0.1188$; Fail to reject H_0 since $P = 0.1188 > 0.05 = \alpha$.

7. c **9.** e **11.** b

13. (a) Fail to reject H_0.

(b) Reject H_0 ($P = 0.0461 < 0.05 = \alpha$).

15. 1.645   **17.** -1.88 **19.** ± 2.33

21. Right-tailed ($\alpha = 0.01$)   **23.** Two-tailed ($\alpha = 0.10$)

25. (a) Fail to reject H_0 since $-1.645 < z < 1.645$.

(b) Reject H_0 since $z > 1.645$.

(c) Fail to reject H_0 since $-1.645 < z < 1.645$.

(d) Reject H_0 since $z < -1.645$.

27. (a) Fail to reject H_0 since $z < 1.285$.

(b) Fail to reject H_0 since $z < 1.285$.

(c) Fail to reject H_0 since $z < 1.285$.

(d) Reject H_0 since $z > 1.285$.

29. $H_0: \mu = 40$; $H_a: \mu \neq 40$

$\mu = 0.05 \rightarrow z_0 = \pm 1.96$

$z = \dfrac{\bar{x} - \mu}{\frac{s}{\sqrt{n}}} = \dfrac{39.2 - 40}{\frac{3.23}{\sqrt{75}}} = \dfrac{-0.8}{0.373} \approx -2.145$

Reject H_0. There is enough evidence to reject the claim.

104 CHAPTER 7 | HYPOTHESIS TESTING WITH ONE SAMPLE

31. $H_0: \mu = 6000$; $H_a: \mu \neq 6000$

$\alpha = 0.01 \rightarrow z_0 = \pm 2.575$

$z = \dfrac{\bar{x} - \mu}{\dfrac{s}{\sqrt{n}}} = \dfrac{5800 - 6000}{\dfrac{350}{\sqrt{35}}} = \dfrac{-200}{59.161} \approx -3.381$

Reject H_0. There is enough evidence to support the claim.

33. (a) $H_0: \mu \leq 260$; $H_a: \mu > 260$ (claim)

(b) $z = \dfrac{\bar{x} - \mu}{\dfrac{s}{\sqrt{n}}} = \dfrac{265 - 260}{\dfrac{55}{\sqrt{85}}} = \dfrac{5}{5.966} \approx 0.838$ Area = 0.7995

(c) P-value = {Area to right of $z = 0.84$} = 0.2005

(d) Fail to reject H_0. There is insufficient evidence at the 4% level of significance to support the claim that the mean score for Illinois' eighth grades is more than 260.

35. (a) $H_0: \mu \leq 7$; $H_a: \mu > 7$ (claim)

(b) $z = \dfrac{\bar{x} - \mu}{\dfrac{s}{\sqrt{n}}} = \dfrac{7.8 - 7}{\dfrac{2.67}{\sqrt{100}}} = \dfrac{0.8}{0.267} \approx 2.996$ Area = 0.9987

(c) P-value = {Area to right of $z = 3.00$} = 0.0013

(d) Reject H_0.

(e) There is sufficient evidence at the 7% level of significance to support the claim that the mean consumption of tea by a person in the U.S. is more than 7 gallons per year.

37. (a) $H_0: \mu = 15$ (claim); $H_a: \mu \neq 15$

(b) $\bar{x} \approx 14.834$ $s \approx 4.288$

$z = \dfrac{\bar{x} - \mu}{\dfrac{s}{\sqrt{n}}} = \dfrac{14.834 - 15}{\dfrac{4.288}{\sqrt{32}}} = \dfrac{-0.166}{0.758} \approx -0.219$ Area = 0.4129

(c) P-value = 2{Area to left of $z = -0.22$} = 2{0.4129} = 0.8258

(d) Fail to reject H_0.

(e) There is insufficient evidence at the 5% level of significance to reject the claim that the mean time it takes smokers to quit smoking permanently is 15 years.

39. (a) $H_0: \mu = 40$ (claim); $H_a: \mu \neq 40$

(b) $z_0 = \pm 2.575$; Rejection regions: $z < -2.575$ and $z > 2.575$

(c) $z = \dfrac{\bar{x} - \mu}{\dfrac{s}{\sqrt{n}}} = \dfrac{39.2 - 40}{\dfrac{7.5}{\sqrt{30}}} = \dfrac{-0.8}{1.369} \approx -0.584$

(d) Fail to reject H_0.

(e) There is insufficient evidence at the 1% level of significance to reject the claim that the mean caffeine content per one twelve-ounce bottle of cola is 40 milligrams.

41. (a) $H_0: \mu \geq 750$ (claim); $H_a: \mu < 750$

(b) $z_0 = -2.05$; Rejection region: $z < -2.05$

(c) $z = \dfrac{\bar{x} - \mu}{\dfrac{s}{\sqrt{n}}} = \dfrac{745 - 750}{\dfrac{60}{\sqrt{36}}} = \dfrac{-5}{10} \approx -0.500$

(d) Fail to reject H_0.

(e) There is insufficient evidence at the 2% level of significance to reject the claim that the mean life of the bulb is at least 750 hours.

43. (a) $H_0: \mu \leq 28$; $H_a: \mu > 28$ (claim)

(b) $z_0 = 1.55$; Rejection region: $z > 1.55$

(c) $\bar{x} \approx 32.861 \quad s \approx 22.128$

$z = \dfrac{\bar{x} - \mu}{\dfrac{s}{\sqrt{n}}} = \dfrac{32.861 - 28}{\dfrac{22.128}{\sqrt{36}}} = \dfrac{4.861}{3.688} \approx 1.318$

(d) Fail to reject H_0.

(e) There is insufficient evidence at the 6% level of significance to support the claim that the mean nitrogen dioxide level in West London is greater than 28 parts per billion.

45. (a) $H_0: \mu \geq 10$ (claim); $H_a: \mu < 10$

(b) $z_0 = 1.88$; Rejection region: $z < 1.88$

(c) $\bar{x} \approx 9.780, s \approx 2.362$

$x = \dfrac{\bar{x} - \mu}{\dfrac{s}{\sqrt{n}}} = \dfrac{9.780 - 10}{\dfrac{2.362}{\sqrt{30}}} = \dfrac{-0.22}{0.431} \approx -0.510$

(d) Fail to reject H_0.

(e) There is insufficient evidence at the 3% level to reject the claim that the mean weight loss after 1 month is at least 10 pounds.

47. $z = \dfrac{\bar{x} - \mu}{\dfrac{s}{\sqrt{n}}} = \dfrac{10{,}700 - 10{,}800}{\dfrac{280}{\sqrt{30}}} = \dfrac{-100}{51.121} \approx -1.956$

P-value = {Area left of $z = -1.96$} = 0.025
Fail to reject H_0 because the standardized test statistic $z = -1.96$ is greater than the critical value $z_0 = -2.33$.

49. (a) $\alpha = 0.02$; Fail to reject H_0.

(b) $\alpha = 0.03$; Reject H_0.

(c) $z = \dfrac{\bar{x} - \mu}{\dfrac{s}{\sqrt{n}}} = \dfrac{9900 - 10{,}000}{\dfrac{280}{\sqrt{50}}} = \dfrac{-100}{39.598} \approx -2.525$

P-value = {Area left of $z = -2.53$} = 0.0057 → Reject H_0.

(d) $z = \dfrac{\bar{x} - \mu}{\frac{s}{\sqrt{n}}} = \dfrac{9900 - 10{,}000}{\frac{280}{\sqrt{100}}} = \dfrac{-100}{28} \approx -3.571$

P-value = {Area left of $z = -3.57$} $< 0.0001 \rightarrow$ Reject H_0.

51. Using the classical z-test, the test statistic is compared to critical values. The z-test using a P-value compares the P-value to the level of significance α.

7.3 HYPOTHESIS TESTING FOR THE MEAN (SMALL SAMPLES)

7.3 Try It Yourself Solutions

1a. 2.650   **b.** $t_0 = -2.650$

2a. 1.860   **b.** $t_0 = +1.860$

3a. 2.947   **b.** $t_0 = \pm 2.947$

4a. $H_0: \mu \geq \$1240$ (claim); $H_a: \mu < \$1240$

b. $\alpha = 0.01$ and d.f. $= n - 1 = 8$

c. $t_0 = -2.896$; Reject H_0 if $t \leq -2.896$.

d. $t = \dfrac{\bar{x} - \mu}{\frac{s}{\sqrt{n}}} = \dfrac{1190 - 1240}{\frac{62}{\sqrt{9}}} = \dfrac{-50}{20.667} \approx -2.419$

e. Fail to reject H_0.

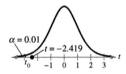

f. There is not enough evidence to reject the claim.

5a. $H_0: \mu = 1890$ (claim); $H_a: \mu \neq 1890$

b. $\alpha = 0.01$ and d.f. $= n - 1 = 18$

c. $t_0 = \pm 2.878$; Reject H_0 if $t < -2.878$ or $t > 2.878$.

d. $t = \dfrac{\bar{x} - \mu}{\frac{s}{\sqrt{n}}} = \dfrac{2500 - 1890}{\frac{700}{\sqrt{19}}} = \dfrac{610}{160.591} \approx 3.798$

e. Reject H_0.

f. There is enough evidence to reject the company's claim.

6a. $t = \dfrac{\bar{x} - \mu}{\frac{s}{\sqrt{n}}} = \dfrac{126 - 136}{\frac{12}{\sqrt{6}}} = \dfrac{-10}{4.899} \approx -2.041$

P-value = {Area left of $t = -2.04$} ≈ 0.0484

b. P-value $= 0.0484 < 0.05 = \alpha$

c. Reject H_0.

d. There is enough evidence to reject the claim.

7.3 EXERCISE SOLUTIONS

1. Identify the level of significance α and the degrees of freedom, d.f. $= n - 1$. Find the critical value(s) using the t-distribution table in the row with $n - 1$ d.f. If the hypothesis test is:

    (1) Left-tailed, use "One Tail, α" column with a negative sign.

    (2) Right-tailed, use "One Tail, α" column with a positive sign.

    (3) Two-tailed, use "Two Tail, α" column with a negative and a positive sign.

3. $t_0 = 1.717$

5. $t_0 = -2.101$

7. $t_0 = \pm 2.779$

9. 1.328

11. -2.473

13. ± 3.747

15. (a) Fail to reject H_0 since $t > -2.086$.

    (b) Fail to reject H_0 since $t > -2.086$.

    (c) Fail to reject H_0 since $t > -2.086$.

    (d) Reject H_0 since $t < -2.086$.

17. (a) Fail to reject H_0 since $-2.602 < t < 2.602$.

    (b) Fail to reject H_0 since $-2.602 < t < 2.602$.

    (c) Reject H_0 since $t > 2.602$.

    (d) Reject H_0 since $t < -2.602$.

19. $H_0: \mu = 15$ (claim); $H_a: \mu \neq 15$

    $\alpha = 0.01$ and d.f. $= n - 1 = 5$

    $t_0 = \pm 4.032$

    $t = \dfrac{\bar{x} - \mu}{\dfrac{s}{\sqrt{n}}} = \dfrac{13.9 - 15}{\dfrac{3.23}{\sqrt{6}}} = \dfrac{-1.1}{1.319} \approx -0.834$

    Fail to reject H_0. There is not enough evidence to reject the claim.

21. $H_0: \mu \geq 8000$ (claim); $H_a: \mu < 8000$

    $\alpha = 0.01$ and d.f. $= n - 1 = 24$

    $t_0 = -2.492$

    $t = \dfrac{\bar{x} - \mu}{\dfrac{s}{\sqrt{n}}} = \dfrac{7700 - 8000}{\dfrac{450}{\sqrt{25}}} = \dfrac{-300}{90} \approx -3.333$

    Reject H_0. There is enough evidence to reject the claim.

108 CHAPTER 7 | HYPOTHESIS TESTING WITH ONE SAMPLE

23. (a) $H_0: \mu \geq 100$; $H_a: \mu < 100$ (claim)

 (b) $t_0 = -3.747$; Reject H_0 if $t < -3.747$.

 (c) $t = \dfrac{\bar{x} - \mu}{\frac{s}{\sqrt{n}}} = \dfrac{75 - 100}{\frac{12.50}{\sqrt{5}}} = \dfrac{-25}{5.590} \approx -4.472$

 (d) Reject H_0.

 (e) There is sufficient evidence at the 1% level to support the claim that the mean repair cost for damaged microwave ovens is less than $100.

25. (a) $H_0: \mu \leq 1$; $H_a: \mu > 1$ (claim)

 (b) $t_0 = 1.796$; Reject H_0 if $t > 1.796$.

 (c) $t = \dfrac{\bar{x} - \mu}{\frac{s}{\sqrt{n}}} = \dfrac{1.2 - 1}{\frac{0.3}{\sqrt{12}}} = \dfrac{0.2}{0.866} \approx 2.309$

 (d) Reject H_0.

 (e) There is sufficient evidence at the 5% level to support the claim that the mean waste recycled by adults in the United States is more than 1 pound per person per day.

27. (a) $H_0: \mu = \$22{,}300$ (claim); $H_a: \mu \neq \$22{,}300$

 (b) $t_0 = 2.262$; Reject H_0 if $t < -2.262$ or $t > 2.262$.

 (c) $\bar{x} \approx 21{,}824.3$ $s \approx \$2628.9$

 $t = \dfrac{\bar{x} - \mu}{\frac{s}{\sqrt{n}}} = \dfrac{21{,}824.3 - 22{,}300}{\frac{2628.9}{\sqrt{10}}} = \dfrac{-475.7}{831.3} \approx -0.572$

 (d) Fail to reject H_0.

 (e) There is insufficient evidence at the 5% level to reject the claim that the mean salary for full-time male workers over age 25 without a high school diploma is $22,300.

29. (a) $H_0: \mu \geq 3.0$; $H_a: \mu < 3.0$ (claim)

 (b) $\bar{x} = 2.785$ $x = 0.828$

 $t = \dfrac{\bar{x} - \mu}{\frac{s}{\sqrt{n}}} = \dfrac{2.785 - 3.0}{\frac{0.828}{\sqrt{20}}} = \dfrac{-0.215}{0.185} \approx -1.161$

 P-value = {Area left of $t = -1.161$} = 0.130

 (c) Fail to reject H_0.

 (e) There is insufficient evidence at the 5% level to reject the claim that teenage males drink less than three 12-ounce servings of soda per day.

31. (a) $H_0: \mu \geq 32$; $H_a: \mu < 32$ (claim)

 (b) $\bar{x} = 30.167$ $s = 4.004$

 $t = \dfrac{\bar{x} - \mu}{\frac{s}{\sqrt{n}}} = \dfrac{30.167 - 32}{\frac{4.004}{\sqrt{18}}} = \dfrac{-1.833}{0.944} \approx -1.942$

 P-value = {Area left of $t = -1.942$} ≈ 0.034

(c) Fail to reject H_0.

(e) There is insufficient evidence at the 1% level to support the claim that the mean class size for full-time faculty is less than 32.

33. (a) H_0: $\mu = \$2116$ (claim); H_a: $\mu \neq \$2116$

(b) $\bar{x} = \$2423.75 \quad s = \$800.566$

$$t = \frac{\bar{x} - \mu}{\frac{s}{\sqrt{n}}} = \frac{2423.75 - 2116}{\frac{800.566}{\sqrt{12}}} = \frac{307.75}{231.103} = 1.332$$

P-value = 2{Area right of $t = 1.332$} = 2{0.105} = 0.210

(c) Fail to reject H_0.

(e) There is insufficient evidence at the 2% level to reject the claim that the typical household in the U.S. spends a mean amount of $2116 per year on food away from home.

35. H_0: $\mu \leq \$2294$; H_a: $\mu > 2294$ (claim)

$$t = \frac{\bar{x} - \mu}{\frac{s}{\sqrt{n}}} = \frac{2494 - 2294}{\frac{325}{\sqrt{6}}} = \frac{200}{132.681} \approx 1.507$$

P-value = {Area right of $t = 1.507$} ≈ 0.096

Since $0.096 > 0.01 = \alpha$, fail to reject H_0.

37. Since σ is unknown, $n < 30$, and the gas mileage is normally distributed, use the t-distribution.

H_0: $\mu \geq 21$ (claim); H_a: $\mu < 21$

$$t = \frac{\bar{x} - \mu}{\frac{s}{\sqrt{n}}} = \frac{19 - 21}{\frac{4}{\sqrt{5}}} = \frac{-2}{1.789} \approx 1.118$$

P-value = {Area left of $t = -1.118$} = 0.163

Fail to reject H_0. There is insufficient evidence at the 5% level to reject the claim that the mean gas mileage for the luxury sedan is at least 21 mpg.

7.4 HYPOTHESIS TESTING FOR PROPORTIONS

7.4 Try It Yourself Solutions

1a. $np = (86)(0.30) = 25.8 > 5$, $nq = (86)(0.70) = 60.2 > 5$

b. H_0: $p \geq 0.30$; H_a: $p < 0.30$ (claim)

c. $\alpha = 0.05$

d. $z_0 = -1.645$; Reject H_0 if $z < -1.645$.

e. $z = \frac{\hat{p} - p}{\sqrt{\frac{pq}{n}}} = \frac{0.20 - 0.30}{\sqrt{\frac{(0.30)(0.70)}{86}}} = \frac{-0.1}{0.0494} \approx -2.024$

f. Reject H_0.

g. There is enough evidence to support the claim.

2a. $np = (250)(0.05) = 12.5 > 5$, $nq = (250)(0.95) = 237.5 > 5$

b. $H_0: p = 0.05$ (claim); $H_a: p \neq 0.05$

c. $\alpha = 0.01$

d. $z_0 = \pm 2.575$; Reject H_0 if $z < -2.575$ or $z > 2.575$.

e. $z = \dfrac{\hat{p} - p}{\sqrt{\dfrac{pq}{n}}} = \dfrac{0.08 - 0.05}{\sqrt{\dfrac{(0.05)(0.95)}{250}}} = \dfrac{0.03}{0.0138} \approx 2.176$

f. Fail to reject H_0.

g. There is not enough evidence to reject the claim.

3a. $np = (75)(0.38) = 28.5 > 5$, $nq = (75)(0.62) = 46.5 > 5$

b. $H_0: p \leq 0.38$; $H_a: p > 0.38$ (claim)

c. $\alpha = 0.01$

d. $z_0 = 2.33$; Reject H_0 if $z > 2.33$.

e. $\hat{p} = \dfrac{x}{n} = \dfrac{33}{75} = 0.440$

$z = \dfrac{\hat{p} - p}{\sqrt{\dfrac{pq}{n}}} = \dfrac{0.440 - 0.38}{\sqrt{\dfrac{(0.38)(0.62)}{75}}} = \dfrac{0.06}{0.056} \approx 1.071$

f. Fail to reject H_0.

g. There is not enough evidence to support the claim.

7.4 EXERCISE SOLUTIONS

1. Verify that $np \geq 5$ and $nq \geq 5$. State H_0 and H_a. Specify the level of significance α. Determine the critical value(s) and rejection region(s). Find the standardized test statistic. Make a decision and interpret in the context of the original claim.

3. $np = (105)(0.25) = 26.25 > 5$

   $nq = (105)(0.75) = 78.75 > 5 \rightarrow$ use normal distribution

   $H_0: p = 0.25$; $H_a: p \neq 0.25$ (claim)

   $z_0 \pm 1.96$

   $z = \dfrac{\hat{p} - p}{\sqrt{\dfrac{pq}{n}}} = \dfrac{0.239 - 0.25}{\sqrt{\dfrac{(0.25)(0.75)}{105}}} = \dfrac{-0.011}{0.0423} \approx -0.260$

   Fail to reject H_0. There is not enough evidence to support the claim.

5. $np = (20)(0.12) = 2.4 < 5$

   $nq = (20)(0.88) = 17.6 \geq 5 \rightarrow$ cannot use normal distribution

7. $np = (70)(0.48) = 33.6 \geq 5$

$nq = (70)(0.52) = 36.4 \geq 5 \rightarrow$ use normal distribution

$H_0: p \geq 0.48$ (claim); $H_a: p < 0.48$

$z_0 = -1.29$

$z = \dfrac{\hat{p} - p}{\sqrt{\dfrac{pz}{n}}} = \dfrac{0.40 - 0.48}{\sqrt{\dfrac{(0.48)(0.52)}{70}}} = \dfrac{-0.08}{0.060} \approx -1.34$

Reject H_0. There is not enough evidence to support the claim.

9. (a) $H_0: p \geq 0.23$ (claim); $H_a: p < 0.23$

(b) $z_0 = -2.33$; Reject H_0 if $z < -2.33$.

(c) $z = \dfrac{\hat{p} - p}{\sqrt{\dfrac{pq}{n}}} = \dfrac{0.225 - 0.23}{\sqrt{\dfrac{(0.23)(0.77)}{200}}} = \dfrac{-0.005}{0.0298} \approx -0.168$

(d) Fail to reject H_0.

(e) There is insufficient evidence at the 1% level to reject the claim that at least 23% of U.S. adults are smokers.

11. (a) $H_0: p \leq 0.30$; $H_a: p > 0.30$ (claim)

(b) $z_0 = 1.88$; Reject H_0 if $z > 1.88$.

(c) $z = \dfrac{\hat{p} - p}{\sqrt{\dfrac{pq}{n}}} = \dfrac{0.32 - 0.3}{\sqrt{\dfrac{(0.30)(0.70)}{1050}}} = \dfrac{0.02}{0.0141} \approx 1.414$

(d) Fail to reject H_0.

(e) There is insufficient evidence at the 3% level to support the claim that more than 30% of U.S. consumers have stopped buying the product because the manufacturing of the product pollutes the environment.

13. (a) $H_0: p = 0.44$ (claim); $H_a: p \neq 0.44$

(b) $z_0 = \pm 2.33$; Reject H_0 if $z < -2.33$ or $z > 2.33$.

(c) $\hat{p} = \dfrac{722}{1762} \approx 0.410$

$z = \dfrac{\hat{p} - p}{\sqrt{\dfrac{pq}{n}}} = \dfrac{0.410 - 0.44}{\sqrt{\dfrac{(0.44)(0.56)}{1762}}} = \dfrac{-0.03}{0.01} \approx -2.537$

(d) Reject H_0.

(e) There is sufficient evidence at the 2% level to reject the claim that 44% of home buyers find their real estate agent through a friend.

15. $H_0: p \geq 0.52$ (claim); $H_a: p < 0.52$

$z_0 = -1.645$; Rejection region: $z < -1.645$

$z = \dfrac{\hat{p} - p}{\sqrt{\dfrac{pq}{n}}} = \dfrac{0.48 - 0.52}{\sqrt{\dfrac{(0.52)(0.48)}{50}}} = \dfrac{-0.04}{0.0707} \approx -0.566$

Fail to reject H_0. There is insufficient evidence to reject the claim.

17. $H_0: p = 0.44$ (claim); $H_a: p \neq 0.44$

$$z = \frac{x - np}{\sqrt{npq}} = \frac{722 - (1762)(0.44)}{\sqrt{(1762)(0.44)(0.56)}} = \frac{-53.28}{20.836} \approx -2.557$$

Reject H_0. The results are the same.

7.5 HYPOTHESIS TESTING FOR VARIANCE AND STANDARD DEVIATION

7.5 Try It Yourself Solutions

1a. $\chi_0^2 = 33.409$ **2a.** $\chi_0^2 = 17.708$ **3a.** $\chi_R^2 = 31.526$

b. $\chi_L^2 = 8.231$

4a. $H_0: \sigma^2 \leq 0.40$ (claim); $H_a: \sigma^2 > 0.40$

b. $\alpha = 0.01$ and d.f. $= n - 1 = 30$

c. $\chi_0^2 = 50.892$; Reject H_0 if $\chi^2 > 50.892$.

d. $\chi^2 = \frac{(n-1)s^2}{\sigma^2} = \frac{(30)(0.75)}{0.40} = 56.250$

e. Reject H_0.

f. There is enough evidence to reject the claim.

5a. $H_0: \sigma \geq 3.7$; $H_a: \sigma < 3.7$ (claim)

b. $\alpha = 0.05$ and d.f. $= n - 1 = 8$

c. $\chi_0^2 = 2.733$; Reject H_0 if $\chi^2 < 2.733$.

d. $\chi^2 = \frac{(n-1)s^2}{\sigma^2} = \frac{(8)(3.0)^2}{(3.7)^2} \approx 5.259$

e. Fail to reject H_0.

f. There is not enough evidence to support the claim.

6a. $H_0: \sigma^2 = 8.6$ (claim); $H_a: \sigma^2 \neq 8.6$

b. $\alpha = 0.01$ and d.f. $= n - 1 = 9$

c. $\chi_L^2 = 1.735$ and $\chi_R^2 = 23.589$

Reject H_0 if $\chi^2 > 23.589$ or $\chi^2 < 1.735$.

d. $\chi^2 = \frac{(n-1)s^2}{\sigma^2} = \frac{(9)(4.3)}{(8.6)} = 4.50$

e. Fail to reject H_0.

f. There is not enough evidence to reject the claim.

CHAPTER 7 | HYPOTHESIS TESTING WITH ONE SAMPLE

7.5 EXERCISE SOLUTIONS

1. Specify the level of significance α. Determine the degrees of freedom. Determine the critical values using the χ^2 distribution. If (a) right-tailed test, use the value that corresponds to d.f. and α. (b) left-tailed test, use the value that corresponds to d.f. and $1 - \alpha$; and (c) two-tailed test, use the value that corresponds to d.f. and $\frac{1}{2}\alpha$ and $1 - \frac{1}{2}\alpha$.

3. $\chi_0^2 = 38.885$ 5. $\chi_0^2 = 0.872$ 7. $\chi_L^2 = 7.261, \chi_R^2 = 24.996$

9. (a) Fail to reject H_0.   (b) Fail to reject H_0.
   (c) Fail to reject H_0.   (d) Reject H_0.

11. (a) Fail to reject H_0.   (b) Reject H_0.
    (c) Reject H_0.   (d) Fail to reject H_0.

13. $H_0: \sigma^2 = 0.52$ (claim); $H_a: \sigma^2 \neq 0.52$

    $\chi_L^2 = 7.564, \chi_R^2 = 30.191$

    $\chi^2 = \dfrac{(n-1)s^2}{\sigma^2} = \dfrac{(17)(0.508)^2}{(0.52)} \approx 16.608$

    Fail to reject H_0. There is insufficient evidence to reject the claim.

15. (a) $H_0: \sigma^2 = 3$ (claim); $H_a: \sigma^2 \neq 3$

    (b) $\chi_L^2 = 13.844, \chi_R^2 = 41.923$; Reject H_0 if $\chi^2 > 41.923$ or $\chi^2 < 13.844$.

    (c) $\chi^2 = \dfrac{(n-1)s^2}{\sigma^2} = \dfrac{(26)(2.8)}{3} \approx 24.267$

    (d) Fail to reject H_0.

    (e) There is insufficient evidence at the 5% level of significance to reject the claim that the variance of the life of the appliances is 3.

17. (a) $H_0: \sigma \geq 29$; $H_a: \sigma < 29$ (claim)

    (b) $\chi_0^2 = 13.240$; Reject H_0 if $\chi^2 < 13.240$.

    (c) $\chi^2 = \dfrac{(n-1)s^2}{\sigma^2} = \dfrac{(21)(27.7)^2}{(29)^2} \approx 19.159$

    (d) Fail to reject H_0.

    (e) There is insufficient evidence at the 10% level of significance to support the claim that the standard deviation for eighth graders on the examination is less than 29.

19. (a) $H_0: \sigma \leq 0.5$ (claim); $H_a: \sigma > 0.5$

    (b) $\chi_0^2 = 33.196$; Reject H_0 if $\chi^2 > 33.196$.

    (c) $\chi^2 = \dfrac{(n-1)s^2}{\sigma^2} = \dfrac{(24)(0.7)^2}{(0.5)^2} = 47.040$

    (d) Reject H_0.

    (e) There is sufficient evidence at the 10% level to reject the claim that the standard deviation of waiting times is no more than 0.5 minute.

21. (a) $H_0: \sigma > \$18{,}000$; $H_a: \sigma < \$18{,}000$ (claim)

    (b) $\chi_0^2 = 18.114$; Reject H_0 if $\chi^2 < 18.114$.

(c) $\chi^2 = \dfrac{(n-1)s^2}{\sigma^2} = \dfrac{(27)(18{,}500)^2}{(18{,}000)^2} \approx 28.521$

(d) Fail to reject H_0.

(e) There is insufficient evidence at the 10% level to support the claim that the standard deviation of the total charge for patients involved in a crash where the vehicle struck a wall is less than $18,000.

23. (a) H_0: $\sigma \leq 20{,}000$; H_a: $\sigma > 20{,}000$ (claim)

(b) $\chi_0^2 = 24.996$; Reject H_0 if $\chi^2 > 24.996$.

(c) $s = 20{,}662.992$

$\chi^2 = \dfrac{(n-1)s^2}{\sigma^2} = \dfrac{(15)(20{,}662.992)^2}{(20{,}000)^2} \approx 16.011$

(d) Fail to reject H_0.

(e) There is insufficient evidence at the 5% level to support the claim that the standard deviation of the annual salaries for actuaries is more than $20,000.

25. $\chi^2 = 28.521$

P-value = {Area left of $\chi^2 = 28.521$} = 0.615

Fail to reject H_0 because P-value = $0.385 > 0.10 = \alpha$.

27. $\chi^2 = 16.011$

P-value = {Area right of $\chi^2 = 16.011$} = 0.381

Fail to reject H_0 because P-value = $0.381 > 0.05 = \alpha$.

CHAPTER 7 REVIEW EXERCISE SOLUTIONS

1. H_0: $\mu \leq 1593$ (claim); H_a: $\mu > 1593$

3. H_0: $p \geq 0.205$; H_a: $p < 0.205$ (claim)

5. H_0: $\sigma \leq 4.5$; H_a: $\sigma > 4.5$ (claim)

7. (a) H_0: $p = 0.73$ (claim); H_a: $p \neq 0.73$

(b) Type I error will occur if H_0 is rejected when the actual proportion of American adults who use nonprescription pain relievers is 0.73.

Type II error if H_0 is not rejected when the actual proportion of American adults who use nonprescription pain relievers is not 0.73.

(c) Two-tailed, since hypothesis compares "= vs $\neq$".

(d) There is enough evidence to reject the claim.

(e) There is not enough evidence to reject the claim.

9. (a) H_0: $\mu \leq 50$ (claim); H_a: $\mu > 50$

(b) Type I error will occur if H_0 is rejected when the actual standard deviation sodium content is no more than 50 mg.

Type II error if H_0 is not rejected when the actual standard deviation sodium content is more than 50 mg.

(c) Right-tailed, since hypothesis compares "≤ vs >".

(d) There is enough evidence to reject the claim.

(e) There is not enough evidence to reject the claim.

11. $z_0 \approx -2.05$ **13.** $z_0 = 1.96$

15. $H_0: \mu \leq 45$ (claim); $H_a: \mu > 45$

$z_0 = 1.645$

$$z = \frac{\bar{x} - \mu}{\frac{s}{\sqrt{n}}} = \frac{47.2 - 45}{\frac{6.7}{\sqrt{42}}} = \frac{2.2}{1.0338} \approx 2.128$$

Reject H_0. There is enough evidence to reject the claim.

17. $H_0: \mu \geq 5.500$; $H_a: \mu < 5.500$ (claim)

$z_0 = -2.33$

$$z = \frac{\bar{x} - \mu}{\frac{s}{\sqrt{n}}} = \frac{5.497 - 5.500}{\frac{0.011}{\sqrt{36}}} = \frac{-0.003}{0.00183} \approx -1.636$$

Fail to reject H_0. There is not enough evidence to support the claim.

19. $H_0: \mu \leq 0.05$ (claim); $H_a: \mu > 0.05$

$$z = \frac{\bar{x} - \mu}{\frac{s}{\sqrt{n}}} = \frac{0.057 - 0.05}{\frac{0.018}{\sqrt{32}}} = \frac{0.007}{0.00318} \approx 2.200$$

P-value = {Area right of $z = 2.20$} = 0.0139

$\alpha = 0.10$; Reject H_0.

$\alpha = 0.05$; Reject H_0.

$\alpha = 0.01$; Fail to reject H_0.

21. $H_0: \mu = 284$ (claim); $H_a: \mu \neq 284$

$$z = \frac{\bar{x} - \mu}{\frac{s}{\sqrt{n}}} = \frac{292 - 284}{\frac{25}{\sqrt{50}}} = \frac{8}{3.536} \approx 2.263$$

P-value = 2{Area right of $z = 2.263$} = 2{0.012} = 0.024

Reject H_0. There is sufficient evidence to reject the claim.

23. $t_0 = \pm 2.093$ **25.** $t_0 = -1.345$

27. $H_0: \mu = 95$; $H_a: \mu \neq 95$ (claim)

$t_0 = \pm 2.201$

$$t = \frac{\bar{x} - \mu}{\frac{s}{\sqrt{n}}} = \frac{94.1 - 95}{\frac{1.53}{\sqrt{12}}} = \frac{-0.9}{0.442} \approx -2.038$$

Fail to reject H_0. There is not enough evidence to support the claim.

29. $H_0: \mu \geq 0$ (claim); $H_a: \mu < 0$

$t_0 = -1.341$

$t = \dfrac{\bar{x} - \mu}{\dfrac{s}{\sqrt{n}}} = \dfrac{-0.45 - 0}{\dfrac{1.38}{\sqrt{16}}} = \dfrac{-0.45}{0.345} \approx -1.304$

Fail to reject H_0. There is not enough evidence to reject the claim.

31. $H_0: \mu \leq 48$ (claim); $H_a: \mu > 48$

$t_0 = 3.148$

$t = \dfrac{\bar{x} - \mu}{\dfrac{s}{\sqrt{n}}} = \dfrac{52 - 48}{\dfrac{2.5}{\sqrt{7}}} = \dfrac{4}{0.945} \approx 4.233$

Reject H_0. There is enough evidence to reject the claim.

33. $H_0: \mu = \$25$ (claim); $H_a: \mu \neq \$25$

$t_0 = \pm 1.740$

$t = \dfrac{\bar{x} - \mu}{\dfrac{s}{\sqrt{n}}} = \dfrac{26.25 - 25}{\dfrac{3.23}{\sqrt{18}}} = \dfrac{1.25}{0.761} \approx 1.642$

Fail to reject H_0. There is not enough evidence to reject the claim.

35. $H_0: \mu \geq 6$ (claim); $H_a: \mu < 6$

$t_0 = -2.539$

$\bar{x} = 5.765 \quad s = 1.709$

$t = \dfrac{\bar{x} - \mu}{\dfrac{s}{\sqrt{n}}} = \dfrac{5.765 - 6}{\dfrac{1.709}{\sqrt{20}}} = \dfrac{-0.235}{0.382} \approx -0.615$

P-value ≈ 0.273

Fail to reject H_0. There is not enough evidence to reject the claim.

37. $H_0: p = 0.15$ (claim); $H_a: p \neq 0.15$

$z_0 = \pm 1.96$

$z = \dfrac{\hat{p} - p}{\sqrt{\dfrac{pq}{n}}} = \dfrac{0.09 - 0.15}{\sqrt{\dfrac{(0.15)(0.85)}{40}}} = \dfrac{-0.06}{0.0565} \approx -1.063$

Fail to reject H_0. There is not enough evidence to reject the claim.

39. Because $np = 3.6$ is less than 5, the normal distribution cannot be used to approximate the binomial distribution.

41. Because $np = 1.2 < 5$, the normal distribution cannot be used to approximate the binomial distribution.

43. $H_0: p = 0.20$; $H_a: p \neq 0.20$ (claim)

$z_0 \pm 2.575$

$z = \dfrac{\hat{p} - p}{\sqrt{\dfrac{pq}{n}}} = \dfrac{0.23 - 0.20}{\sqrt{\dfrac{(0.20)(0.80)}{56}}} = \dfrac{0.03}{0.0534} \approx 0.561$

Fail to reject H_0. There is not enough evidence to support the claim.

45. $H_0: p \leq 0.75$; $H_a: p > 0.75$ (claim)

$z_0 = 1.28$

$\hat{p} = \dfrac{x}{n} = \dfrac{818}{1036} \approx 0.790$

$z = \dfrac{\hat{p} - p}{\sqrt{\dfrac{pq}{n}}} = \dfrac{0.790 - 0.75}{\sqrt{\dfrac{(0.75)(0.25)}{1036}}} \approx \dfrac{0.04}{0.0134} \approx 2.973$

Reject H_0. There is enough evidence to support the claim.

47. $\chi_R^2 = 30.144$ **49.** $\chi_R^2 = 33.196$

51. $H_0: \sigma^2 \leq 2$; $H_a: \sigma^2 > 2$ (claim)

$\chi_0^2 = 24.769$

$\chi^2 = \dfrac{(n-1)s^2}{\sigma^2} = \dfrac{(17)(2.38)}{(2)} = 20.230$

Fail to reject H_0. There is not enough evidence to support the claim.

53. $H_0: \sigma^2 = 1.25$ (claim); $H_a: \sigma^2 \neq 1.25$

$\chi_L^2 = 0.831$, $\chi_R^2 = 12.833$

$\chi^2 = \dfrac{(n-1)s^2}{\sigma^2} = \dfrac{(5)(1.03)^2}{(1.25)^2} \approx 3.395$

Fail to reject H_0. There is not enough evidence to reject the claim.

55. $H_0: \sigma^2 \leq 0.01$ (claim); $H_a: \sigma^2 > 0.01$

$\chi_0^2 = 49.645$

$\chi^2 = \dfrac{(n-1)s^2}{\sigma^2} = \dfrac{(27)(0.064)}{(0.01)} = 172.800$

Reject H_0. There is not enough evidence to reject the claim.

CHAPTER 7 QUIZ SOLUTIONS

1. (a) $H_0: \mu \geq 94$ (claim); $H_a: \mu < 94$

(b) "$\geq$ vs $<$" $\rightarrow$ Left-tailed

σ is unknown and $n \geq 30 \rightarrow z$-test.

(c) $z_0 = -2.05$; Reject H_0 if $z < -2.05$.

(d) $z = \dfrac{\bar{x} - \mu}{\dfrac{s}{\sqrt{n}}} = \dfrac{93.5 - 94}{\dfrac{30}{\sqrt{103}}} = \dfrac{-0.5}{2.956} \approx -0.169$

(e) Fail to reject H_0. There is insufficient evidence at the 2% level to reject the claim that the mean consumption of fresh citrus fruits by people in the U.S. is at least 94 pounds per year.

2. (a) $H_0: \mu \geq 25$ (claim); $H_a: \mu < 25$

(b) "$\geq$ vs $<$" $\rightarrow$ Left-tailed

σ is unknown, the population is normal, and $n < 30 \rightarrow t$-test.

(c) $t_0 = -1.895$; Reject H_0 if $t < -1.895$.

(d) $z = \dfrac{\bar{x} - \mu}{\dfrac{s}{\sqrt{n}}} = \dfrac{23 - 25}{\dfrac{5}{\sqrt{8}}} = \dfrac{-2}{1.768} \approx -1.131$

118 CHAPTER 7 | HYPOTHESIS TESTING WITH ONE SAMPLE

(e) Fail to reject H_0. There is insufficient evidence at the 5% level to reject the claim that the mean gas mileage is at least 25 miles per gallon.

3. (a) $H_0: p \le 0.10$ (claim); $H_a: p > 0.10$

(b) "$\le$ vs $>$" $\rightarrow$ Right-tailed
$np \ge 5$ and $nq \ge 5 \rightarrow z$-test

(c) $z_0 = 1.75$; Reject H_0 if $z > 1.75$.

(d) $z = \dfrac{\hat{p} - p}{\sqrt{\dfrac{pq}{n}}} = \dfrac{0.13 - 0.10}{\sqrt{\dfrac{(0.10)(0.90)}{57}}} = \dfrac{0.03}{0.0397} \approx 0.755$

(e) Fail to reject H_0. There is insufficient evidence at the 4% level to reject the claim that no more than 10% of microwaves need repair during the first five years of use.

4. (a) $H_0: \sigma = 105$ (claim); $H_a: \sigma \ne 105$

(b) "= vs $\ne$" $\rightarrow$ Two-tailed
Assuming the scores are normally distributed and you are testing the hypothesized standard deviation $\rightarrow \chi^2$ test.

(c) $\chi_L^2 = 3.565$, $\chi_R^2 = 29.819$; Reject H_0 if $\chi^2 < 3.565$ or if $\chi^2 > 29.819$.

(d) $\chi^2 = \dfrac{(n-1)s^2}{\sigma^2} = \dfrac{(13)(113)^2}{(105)^2} \approx 15.056$

(e) Fail to reject H_0. There is insufficient evidence at the 1% level to reject the claim that the standard deviation of the SAT verbal scores for the state is 105.

5. (a) $H_0: \mu = \$59{,}482$ (claim); $H_a: \mu \ne \$59{,}482$

(b) "= vs $\ne$" $\rightarrow$ Two-tailed
σ is unknown, $n < 30$, and assuming the salaries are normally distributed $\rightarrow$ t-test.

(c) not applicable

(d) $t = \dfrac{\bar{x} - \mu}{\dfrac{s}{\sqrt{n}}} = \dfrac{58{,}581 - 59{,}482}{\dfrac{6500}{\sqrt{12}}} = \dfrac{-901}{1876.388} \approx -0.480$

P-value = 2{Area left of $t = -0.480$} = 2(0.320) = 0.640

(e) Fail to reject H_0. There is insufficient evidence at the 5% level to reject the claim that the mean annual salary for full-time male workers age 25 to 34 with a bachelor's degree is $59,482.

6. (a) $H_0: \mu = \$276$ (claim); $H_a: \mu \ne \$276$

(b) "= vs $\ne$" $\rightarrow$ Two-tailed
σ is unknown, $n \ge 30$ $\rightarrow$ z-test.

(c) not applicable

(d) $z = \dfrac{\bar{x} - \mu}{\dfrac{s}{\sqrt{n}}} = \dfrac{285 - 276}{\dfrac{30}{\sqrt{35}}} = \dfrac{9}{5.071} \approx 1.775$

P-value = 2{Area right of $z = 1.775$} = 2{0.038} = 0.0759

(e) Fail to reject H_0. There is insufficient evidence at the 5% level to reject the claim that the mean daily cost of meals and lodging for a family of four traveling in Massachusetts is $276.

Hypothesis Testing with Two Samples

8.1 TESTING THE DIFFERENCE BETWEEN MEANS (LARGE INDEPENDENT SAMPLES)

8.1 Try It Yourself Solutions

1a. $H_0: \mu_1 = \mu_2; H_a: \mu_1 \neq \mu_2$ (claim)

b. $\alpha = 0.01$

c. $z_0 = \pm 2.575$; Reject H_0 if $z > 2.575$ or $z < -2.575$.

d. $z = \dfrac{(\bar{x}_1 - \bar{x}_2) - (\mu_1 - \mu_2)}{\sqrt{\dfrac{s_1^2}{n_1} + \dfrac{s_2^2}{n_2}}} = \dfrac{(3900 - 3500) - (0)}{\sqrt{\dfrac{(900)^2}{50} + \dfrac{(500)^2}{50}}} = \dfrac{400}{\sqrt{21200}} \approx 2.747$

e. Reject H_0.

f. There is enough evidence to support the claim.

2a. $z = \dfrac{(\bar{x}_1 - \bar{x}_2) - (\mu_1 - \mu_2)}{\sqrt{\dfrac{s_1^2}{n_1} + \dfrac{s_2^2}{n_2}}} = \dfrac{(252 - 242) - (0)}{\sqrt{\dfrac{(22)^2}{150} + \dfrac{(18)^2}{200}}} = \dfrac{10}{\sqrt{4.847}} \approx 4.542$

$\rightarrow$ P-value = {area right of $z = 4.542$} = 0.00000278

b. Reject H_0.

8.1 EXERCISE SOLUTIONS

1. State the hypotheses and identify the claim. Specify the level of significance and find the critical value(s). Identify the rejection regions. Find the standardized test statistic. Make a decision and interpret in the context of the claim.

3. $H_0: \mu_1 = \mu_2$ (claim); $H_a: \mu_1 \neq \mu_2$

   Rejection regions: $z_0 < -1.96$ and $z_0 > 1.96$ (Two-tailed test)

   (a) $\bar{x}_1 - \bar{x}_2 = 16 - 14 = 2$

   (b) $z = \dfrac{(\bar{x}_1 - \bar{x}_2) - (\mu_1 - \mu_2)}{\sqrt{\dfrac{s_1^2}{n_1} + \dfrac{s_2^2}{n_2}}} = \dfrac{(16 - 14) - (0)}{\sqrt{\dfrac{(1.1)^2}{50} + \dfrac{(1.5)^2}{50}}} = \dfrac{2}{\sqrt{0.0692}} \approx 7.603$

   (c) z is in the rejection region because $7.603 > 1.96$.

   (d) Reject H_0. There is enough evidence to reject the claim.

5. $H_0: \mu_1 \geq \mu_2; H_a: \mu_1 < \mu_2$ (claim)

   Rejection region: $z_0 < -2.33$ (Left-tailed test)

   (a) $\bar{x}_1 - \bar{x}_2 = 1225 - 1195 = 30$

(b) $z = \dfrac{(\bar{x}_1 - \bar{x}_2) - (\mu_1 - \mu_2)}{\sqrt{\dfrac{s_1^2}{n_1} + \dfrac{s_2^2}{n_2}}} = \dfrac{(1225 - 1195) - (0)}{\sqrt{\dfrac{(75)^2}{35} + \dfrac{(105)^2}{105}}} = \dfrac{30}{\sqrt{265.714}} \approx 1.84$

(c) z is not in the rejection region because $1.84 > -2.330$.

(d) Fail to reject H_0. There is not enough evidence to support the claim.

7. $H_0: \mu_1 \leq \mu_2; H_a: \mu_1 > \mu_2$ (claim)

$z_0 = 2.33$; Reject H_0 if $z > 2.33$.

$z = \dfrac{(\bar{x}_1 - \bar{x}_2) - (\mu_1 - \mu_2)}{\sqrt{\dfrac{s_1^2}{n_1} + \dfrac{s_2^2}{n_2}}} = \dfrac{(5.2 - 5.5) - (0)}{\sqrt{\dfrac{(0.2)^2}{45} + \dfrac{(0.3)^2}{37}}} = \dfrac{-.30}{\sqrt{0.00332}} \approx -5.207$

Reject H_0. There is enough evidence to reject the claim.

9. (a) $H_0: \mu_1 = \mu_2; H_a: \mu_1 \neq \mu_2$ (claim)

(b) $z_0 = \pm 1.645$; Reject H_0 if $z < -1.645$ or $z > 1.645$.

(c) $z = \dfrac{(\bar{x}_1 - \bar{x}_2) - (\mu_1 - \mu_2)}{\sqrt{\dfrac{s_1^2}{n_1} + \dfrac{s_2^2}{n_2}}} = \dfrac{(42 - 45) - (0)}{\sqrt{\dfrac{(4.7)^2}{35} + \dfrac{(4.3)^2}{35}}} = \dfrac{-3}{\sqrt{1.159}} \approx -2.786$

(d) Reject H_0.

(e) There is sufficient evidence at the 10% level to support the claim that the mean braking distance is different for both types of tires.

11. (a) $H_0: \mu_1 \geq \mu_2; H_a: \mu_1 < \mu_2$ (claim)

(b) $z_0 = -2.33$; Reject H_0 if $z < -2.33$.

(c) $z = \dfrac{(\bar{x}_1 - \bar{x}_2) - (\mu_1 - \mu_2)}{\sqrt{\dfrac{s_1^2}{n_1} + \dfrac{s_2^2}{n_2}}} = \dfrac{(75 - 80) - (0)}{\sqrt{\dfrac{(12.50)^2}{47} + \dfrac{(20)^2}{55}}} = \dfrac{-5}{\sqrt{10.597}} \approx -1.536$

(d) Fail to reject H_0.

(e) There is insufficient evidence at the 1% level to conclude that the repair costs for Model A are lower than for Model B.

13. (a) $H_0: \mu_1 = \mu_2$ (claim); $H_a: \mu_1 \neq \mu_2$

(b) $z_0 = \pm 2.575$; Reject H_0 if $z < -2.575$ or $z > 2.575$.

(c) $z = \dfrac{(\bar{x}_1 - \bar{x}_2) - (\mu_1 - \mu_2)}{\sqrt{\dfrac{s_1^2}{n_1} + \dfrac{s_2^2}{n_2}}} = \dfrac{(21.0 - 20.8) - (0)}{\sqrt{\dfrac{(5.0)^2}{43} + \dfrac{(4.7)^2}{56}}} = \dfrac{0.2}{\sqrt{0.976}} \approx 0.202$

(d) Fail to reject H_0.

(e) There is insufficient evidence at the 1% level to reject the claim that the male and female high school students have equal ACT scores.

15. (a) $H_0: \mu_1 = \mu_2$ (claim); $H_a: \mu_1 \neq \mu_2$

(b) $z_0 = \pm 1.645$; Reject H_0 if $z < -1.645$ or $z > 1.645$.

(c) $z = \dfrac{(\bar{x}_1 - \bar{x}_2) - (\mu_1 - \mu_2)}{\sqrt{\dfrac{s_1^2}{n_1} + \dfrac{s_2^2}{n_2}}} = \dfrac{(136 - 140) - (0)}{\sqrt{\dfrac{(25)^2}{35} + \dfrac{(30)^2}{35}}} = \dfrac{-4}{\sqrt{43.571}} \approx 0.606$

(d) Fail to reject H_0.

(e) There is insufficient evidence at the 10% level to reject the claim that the lodging cost for a family traveling in California is the same as Florida.

17. (a) $H_0: \mu_1 = \mu_2$ (claim); $H_a: \mu_1 \neq \mu_2$

(b) $z_0 = \pm 1.645$; Reject H_0 if $z < -1.645$ or $z > 1.645$.

(c) $z = \dfrac{(\bar{x}_1 - \bar{x}_2) - (\mu_1 - \mu_2)}{\sqrt{\dfrac{s_1^2}{n_1} + \dfrac{s_2^2}{n_2}}} = \dfrac{(146 - 142) - (0)}{\sqrt{\dfrac{(30)^2}{50} + \dfrac{(32)^2}{50}}} = \dfrac{4}{\sqrt{38.48}} \approx 0.645$

(d) Fail to reject H_0.

(e) There is insufficient evidence at the 10% level to reject the claim that the lodging cost for a family traveling in California is the same as Florida. The new samples do not lead to a different conclusion.

19. (a) $H_0: \mu_1 \leq \mu_2$; $H_a: \mu_1 > \mu_2$ (claim)

(b) $z_0 = 1.96$; Reject H_0 if $z > 1.96$.

(c) $\bar{x}_1 \approx 2.130$, $s_1 \approx 0.490$, $n_1 = 30$
$\bar{x}_2 \approx 1.593$, $s_2 \approx 0.328$, $n_2 = 30$

$z = \dfrac{(\bar{x}_1 - \bar{x}_2) - (\mu_1 - \mu_2)}{\sqrt{\dfrac{s_1^2}{n_1} + \dfrac{s_2^2}{n_2}}} = \dfrac{(2.130 - 1.593) - (0)}{\sqrt{\dfrac{(0.490)^2}{30} + \dfrac{(0.328)^2}{30}}} = \dfrac{0.537}{\sqrt{0.0116}} \approx 4.988$

(d) Reject H_0.

(e) At the 2.5% level of significance, there is sufficient evidence to support the claim.

21. (a) $H_0: \mu_1 = \mu_2$ (claim); $H_a: \mu_1 \neq \mu_2$

(b) $z_0 = \pm 2.575$; Reject H_0 if $z < -2.575$ or $z > 2.575$.

(c) $\bar{x}_1 \approx 0.875$, $s_1 \approx 0.011$, $n_1 = 35$
$\bar{x}_2 \approx 0.701$, $s_2 \approx 0.011$, $n_2 = 35$

$z = \dfrac{(\bar{x}_1 - \bar{x}_2) - (\mu_1 - \mu_2)}{\sqrt{\dfrac{s_1^2}{n_1} + \dfrac{s_2^2}{n_2}}} = \dfrac{(0.875 - 0.701) - (0)}{\sqrt{\dfrac{(0.011)^2}{35} + \dfrac{(0.011)^2}{35}}} = \dfrac{0.174}{\sqrt{0.0000006914}} \approx 66.172$

(d) Reject H_0.

(e) At the 1% level of significance, there is sufficient evidence to reject the claim.

23. They are equivalent through algebraic manipulation of the equation.
$\mu_1 = \mu_2 \rightarrow \mu_1 - \mu_2 = 0$

122 CHAPTER 8 | HYPOTHESIS TESTING WITH TWO SAMPLES

25. $H_0: \mu_1 - \mu_2 = -9$ (claim); $H_a: \mu_1 - \mu_2 \neq -9$

$z_0 = \pm 2.575$; Reject H_0 if $z < -2.575$ or $z > 2.575$.

$$z = \frac{(\bar{x}_1 - \bar{x}_2) - (\mu_1 - \mu_2)}{\sqrt{\frac{s_1^2}{n_1} + \frac{s_2^2}{n_2}}} = \frac{(11.5 - 20) - (-9)}{\sqrt{\frac{(3.8)^2}{70} + \frac{(6.7)^2}{65}}} = \frac{0.5}{\sqrt{0.897}} \approx 0.528$$

Fail to reject H_0. There is not enough evidence to reject the claim.

27. $H_0: \mu_1 - \mu_2 \leq 6000$; $H_a: \mu_1 - \mu_2 > 6000$ (claim)

$z_0 = 1.28$; Reject H_0 if $z > 1.28$.

$$z = \frac{(\bar{x}_1 - \bar{x}_2) - (\mu_1 - \mu_2)}{\sqrt{\frac{s_1^2}{n_1} + \frac{s_2^2}{n_2}}} = \frac{(55{,}900 - 54{,}200) - (6000)}{\sqrt{\frac{(8875)^2}{45} + \frac{(9175)^2}{37}}} = \frac{-4300}{\sqrt{4{,}025{,}499.249}} \approx -2.143$$

Fail to reject H_0. There is not enough evidence to support the claim.

29. $(\bar{x}_1 - \bar{x}_2) - z_c \sqrt{\frac{s_1^2}{n_1} + \frac{s_2^2}{n_2}} < \mu_1 - \mu_2 < (\bar{x}_1 - \bar{x}_2) + z_c \sqrt{\frac{s_1^2}{n_1} + \frac{s_2^2}{n_2}}$

$(3.2 - 4.1) - 1.96 \sqrt{\frac{(3.3)^2}{42} + \frac{(3.9)^2}{42}} < \mu_1 - \mu_2 < (3.2 - 4.1) + 1.96 \sqrt{\frac{(3.3)^2}{42} + \frac{(3.9)^2}{42}}$

$-0.9 - 1.96\sqrt{0.621} < \mu_1 - \mu_2 < -0.9 + 1.96\sqrt{0.621}$

$-2.45 < \mu_1 - \mu_2 < 0.65$

31. $H_0: \mu_1 - \mu_2 \leq 0$; $H_a: \mu_1 - \mu_2 > 0$ (claim)

$z_0 = 1.645$; Reject H_0 if $z > 1.645$.

$$z = \frac{(\bar{x}_1 - \bar{x}_2) - (\mu_1 - \mu_2)}{\sqrt{\frac{s_1^2}{n_1} + \frac{s_2^2}{n_2}}} = \frac{(3.2 - 4.1) - (0)}{\sqrt{\frac{(3.3)^2}{42} + \frac{(3.9)^2}{42}}} = \frac{-0.9}{\sqrt{0.621}} \approx -1.14$$

Fail to reject H_0. There is not enough evidence to support the claim. I would not recommend using the herbal supplement with a high-fiber, low-calorie diet to lose weight because there was not a significant difference between the weight loss it produced and that of the placebo.

33. $H_0: \mu_1 - \mu_2 \leq 0$; $H_a: \mu_1 - \mu_2 > 0$ (claim)

The 95% CI for $\mu_1 - \mu_2$ in Exercise 29 contained values less than or equal to zero and, as found in Exercise 31, there was not enough evidence at the 5% level of significance to support the claim. If zero is contained in the CI for $\mu_1 - \mu_2$, you fail to reject H_0 because the null hypothesis states that $\mu_1 - \mu_2$ is less than or equal to zero.

8.2 TESTING THE DIFFERENCE BETWEEN MEANS (SMALL INDEPENDENT SAMPLES)

8.2 Try It Yourself Solutions

1a. $H_0: \mu_1 = \mu_2$; $H_a: \mu_1 \neq \mu_2$ (claim)   **b.** $\alpha = 0.05$

c. d.f. $= \min\{n_1 - 1, n_2 - 1\} = \min\{10 - 1, 12 - 1\} = 9$

d. $t_0 = \pm 2.262$; Reject H_0 if $t < -2.262$ or $t > 2.262$.

e. $z = \dfrac{(\bar{x}_1 - \bar{x}_2) - (\mu_1 - \mu_2)}{\sqrt{\dfrac{s_1^2}{n_1} + \dfrac{s_2^2}{n_2}}} = \dfrac{(102 - 94) - (0)}{\sqrt{\dfrac{(10)^2}{10} + \dfrac{(4)^2}{12}}} = \dfrac{8}{\sqrt{11.333}} \approx 2.376$

f. Reject H_0.

g. There is enough evidence to support the claim.

2a. $H_0: \mu_1 \geq \mu_2$; $H_a: \mu_1 < \mu_2$ (claim)   **b.** $\alpha = 0.10$

c. d.f. $= n_1 = n_2 - 2 = 12 + 15 - 2 = 25$ **d.** $t_0 = -1.316$; Reject H_0 if $t < -1.316$.

e. $t = \dfrac{(\bar{x}_1 - \bar{x}_2) - (\mu_1 - \mu_2)}{\sqrt{\dfrac{(n_1 - 1)s_1^2 + (n_2 - 1)s_2^2}{n_1 + n_2 - 2}} \sqrt{\dfrac{1}{n_1} + \dfrac{1}{n_2}}} = \dfrac{(73 - 74) - (0)}{\sqrt{\dfrac{(12 - 1)(2.4)^2 + (15 - 1)(3.2)^2}{12 + 15 - 2}} \sqrt{\dfrac{1}{12} + \dfrac{1}{15}}}$

$= \dfrac{-1}{\sqrt{8.269}\sqrt{0.150}} \approx -0.898$

f. Fail to reject H_0.

g. There is not enough evidence to support the claim.

8.2 EXERCISE SOLUTIONS

1. State hypotheses and identify the claim. Specify the level of significance. Determine the degrees of freedom. Find the critical value(s) and identify the rejection region(s). Find the standardized test statistic. Make a decision and interpret in the context of the original claim.

3. (a) d.f. $= n_1 + n_2 - 2 = 20$
$t_0 = \pm 1.725$

    (b) d.f. $= \min\{n_1 - 1, n_2 - 1\} = 9$
$t_0 = \pm 1.833$

5. (a) d.f. $= n_1 + n_2 - 2 = 22$
$t_0 = -2.074$

    (b) d.f. $= \min\{n_1 - 1, n_2 - 1\} = 8$
$t_0 = -2.306$

7. (a) d.f. $= n_1 + n_2 - 2 = 19$
$t_0 = 1.729$

    (b) d.f. $= \min\{n_1 - 1, n_2 - 1\} = 7$
$t_0 = 1.895$

9. (a) d.f. $= n_1 + n_2 - 2 = 27$
$t_0 = \pm 2.771$

    (b) d.f. $= \min\{n_1 - 1, n_2 - 1\} = 11$
$t_0 = \pm 3.106$

11. $H_0: \mu_1 = \mu_2$ (claim); $H_a: \mu_1 \neq \mu_2$
d.f. $= n_1 + n_2 - 2 = 15$
$t_0 = \pm 2.947$ (Two-tailed test)

    (a) $\bar{x}_1 - \bar{x}_2 = 33.7 - 35.5 = -1.8$

    (b) $t = \dfrac{(\bar{x}_1 - \bar{x}_2) - (\mu_1 - \mu_2)}{\sqrt{\dfrac{(n_1 - 1)s_1^2 + (n_2 - 1)s_2^2}{n_1 + n_2 - 2}} \sqrt{\dfrac{1}{n_1} + \dfrac{1}{n_2}}} = \dfrac{(33.7 - 35.5) - (0)}{\sqrt{\dfrac{(10 - 1)(3.5)^2 + (7 - 1)(2.2)^2}{10 + 7 - 2}} \sqrt{\dfrac{1}{10} + \dfrac{1}{7}}}$

    $= \dfrac{-1.8}{\sqrt{9.286}\sqrt{0.243}} \approx -1.199$

    (c) t is not in the rejection region.

    (d) Fail to reject H_0. There is not enough evidence to reject the claim.

13. There is no need to run the test since this is a right-tailed test and the test statistic is negative. It is obvious that the standardized test statistic will also be negative and fall outside of the rejection region. So, the decision is to fail to reject H_0.

15. (a) $H_0: \mu_1 = \mu_2$ (claim); $H_a: \mu_1 \neq \mu_2$

(b) d.f. $= n_1 + n_2 - 2 = 13 + 15 - 2 = 26$
$t_0 = \pm 1.706$; Reject H_0 if $t < -1.706$ or $t > 1.706$.

(c) $t = \dfrac{(\bar{x}_1 - \bar{x}_2) - (\mu_1 - \mu_2)}{\sqrt{\dfrac{(n_1 - 1)s_1^2 + (n_2 - 1)s_2^2}{n_1 + n_2 - 2}}\sqrt{\dfrac{1}{n_1} + \dfrac{1}{n_2}}}$

$= \dfrac{(15.1 - 12.8) - (0)}{\sqrt{\dfrac{(13 - 1)(7.55)^2 + (15 - 1)(6.35)^2}{13 + 15 - 2}}\sqrt{\dfrac{1}{13} + \dfrac{1}{15}}} = \dfrac{2.3}{\sqrt{48.021}\sqrt{0.144}} \approx 0.876$

(d) Fail to reject H_0.

(e) There is not enough evidence to reject the claim.

17. (a) $H_0: \mu_1 \geq \mu_2$; $H_a: \mu_1 < \mu_2$ (claim)

(b) d.f. $= \min\{n_1 - 1, n_2 - 1\} = 14$
$t_0 = -1.345$; Reject H_0 if $t < -1.345$.

(c) $t = \dfrac{(\bar{x}_1 - \bar{x}_2) - (\mu_1 - \mu_2)}{\sqrt{\dfrac{s_1^2}{n_1} + \dfrac{s_2^2}{n_2}}} = \dfrac{(473 - 741) - (0)}{\sqrt{\dfrac{(190)^2}{14} + \dfrac{(205)^2}{23}}} = \dfrac{-268}{\sqrt{4405.745}} = -4.038$

(d) Reject H_0.

(e) There is enough evidence to support the claim.

19. (a) $H_0: \mu_1 \leq \mu_2$; $H_a: \mu_1 > \mu_2$ (claim)

(b) d.f. $= \min\{n_1 - 1, n_2 - 1\} = 14$
$t_0 = 1.345$; Reject H_0 if $t > 1.345$.

(c) $z = \dfrac{(\bar{x}_1 - \bar{x}_2) - (\mu_1 - \mu_2)}{\sqrt{\dfrac{s_1^2}{n_1} + \dfrac{s_2^2}{n_2}}} = \dfrac{(41{,}300 - 37{,}800) - (0)}{\sqrt{\dfrac{(8600)^2}{19} + \dfrac{(5500)^2}{15}}} = \dfrac{3500}{\sqrt{5{,}909{,}298.246}} \approx 1.440$

(d) Reject H_0.

(e) There is enough evidence to support the claim.

21. (a) $H_0: \mu_1 = \mu_2$; $H_a: \mu_1 \neq \mu_2$ (claim)

(b) d.f. $= n_1 + n_2 - 2 = 21$
$t_0 \pm 2.831$; Reject H_0 if $t < -2.831$ or $t > 2.831$.

(c) $\bar{x}_1 = 340.300$, $s_1 = 22.301$, $n_1 = 10$
$\bar{x}_2 = 389.538$, $s_2 = 14.512$, $n_2 = 13$

$t = \dfrac{(\bar{x}_1 - \bar{x}_2) - (\mu_1 - \mu_2)}{\sqrt{\dfrac{(n_1 - 1)s_1^2 + (n_2 - 1)s_2^2}{n_1 + n_2 - 2}}\sqrt{\dfrac{1}{n_1} + \dfrac{1}{n_2}}}$

$= \dfrac{(340.300 - 389.538) - (0)}{\sqrt{\dfrac{(10 - 1)(22.301)^2 + (13 - 1)(14.512)^2}{10 + 13 - 2}}\sqrt{\dfrac{1}{10} + \dfrac{1}{13}}} = \dfrac{-49.238}{\sqrt{333.485}\sqrt{0.177}} \approx -6.410$

(d) Reject H_0.

(e) There is enough evidence to support the claim.

23. (a) H_0: $\mu_1 \geq \mu_2$; H_a: $\mu_1 < \mu_2$ (claim)

(b) d.f. $= n_1 + n_2 - 2 = 42$
$t_0 = -1.282 \to t < -1.282$

(c) $\bar{x}_1 = 56.684$, $s_1 = 6.961$, $n_1 = 19$
$\bar{x}_2 = 67.400$, $s_2 = 9.014$, $n_2 = 25$

$$t = \frac{(\bar{x}_1 - \bar{x}_2) - (\mu_1 - \mu_2)}{\sqrt{\frac{(n_1-1)s_1^2 + (n_2-1)s_2^2}{n_1+n_2-2}}\sqrt{\frac{1}{n_1}+\frac{1}{n_2}}}$$

$$= \frac{(56.684 - 67.400) - (0)}{\sqrt{\frac{(19-1)(6.961)^2 + (25-1)(9.014)^2}{19+25-2}}\sqrt{\frac{1}{19}+\frac{1}{25}}} = \frac{-10.716}{\sqrt{67.196}\sqrt{0.0926}} \approx -4.295$$

(d) Reject H_0.

(e) There is enough evidence to support the claim and to recommend changing to the new method.

25. $\hat{\sigma} = \sqrt{\frac{(n_1-1)s_1^2 + (n_2-1)s_2^2}{n_1+n_2-2}} = \sqrt{\frac{(15-1)(6.2)^2 + (12-1)(8.1)^2}{15+12-2}} \approx 7.099$

$(\bar{x}_1 - \bar{x}_2) \pm t_c \hat{\sigma}\sqrt{\frac{1}{n_1}+\frac{1}{n_2}} \to (410 - 400) \pm 2.060 \cdot 7.099 \sqrt{\frac{1}{15}+\frac{1}{12}}$

$\to 10 \pm 5.664 \to 4.336 < \mu_1 - \mu_2 < 15.644$

27. $(\bar{x}_1 - \bar{x}_2) \pm t_c \sqrt{\frac{s_1^2}{n_1}+\frac{s_2^2}{n_2}} \to (60 - 70) \pm 1.363 \sqrt{\frac{(3.59)^2}{15}+\frac{(2.41)^2}{12}}$

$\to -10 \pm 1.580 \to -11.58 < \mu_1 - \mu_2 < -8.42$

8.3 TESTING THE DIFFERENCE BETWEEN MEANS (DEPENDENT SAMPLES)

8.3 Try It Yourself Solutions

1a. (1) Independent   (2) Dependent

2.

Before	After	d	d^2
72	73	−1	1
81	80	1	1
76	79	−3	9
74	76	−2	4
75	76	−1	1
80	80	0	0
68	74	−6	36
75	77	−2	4
78	75	3	9
76	74	2	4
74	76	−2	4
77	78	−1	1
		$\Sigma d = -12$	$\Sigma d^2 = 74$

a. H_0: $\mu_d \geq 0$; H_a: $\mu_d < 0$ (claim)

b. $\alpha = 0.05$ and d.f. $= n - 1 = 11$

c. $t_0 \approx -1.796$; Reject H_0 if $t < -1.796$.

d. $\bar{d} = \frac{\Sigma d}{n} = \frac{-12}{12} = -1$

$s_d = \sqrt{\frac{n(\Sigma d^2) - (\Sigma d)^2}{n(n-1)}} = \sqrt{\frac{12(74) - (-12)^2}{12(11)}} \approx 2.374$

e. $t = \frac{\bar{d} - \mu_d}{\frac{s_d}{\sqrt{n}}} = \frac{-1 - 0}{\frac{2.374}{\sqrt{12}}} \approx -1.459$

f. Fail to reject H_0.

g. There is not enough evidence to support the claim.

3.

Before	After	d	d^2
101.8	99.2	2.6	6.76
98.5	98.4	0.1	0.01
98.1	98.2	−0.1	0.01
99.4	99	0.4	0.16
98.9	98.6	0.3	0.09
100.2	99.7	0.5	0.25
97.9	97.8	0.1	0.01
		$\Sigma d = 3.9$	$\Sigma d^2 = 7.29$

a. $H_0: \mu_d = 0$; $H_a: \mu_d \neq 0$ (claim)

b. $\alpha = 0.05$ and d.f. $= n - 1 = 6$

c. $t_0 = \pm 2.447$; Reject H_0 if $t < -2.447$ or $t > 2.447$.

d. $\bar{d} = \dfrac{\Sigma d}{n} = \dfrac{3.9}{7} \approx 0.557$

$s_d = \sqrt{\dfrac{n(\Sigma d^2) - (\Sigma d)^2}{n(n-1)}} = \sqrt{\dfrac{7(7.29) - (3.9)^2}{7(6)}} \approx 0.924$

e. $t = \dfrac{\bar{d} - \mu_d}{\dfrac{s_d}{\sqrt{n}}} = \dfrac{0.557 - 0}{\dfrac{0.924}{\sqrt{7}}} \approx 1.595$

f. Fail to reject H_0.

g. There is not enough evidence to conclude that the drug changes the body's temperature at the specified level of significance.

8.3 EXERCISE SOLUTIONS

1. Two samples are dependent if each member of one sample corresponds to a member of the other sample. Example: The weights of 22 people before starting an exercise program and the weights of the same 22 people six weeks after starting the exercise program. Two samples are independent if the sample selected from one population is not related to the sample selected from the second population. Example: The weights of 25 cats and the weights of 20 dogs.

3. Independent because the scores are from different students.

5. Dependent because the same adults were sampled.

7. Independent because different boats were sampled.

9. Dependent because the same tire sets were sampled.

11. $H_0: \mu_d \geq 0$; $H_a: \mu_d < 0$ (claim)

    $\alpha = 0.05$ and d.f. $= n - 1 = 9$

    $t_0 = -1.833$ (Left-tailed)

    $t = \dfrac{\bar{d} - \mu_d}{\dfrac{s_d}{\sqrt{n}}} = \dfrac{10 - 0}{\dfrac{1.5}{\sqrt{10}}} = \dfrac{10}{0.474} \approx 21.082$

    Fail to reject H_0. There is not enough evidence to support the claim.

13. $H_0: \mu_d \leq 0$ (claim); $H_a: \mu_d > 0$

$\alpha = 0.10$ and d.f. $= n - 1 = 15$

$t_0 = 1.341$ (Right-tailed)

$t = \dfrac{\bar{d} - \mu_d}{\dfrac{s_d}{\sqrt{n}}} = \dfrac{6.1 - 0}{\dfrac{0.36}{\sqrt{16}}} = \dfrac{6.1}{0.09} \approx 67.778$

Reject H_0. There is enough evidence to reject the claim.

15. $H_0: \mu_d \geq 0$ (claim); $H_a: \mu_d < 0$

$\alpha = 0.01$ and d.f. $= n - 1 = 14$

$t_0 = -2.624$ (Left-tailed)

$t = \dfrac{\bar{d} - \mu_d}{\dfrac{s_d}{\sqrt{n}}} = \dfrac{-2.3 - 0}{\dfrac{1.2}{\sqrt{15}}} = \dfrac{-2.3}{0.3098} \approx -7.423$

Reject H_0. There is enough evidence to reject the claim.

17. (a) $H_0: \mu_d \geq 0$; $H_a: \mu_d < 0$ (claim)

(b) $t_0 = -2.650$; Reject H_0 if $t < -2.650$.

(c) $\bar{d} \approx -33.714$ and $s_d \approx 42.034$

(d) $t = \dfrac{\bar{d} - \mu_d}{\dfrac{s_d}{\sqrt{n}}} = \dfrac{-33.714 - 0}{\dfrac{42.034}{\sqrt{14}}} = \dfrac{-33.714}{11.234} \approx -3.001$

(e) Reject H_0.

(f) There is enough evidence to support the claim that the second SAT scores are improved.

19. (a) $H_0: \mu_d \geq 0$; $H_a: \mu_d < 0$ (claim)

(b) $t_0 = -1.415$; Reject H_0 if $t > 1.415$.

(c) $\bar{d} \approx -1.125$ and $s_d \approx 0.871$

(d) $t = \dfrac{\bar{d} - \mu_d}{\dfrac{s_d}{\sqrt{n}}} = \dfrac{-1.125 - 0}{\dfrac{0.871}{\sqrt{8}}} = \dfrac{-1.125}{0.308} = -3.653$

(e) Reject H_0.

(f) There is enough evidence to support the fuel additive improved gas mileage.

21. (a) $H_0: \mu_d \leq 0$; $H_a: \mu > 0$ (claim)

(b) $t_0 = 1.363$; Reject H_0 if $t > 1.363$.

(c) $\bar{d} = 3.75$ and $s_d \approx 7.84$

(d) $t = \dfrac{\bar{d} - \mu_d}{\dfrac{s_d}{\sqrt{n}}} = \dfrac{3.75 - 0}{\dfrac{7.84}{\sqrt{12}}} = \dfrac{3.75}{2.26} = 1.657$

(e) Reject H_0.

(f) There is enough evidence to support the claim that the exercise program helps participants lose weight.

128 CHAPTER 8 | HYPOTHESIS TESTING WITH TWO SAMPLES

23. (a) $H_0: \mu_d \leq 0$; $H_a: \mu_d > 0$ (claim)

    (b) $t_0 = 2.764$; Reject H_0 if $t > 2.764$.

    (c) $\bar{d} \approx 1.255$ and $s_d \approx 0.441$

    (d) $t = \dfrac{\bar{d} - \mu_d}{\dfrac{s_d}{\sqrt{n}}} = \dfrac{1.255 - 0}{\dfrac{0.441}{\sqrt{11}}} = \dfrac{1.255}{0.133} \approx 9.438$

    (e) Reject H_0.

    (f) There is enough evidence to support the claim that soft tissue therapy and spinal manipulation help reduce the length of time patients suffer from headaches.

25. (a) $H_0: \mu_d \leq 0$; $H_a: \mu_d > 0$ (claim)

    (b) $t_0 = 1.895$; Reject H_0 if $t > 1.895$.

    (c) $\bar{d} = 14.75$ and $s_d \approx 6.86$

    (d) $t = \dfrac{\bar{d} - \mu_d}{\dfrac{s_d}{\sqrt{n}}} = \dfrac{14.75 - 0}{\dfrac{6.86}{\sqrt{8}}} = \dfrac{14.75}{2.425} \approx 6.082$

    (e) Reject H_0.

    (f) There is enough evidence to support the claim that the new drug reduces systolic blood pressure.

27. (a) $H_0: \mu_d = 0$; $H_a: \mu_d \neq 0$ (claim)

    (b) $t_0 = \pm 2.365$; Reject H_0 if $t < -2.365$ or $t > 2.365$.

    (c) $\bar{d} = -1$ and $s_d = 1.31$

    (d) $t = \dfrac{\bar{d} - \mu_d}{\dfrac{s_d}{\sqrt{n}}} = \dfrac{-1 - 0}{\dfrac{1.31}{\sqrt{8}}} = \dfrac{-1}{0.463} = -2.159$

    (e) Fail to reject H_0.

    (f) There is not enough evidence to support the claim that the product ratings have changed.

29. $\bar{d} \approx -1.525$ and $s_d \approx 0.542$

$$\bar{d} - t_{\alpha/2}\dfrac{s_d}{\sqrt{n}} < \mu_d < \bar{d} - t_{\alpha/2}\dfrac{s_d}{\sqrt{n}}$$

$$-1.525 - 1.753\left(\dfrac{0.542}{\sqrt{16}}\right) < \mu_d < -1.525 + 1.753\left(\dfrac{0.542}{\sqrt{16}}\right)$$

$$-1.525 - 0.238 < \mu_d < -1.525 + 0.238$$

$$-1.763 < \mu_d < -1.287$$

8.4 TESTING THE DIFFERENCE BETWEEN PROPORTIONS

8.4 Try It Yourself Solutions

1a. $H_0: p_1 = p_2; H_a: p_1 \neq p_2$ (claim)   **b.** $\alpha = 0.05$

c. $z_0 = \pm 1.96$; Reject H_0 if $z < -1.96$ or $z > 1.96$.

d. $\bar{p} = \dfrac{x_1 + x_2}{n_1 + n_2} = \dfrac{1666 + 1435}{6771 + 6767} = 0.229$

$\bar{q} = 0.771$

e. $n_1 \bar{p} \approx 1550.559 > 5$, $n_1 \bar{q} \approx 5220.411 > 5$, $n_2 \bar{p} \approx 1549.643 > 5$, and $n_2 \bar{q} \approx 5217.357 > 5$.

f. $z = \dfrac{(\hat{p}_1 - \hat{p}_2) - (p_1 - p_2)}{\sqrt{\bar{p}\bar{q}\left(\dfrac{1}{n_1} + \dfrac{1}{n_2}\right)}} = \dfrac{(0.246 - 0.212) - (0)}{\sqrt{0.229 \cdot 0.771\left(\dfrac{1}{6771} + \dfrac{1}{6767}\right)}} = \dfrac{0.034}{\sqrt{0.000052167}} \approx 4.707$

g. Reject H_0.

h. There is enough evidence to support the claim.

2a. $H_0: p_1 \leq p_2; H_a: p_1 > p_2$ (claim)   **b.** $\alpha = 0.05$

c. $z_0 = 1.645$; Reject H_0 if $z > 1.645$.

d. $\bar{p} = \dfrac{x_1 + x_2}{n_1 + n_2} = \dfrac{1144 + 420}{6771 + 6767} = 0.116$

$\bar{q} = 0.884$

e. $n_1 \bar{p} \approx 785.436 > 5$, $n_1 \bar{q} \approx 5985.564 > 5$, $n_2 \bar{p} \approx 778.205 > 5$, and $n_2 \bar{q} \approx 5982.028 > 5$.

f. $z = \dfrac{(\hat{p}_1 - \hat{p}_2) - (p_1 - p_2)}{\sqrt{\bar{p}\bar{q}\left(\dfrac{1}{n_1} + \dfrac{1}{n_2}\right)}} = \dfrac{(0.169 - 0.062) - (0)}{\sqrt{0.116 \cdot 0.884\left(\dfrac{1}{6771} + \dfrac{1}{6767}\right)}} = \dfrac{0.107}{\sqrt{0.000030298}} \approx 19.44$

g. Reject H_0.

h. There is enough evidence to support the claim.

8.4 EXERCISE SOLUTIONS

1. State the hypotheses and identify the claim. Specify the level of significance. Find the critical value(s) and rejection region(s). Find $\bar{p}$ and $\bar{q}$. Find the standardized test statistic. Make a decision and interpret in the context of the claim.

3. $H_0: p_1 = p_2; H_a: p_1 \neq p_2$ (claim)

    $z_0 = \pm 2.575$ (Two-tailed test)

    $\bar{p} = \dfrac{x_1 + x_2}{n_1 + n_2} = \dfrac{35 + 36}{70 + 60} = 0.546$

    $\bar{q} = 0.454$

    $z = \dfrac{(\hat{p}_1 - \hat{p}_2) - (p_1 - p_2)}{\sqrt{\bar{p}\bar{q}\left(\dfrac{1}{n_1} + \dfrac{1}{n_2}\right)}} = \dfrac{(0.500 - 0.600) - (0)}{\sqrt{0.546 \cdot 0.454\left(\dfrac{1}{70} + \dfrac{1}{60}\right)}} = \dfrac{-0.100}{\sqrt{0.00767}} \approx -1.142$

    Fail to reject H_0. There is not enough evidence to support the claim.

5. $H_0: p_1 \leq p_2$ (claim); $H_a: p_1 > p_2$

$z_0 = 1.282$ (Right-tailed test)

$\bar{p} = \dfrac{x_1 + x_2}{n_1 + n_2} = \dfrac{344 + 304}{860 + 800} = 0.390$

$\bar{q} = 0.610$

$z = \dfrac{(\hat{p}_1 - \hat{p}_2) - (p_1 - p_2)}{\sqrt{\bar{p}\,\bar{q}\left(\dfrac{1}{n_1} + \dfrac{1}{n_2}\right)}} = \dfrac{(0.400 - 0.380) - (0)}{\sqrt{0.390 \cdot 0.610\left(\dfrac{1}{860} + \dfrac{1}{800}\right)}} = \dfrac{0.020}{\sqrt{0.0000574003}} \approx 0.835$

Fail to reject H_0. There is not enough evidence to reject the claim.

7. (a) $H_0: p_1 = p_2$ (claim); $H_a: p_1 \neq p_2$

(b) $z_0 = \pm 1.96$; Reject H_0 if $z < -1.96$ or $z > 1.96$.

$\bar{p} = \dfrac{x_1 + x_2}{n_1 + n_2} = \dfrac{520 + 865}{1539 + 2055} = 0.385$

$\bar{q} = 0.615$

(c) $z = \dfrac{(\hat{p}_1 - \hat{p}_2) - (p_1 - p_2)}{\sqrt{\bar{p}\,\bar{q}\left(\dfrac{1}{n_1} + \dfrac{1}{n_2}\right)}} = \dfrac{(0.338 - 0.421) - (0)}{\sqrt{0.385 \cdot 0.615\left(\dfrac{1}{1539} + \dfrac{1}{2055}\right)}}$

$= \dfrac{-0.083}{\sqrt{0.0000269069}} \approx -5.060$

(d) Reject H_0. There is sufficient evidence at the 5% level to reject the claim that the proportion of adults using alternative medicines has not changed since 1990.

9. (a) $H_0: p_1 = p_2$ (claim); $H_a: p_1 \neq p_2$

(b) $z_0 = \pm 1.645$; Reject H_0 if $z < -1.645$ or $z > 1.645$.

$\bar{p} = \dfrac{x_1 + x_2}{n_1 + n_2} = \dfrac{2201 + 2348}{5240 + 6180} = 0.398$

$\bar{q} = 0.602$

(c) $z = \dfrac{(\hat{p}_1 - \hat{p}_2) - (p_1 - p_2)}{\sqrt{\bar{p}\,\bar{q}\left(\dfrac{1}{n_1} + \dfrac{1}{n_2}\right)}} = \dfrac{(0.420 - 0.380) - 0}{\sqrt{(0.398)(0.602)\left(\dfrac{1}{5240} + \dfrac{1}{6180}\right)}}$

$= \dfrac{0.04}{\sqrt{0.00008449}} = 4.351$

(d) Reject H_0.

(e) There is sufficient evidence at the 10% level to reject the claim that the proportions of male and female senior citizens that eat the daily recommended number of servings of vegetables is not different.

11. (a) $H_0: p_1 \geq p_2$; $H_a: p_1 < p_2$ (claim)

(b) $z_0 = -2.33$; Reject H_0 if $z < -2.33$.

$\bar{p} = \dfrac{x_1 + x_2}{n_1 + n_2} = \dfrac{488 + 532}{2000 + 2000} = 0.255$

$\bar{q} = 0.745$

(c) $z = \dfrac{(\hat{p}_1 - \hat{p}_2) - (p_1 - p_2)}{\sqrt{\bar{p}\bar{q}\left(\dfrac{1}{n_1} + \dfrac{1}{n_2}\right)}} = \dfrac{(0.244 - 0.266) - (0)}{\sqrt{0.255 \cdot 0.745\left(\dfrac{1}{2000} + \dfrac{1}{2000}\right)}} = \dfrac{-0.022}{\sqrt{0.00018997}} \approx -1.596$

(d) Fail to reject H_0.

(e) There is not sufficient evidence at the 1% level to support the claim that the proportion of adults who are smokers is lower in Alabama than in Missouri.

13. (a) $H_0: p_1 \geq p_2$; $H_a: p_1 < p_2$ (claim)

(b) $z_0 = -2.33$; Reject H_0 if $z < -2.33$.

$\bar{p} = \dfrac{x_1 + x_2}{n_1 + n_2} = \dfrac{1589 + 1683}{5800 + 6600} = 0.264$

$\bar{q} = 0.736$

(c) $z = \dfrac{(\hat{p}_1 - \hat{p}_2) - (p_1 - p_2)}{\sqrt{\bar{p}\bar{q}\left(\dfrac{1}{n_1} + \dfrac{1}{n_2}\right)}} = \dfrac{(0.274 - 0.255) - (0)}{\sqrt{0.264 \cdot 0.736\left(\dfrac{1}{5800} + \dfrac{1}{6600}\right)}} = \dfrac{0.019}{\sqrt{0.00006294}} \approx 2.395$

(d) Fail to reject H_0.

(e) There is sufficient evidence at the 1% level to support the claim that the proportion of males who said they had smoked in the last 30 days is less than the proportion of females.

15. (a) $H_0: p_1 = p_2$ (claim); $H_a: p_1 \neq p_2$

(b) $z_0 = \pm 1.96$; Reject H_0 if $z < -1.96$ or $z > 1.96$.

$\bar{p} = \dfrac{x_1 + x_2}{n_1 + n_2} = \dfrac{2496 + 1380}{9600 + 11{,}500} = 0.184$

$\bar{q} = 0.816$

(c) $z = \dfrac{(\hat{p}_1 - \hat{p}_2) - (p_1 - p_2)}{\sqrt{\bar{p}\bar{q}\left(\dfrac{1}{n_1} + \dfrac{1}{n_2}\right)}} = \dfrac{(0.260 - 0.120) - 0}{\sqrt{(0.184)(0.816)\left(\dfrac{1}{9600} + \dfrac{1}{11{,}500}\right)}} = \dfrac{0.14}{\sqrt{0.00002869}} = 26.135$

(d) Reject H_0.

(e) There is insufficient evidence at the 5% level to reject the claim that the proportions of adults that have chronic backs are the same for both groups.

17. $H_0: p_1 \geq p_2$; $H_a: p_1 < p_2$ (claim)

$z_0 = -2.33$

$\bar{p} = \dfrac{x_1 + x_2}{n_1 + n_2} = \dfrac{28 + 35}{700 + 500} = 0.053$

$\bar{q} = 0.947$

$z = \dfrac{(\hat{p}_1 - \hat{p}_2) - (p_1 - p_2)}{\sqrt{\bar{p}\bar{q}\left(\dfrac{1}{n_1} + \dfrac{1}{n_2}\right)}} = \dfrac{(0.04 - 0.07) - (0)}{\sqrt{0.053 \cdot 0.947\left(\dfrac{1}{700} + \dfrac{1}{500}\right)}} = \dfrac{-0.03}{\sqrt{0.0001721}} \approx -2.287$

Fail to reject H_0. There is insufficient evidence at the 1% level to support the claim.

19. $H_0: p_1 = p_2$ (claim); $H_a: p_1 \neq p_2$

$z_0 = \pm 1.96$

$\bar{p} = \dfrac{x_1 + x_2}{n_1 + n_2} = \dfrac{189 + 185}{700 + 500} = 0.312$

$\bar{q} = 0.688$

$z = \dfrac{(\hat{p}_1 - \hat{p}_2) - (p_1 - p_2)}{\sqrt{\bar{p}\bar{q}\left(\dfrac{1}{n_1} + \dfrac{1}{n_2}\right)}} = \dfrac{(0.27 - 0.37) - (0)}{\sqrt{0.312 \cdot 0.688\left(\dfrac{1}{700} + \dfrac{1}{500}\right)}} = \dfrac{-0.10}{\sqrt{0.0007359}} \approx -3.686$

Reject H_0. There is sufficient evidence at the 5% level to reject the advocate's claim.

21. $H_0: p_1 \leq p_2$; $H_a: p_1 > p_2$ (claim)

$z_0 = 1.645$

$\bar{p} = \dfrac{x_1 + x_2}{n_1 + n_2} = \dfrac{348 + 275}{580 + 500} = 0.577$

$\bar{q} = 0.423$

$z = \dfrac{(\hat{p}_1 - \hat{p}_2) - (p_1 - p_2)}{\sqrt{\bar{p}\bar{q}\left(\dfrac{1}{n_1} + \dfrac{1}{n_2}\right)}} = \dfrac{(0.60 - 0.55) - (0)}{\sqrt{(0.577)(0.423)\left(\dfrac{1}{580} + \dfrac{1}{500}\right)}} = \dfrac{0.05}{\sqrt{0.0009089}} \approx 1.658$

Reject H_0. There is sufficient evidence at the 5% level to support the claim.

23. $H_0: p_1 = p_2$ (claim); $H_a: p_1 \neq p_2$

$z_0 = \pm 2.576$

$\bar{p} = \dfrac{x_1 + x_2}{n_1 + n_2} = \dfrac{348 + 240}{580 + 500} = 0.544$

$\bar{q} = 0.456$

$z = \dfrac{(\hat{p}_1 - \hat{p}_2) - (p_1 - p_2)}{\sqrt{\bar{p}\bar{q}\left(\dfrac{1}{n_1} + \dfrac{1}{n_2}\right)}} = \dfrac{(0.60 - 0.48) - (0)}{\sqrt{(0.544)(0.456)\left(\dfrac{1}{580} + \dfrac{1}{500}\right)}} = \dfrac{0.12}{\sqrt{0.0009238}} \approx 3.948$

Reject H_0. There is sufficient evidence at the 1% level to reject the claim.

25. $(\hat{p}_1 - \hat{p}_2) \pm z_c \sqrt{\dfrac{\hat{p}_1\hat{q}_1}{n_1} + \dfrac{\hat{p}_2\hat{q}_2}{n_2}} \rightarrow (0.117 - 0.085) \pm 1.96 \sqrt{\dfrac{0.117 \cdot 0.883}{977,000} + \dfrac{0.085 \cdot 0.915}{1,085,000}}$

$\rightarrow 0.032 \pm 1.96 \sqrt{0.000000177425}$

$\rightarrow 0.0312 < p_1 - p_2 < 0.0328$

CHAPTER 8 REVIEW EXERCISE SOLUTIONS

1. $H_0: \mu_1 \geq \mu_2$ (claim); $H_1: \mu_1 < \mu_2$

$z_0 = -1.645$

$z = \dfrac{(\bar{x}_1 - \bar{x}_2) - (\mu_1 - \mu_2)}{\sqrt{\dfrac{s_1^2}{n_1} + \dfrac{s_2^2}{n_2}}} = \dfrac{(1.28 - 1.36) - (0)}{\sqrt{\dfrac{(0.28)^2}{76} + \dfrac{(0.23)^2}{65}}} = \dfrac{-0.08}{\sqrt{0.00185}} \approx -.1862$

Reject H_0. There is enough evidence to reject the claim.

3. $H_0: \mu_1 \geq \mu_2$; $H_1: \mu_1 < \mu_2$ (claim)

   $z_0 = -1.282$

   $$z = \frac{(\bar{x}_1 - \bar{x}_2) - (\mu_1 - \mu_2)}{\sqrt{\frac{s_1^2}{n_1} + \frac{s_2^2}{n_2}}} = \frac{(0.28 - 0.33) - (0)}{\sqrt{\frac{(0.11)^2}{41} + \frac{(0.10)^2}{34}}} = \frac{-0.50}{\sqrt{0.00058924}} \approx -2.060$$

   Reject H_0. There is enough evidence to support the claim.

5. (a) $H_0: \mu_1 \leq \mu_2$; $H_1: \mu_1 > \mu_2$ (claim)

   (b) $z_0 = 1.645$; Reject H_0 if $z > 1.645$.

   (c) $$z = \frac{(\bar{x}_1 - \bar{x}_2) - (\mu_1 - \mu_2)}{\sqrt{\frac{s_1^2}{n_1} + \frac{s_2^2}{n_2}}} = \frac{(430 - 410) - (0)}{\sqrt{\frac{(43)^2}{36} + \frac{(57)^2}{41}}} = \frac{20}{\sqrt{130.605}} \approx 1.750$$

   (d) Reject H_0.

   (e) There is enough evidence to support the claim.

7. $H_0: \mu_1 = \mu_2$ (claim); $H_a: \mu_1 \neq \mu_2$

   d.f. $= n_1 + n_2 - 2 = 31$

   $t_0 = \pm 1.96$

   $$t = \frac{(\bar{x}_1 - \bar{x}_2) - (\mu_1 - \mu_2)}{\sqrt{\frac{(n_1 - 1)s_1^2 + (n_2 - 1)s_2^2}{n_1 + n_2 - 2}} \sqrt{\frac{1}{n_1} + \frac{1}{n_2}}} = \frac{(250 - 240) - (0)}{\sqrt{\frac{(21 - 1)(26)^2 + (12 - 1)(22)^2}{21 + 12 - 2}} \sqrt{\frac{1}{21} + \frac{1}{12}}}$$

   $$= \frac{10}{\sqrt{607.871} \sqrt{0.13095}} \approx 1.121$$

   Fail to reject H_0. There is not enough evidence to reject the claim.

9. $H_0: \mu_1 \leq \mu_2$ (claim); $H_a: \mu_1 > \mu_2$

   d.f. $= \min\{n_1 - 1, n_2 - 1\} = 24$

   $t_0 = 1.711$

   $$t = \frac{(\bar{x}_1 - \bar{x}_2) - (\mu_1 - \mu_2)}{\sqrt{\frac{s_1^2}{n_1} + \frac{s_2^2}{n_2}}} = \frac{(183.5 - 184.7) - (0)}{\sqrt{\frac{(1.3)^2}{25} + \frac{(3.9)^2}{25}}} = \frac{-1.2}{\sqrt{0.676}} \approx -1.460$$

   Fail to reject H_0. There is not enough evidence to reject the claim.

11. $H_0: \mu_1 = \mu_2$; $H_a: \mu_1 \neq \mu_2$ (claim)

    d.f. $= n_1 + n_2 - 2 = 10$

    $t_0 = \pm 3.169$

    $$t = \frac{(\bar{x}_1 - \bar{x}_2) - (\mu_1 - \mu_2)}{\sqrt{\frac{(n_1 - 1)s_1^2 + (n_2 - 1)s_2^2}{n_1 + n_2 - 2}} \cdot \sqrt{\frac{1}{n_1} + \frac{1}{n_2}}} = \frac{(61 - 55) - (0)}{\sqrt{\frac{(5 - 1)3.3^2 + (7 - 1)1.2^2}{5 + 7 - 2}} \cdot \sqrt{\frac{1}{5} + \frac{1}{7}}}$$

    $$= \frac{6}{\sqrt{5.22} \sqrt{0.343}} \approx 4.484$$

    Reject H_0. There is enough evidence to support the claim.

13. (a) $H_0: \mu_1 \le \mu_2$; $H_a: \mu_1 > \mu_2$ (claim)

    (b) d.f. $= n_1 + n_2 - 2 = 42$
    $t_0 = 1.645$; Reject H_0 if $t > 1.645$.

    (c) $\bar{x}_1 = 51.476$, $s_1 = 11.007$, $n_1 = 21$
    $\bar{x}_2 = 41.522$, $s_2 = 17.149$, $n_2 = 23$

    $$t = \frac{(\bar{x}_1 - \bar{x}_2) - (\mu_1 - \mu_2)}{\sqrt{\frac{(n_1-1)s_1^2 + (n_2-1)s_2^2}{n_1 + n_2 - 2}}\sqrt{\frac{1}{n_1}+\frac{1}{n_2}}} = \frac{(51.476 - 41.522) - (0)}{\sqrt{\frac{(21-1)(11.007)^2 + (23-1)(17.149)^2}{21+23-2}}\sqrt{\frac{1}{21}+\frac{1}{23}}}$$

    $$= \frac{9.954}{\sqrt{211.7386}\sqrt{0.0911}} \approx 2.266$$

    (d) Reject H_0.

    (e) There is sufficient evidence at the 5% level to support the claim that the third graders taught with the direct reading activities scored higher than those taught without the activities.

15. Independent since the two samples of laboratory mice are different groups.

17. $H_0: \mu_d = 0$ (claim); $H_a: \mu_d \ne 0$

    $\alpha = 0.05$ and d.f. $= n - 1 = 99$

    $t_0 = \pm 1.96$ (Two-tailed test)

    $$t = \frac{\bar{d} - \mu_d}{\frac{s_d}{\sqrt{n}}} = \frac{10 - 0}{\frac{12.4}{\sqrt{100}}} = \frac{10}{1.24} \approx 8.065$$

    Reject H_0. There is enough evidence to reject the claim.

19. $H_0: \mu_d \le 6$ (claim); $H_a: \mu_d > 6$

    $\alpha = 0.10$ and d.f. $= n - 1 = 32$

    $t_0 = 1.282$ (Right-tailed test)

    $$t = \frac{\bar{d} - \mu_d}{\frac{s_d}{\sqrt{n}}} = \frac{10.3 - 6}{\frac{1.24}{\sqrt{33}}} = \frac{4.3}{0.21586} \approx 19.921$$

    Reject H_0. There is enough evidence to reject the claim.

21. (a) $H_0: \mu_d \le 0$; $H_a: \mu_d > 0$ (claim)

    (b) $t_0 = 1.383$; Reject H_0 if $t > 1.383$.

    (c) $\bar{d} = 5$ and $s_d \approx 8.743$

    (d) $t = \dfrac{\bar{d} - \mu_d}{\frac{s_d}{\sqrt{n}}} = \dfrac{5 - 0}{\frac{8.743}{\sqrt{10}}} = \dfrac{5}{2.765} \approx 1.808$

    (e) Reject H_0.

    (f) There is enough evidence to support the claim.

23. $H_0: p_1 = p_2$; $H_a: p_1 \neq p_2$ (claim)

$z_0 = \pm 1.96$ (Two-tailed test)

$\bar{p} = \dfrac{x_1 + x_2}{n_1 + n_2} = \dfrac{375 + 365}{720 + 660} = 0.536$

$\bar{q} = 0.464$

$z = \dfrac{(\hat{p}_1 - \hat{p}_2) - (p_1 - p_2)}{\sqrt{\bar{p}\bar{q}\left(\dfrac{1}{n_1} + \dfrac{1}{n_2}\right)}} = \dfrac{(0.521 - 0.553) - (0)}{\sqrt{0.536 \cdot 0.464\left(\dfrac{1}{720} + \dfrac{1}{660}\right)}} = \dfrac{-0.032}{\sqrt{0.000722246}} \approx -1.198$

Fail to reject H_0. There is not enough evidence to reject the claim.

25. $H_0: p_1 \leq p_2$; $H_a: p_1 > p_2$ (claim)

$z_0 = 1.282$ (Right-tailed test)

$\bar{p} = \dfrac{x_1 + x_2}{n_1 + n_2} = \dfrac{227 + 198}{556 + 420} = 0.435$

$\bar{q} = 0.565$

$z = \dfrac{(\hat{p}_1 - \hat{p}_2) - (p_1 - p_2)}{\sqrt{\bar{p}\bar{q}\left(\dfrac{1}{n_1} + \dfrac{1}{n_2}\right)}} = \dfrac{(0.408 - 0.471) - (0)}{\sqrt{0.435 \cdot 0.565\left(\dfrac{1}{556} + \dfrac{1}{420}\right)}} = \dfrac{-0.063}{\sqrt{0.001027}} \approx -1.970$

Fail to reject H_0. There is not enough evidence to support the claim.

27. (a) $H_0: p_1 = p_2$ (claim); $H_a: p_1 \neq p_2$

(b) $z_0 = \pm 1.645$; Reject H_0 if $z < -1.645$ or $z > 1.645$.

$\bar{p} = \dfrac{x_1 + x_2}{n_1 + n_2} = \dfrac{26 + 46}{200 + 300} = 0.144$

$\bar{q} = 0.856$

(c) $z = \dfrac{(\hat{p}_1 - \hat{p}_2) - (p_1 - p_2)}{\sqrt{\bar{p}\bar{q}\left(\dfrac{1}{n_1} + \dfrac{1}{n_2}\right)}} = \dfrac{(0.130 - 0.153) - (0)}{\sqrt{0.144 \cdot 0.856\left(\dfrac{1}{200} + \dfrac{1}{300}\right)}} = \dfrac{-0.023}{\sqrt{0.001027}} \approx -0.718$

(d) Fail to reject H_0.

(e) There is not enough evidence to reject the claim.

CHAPTER 8 QUIZ SOLUTIONS

1. (a) $H_0: \mu_1 \leq \mu_2$; $H_a: \mu_1 > \mu_2$ (claim)

(b) n_1 and $n_2 > 30$ and the samples are independent $\rightarrow$ Right tailed z-test

(c) $z_0 = 1.645$; Reject H_0 if $z > 1.645$.

(d) $z = \dfrac{(\bar{x}_1 - \bar{x}_2) - (\mu_1 - \mu_2)}{\sqrt{\dfrac{s_1^2}{n_1} + \dfrac{s_2^2}{n_2}}} = \dfrac{(300.4 - 290.6) - (0)}{\sqrt{\dfrac{(1.6)^2}{49} + \dfrac{(1.5)^2}{50}}} = \dfrac{9.8}{\sqrt{0.0972}} \approx 31.426$

(e) Reject H_0.

(f) There is sufficient evidence at the 5% level to support the claim that the mean score on the science assessment for male high school students was higher than for the female high school students.

136 CHAPTER 8 | HYPOTHESIS TESTING WITH TWO SAMPLES

2. (a) $H_0: \mu_1 = \mu_2$ (claim); $H_a: \mu_1 \neq \mu_2$

(b) n_1 and $n_2 < 30$, the samples are independent, and the populations are normally distributed.
→ Two-tailed t-test (assume vars are equal)

(c) d.f. $= n_1 + n_2 - 2 = 26$
$t_0 = \pm 2.779$; Reject H_0 if $t < -2.779$ or $t > 2.779$.

(d) $t = \dfrac{(\bar{x}_1 - \bar{x}_2) - (\mu_1 - \mu_2)}{\sqrt{\dfrac{(n_1-1)s_1^2 + (n_2-1)s_2^2}{n_1+n_2-2}}\sqrt{\dfrac{1}{n_1}+\dfrac{1}{n_2}}} = \dfrac{(230.9 - 227.9) - (0)}{\sqrt{\dfrac{(13-1)(1.3)^2 + (15-1)(1.1)^2}{13+15-2}}\sqrt{\dfrac{1}{13}+\dfrac{1}{15}}}$

$= \dfrac{3.0}{\sqrt{1.43154}\sqrt{0.14359}} \approx 6.617$

(e) Reject H_0.

(f) There is sufficient evidence at the 1% level to reject the teacher's suggestion that the mean scores on the science assessment test are the same for nine-year old boys and girls.

3. (a) $H_0: p_1 \leq p_2$; $H_a: p_1 > p_2$ (claim)

(b) Testing 2 proportions, $n_1 \bar{p}$, $n_1 \bar{q}$, $n_2 \bar{p}$, and $n_2 \bar{q} \geq 5$, and the samples are independent → Right-tailed z-test

(c) $z_0 = 1.282$; Reject H_0 if $z > 1.282$.

(d) $\bar{p} = \dfrac{x_1 + x_2}{n_1 + n_2} = \dfrac{64{,}800 + 8560}{1{,}296{,}000 + 856{,}000} = 0.034$

$\bar{q} = 0.966$

$z = \dfrac{(\hat{p}_1 - \hat{p}_2) - (p_1 - p_2)}{\sqrt{\bar{p}\bar{q}\left(\dfrac{1}{n_1}+\dfrac{1}{n_2}\right)}} = \dfrac{(0.05 - 0.01) - (0)}{\sqrt{0.034 \cdot 0.966\left(\dfrac{1}{1{,}296{,}000}+\dfrac{1}{856{,}000}\right)}}$

$= \dfrac{0.04}{\sqrt{0.0000000637118}} \approx 158.471$

(e) Reject H_0.

(f) There is sufficient evidence at the 10% level to support the claim that the proportion of accidents involving alcohol is higher for drivers in the 21 to 24 age group than for drivers aged 65 and older.

4. (a) $H_0: \mu_d \geq 0$; $H_a: \mu_d < 0$ (claim)

(b) Dependent samples and both populations are normally distributed. → Dependent t-test

(c) $t_0 = -1.796$; Reject H_0 if $t < -1.796$.

(d) $t = \dfrac{\bar{d} - \mu_d}{\dfrac{s_d}{\sqrt{n}}} = \dfrac{0 - 68.5}{\dfrac{26.318}{\sqrt{12}}} = \dfrac{-68.5}{7.597} \approx 9.016$

(e) Reject H_0.

(f) There is sufficient evidence at the 5% level to conclude that the students' SAT scores improved on the second test.

Correlation and Regression

CHAPTER 9

9.1 CORRELATION

9.1 Try It Yourself Solutions

1ab.

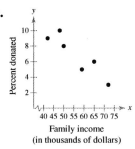

c. Yes, it appears that there is a negative linear correlation. As family income increases, the percent of income donated to charity decreases.

2ab.

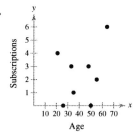

c. No, it appears that there is no correlation between age and subscriptions.

3ab.

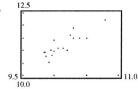

c. Yes, there appears to be a positive linear relationship between men's winning time and women's winning time.

4a. $n = 6$

b.

x	y	xy	x^2	y^2
42	9	378	1764	81
48	10	480	2304	100
50	8	400	2500	64
59	5	295	3481	25
65	6	390	4225	36
72	3	216	5184	9
$\Sigma x = 336$	$\Sigma y = 41$	$\Sigma xy = 2159$	$\Sigma x^2 = 19458$	$\Sigma y^2 = 315$

c. $r = \dfrac{n\Sigma xy - (\Sigma x)(\Sigma y)}{\sqrt{n\Sigma x^2 - (\Sigma x)^2}\sqrt{n\Sigma y^2 - (\Sigma y)^2}} = \dfrac{6(2159) - (336)(41)}{\sqrt{6(19,458) - (336)^2}\sqrt{6(315) - (41)^2}}$

$= \dfrac{-822}{\sqrt{3852}\sqrt{209}} \approx -0.916$

d. Since r is close to -1, there appears to be a strong negative linear correlation between income level and donating percent.

5a. Enter the data.

b. $r \approx 0.849$

c. Since r is close to 1, there appears to be a strong positive linear correlation between men's winning time and women's winning time.

6a. $n = 6$

b. $\alpha = 0.01$

c. $cu = 0.917$

d. Since $|r| \approx 0.916 < 0.917$, the correlation is not significant.

e. There is not enough evidence to conclude that there is a significant correlation between income level and the donating percent.

7a. $H_0: \rho = 0; H_a: \rho \neq 0$

b. $\alpha = 0.01$

c. d.f. $= n - 2 = 16$

d. ± 2.921; Reject H_0 if $t < -2.947$ or $t > 2.947$.

e. $t = \dfrac{r}{\sqrt{\dfrac{1-r^2}{n-2}}} = \dfrac{0.849}{\sqrt{\dfrac{1-(0.849)^2}{18-2}}} = \dfrac{0.849}{\sqrt{0.01777}} \approx 6.427$

f. Reject H_0.

g. There is enough evidence in the sample to conclude that a significant correlation exists.

9.1 EXERCISE SOLUTIONS

1. $r = -0.845$ represents a stronger correlation since it is closer to -1 than $r = 0.731$ is to $+1$.

3. State the null and alternative hypotheses. Specify the level of significance and determine the degrees of freedom. Identify the rejection regions and calculate the standardized test statistic. Make a decision and interpret in the context of the original claim.

5. Positive linear correlation

7. No linear correlation (but there is a nonlinear correlation between the variables)

9. (c), You would expect a positive linear correlation between age and income.

11. (b), You would expect a negative linear correlation between age and balance on student loans.

13. Explanatory variable: Amount of water consumed.
Response variable: Weight loss.

15. (a)

(b)

x	y	xy	x²	y²
16	109	1744	256	11,881
25	122	3050	625	14,884
39	143	5577	1521	20,449
45	132	5940	2025	17,424
49	199	9751	2401	39,601
64	185	11,840	4096	34,225
70	199	13,930	4900	39,601
29	130	3770	841	16,900
57	175	9975	3249	30,625
20	118	2360	400	13,924
$\Sigma x = 414$	$\Sigma y = 1512$	$\Sigma xy = 67{,}937$	$\Sigma x^2 = 20{,}314$	$\Sigma y^2 = 239{,}514$

$$r = \frac{n\Sigma xy - (\Sigma x)(\Sigma y)}{\sqrt{n\Sigma x^2 - (\Sigma x)^2}\sqrt{n\Sigma y^2 - (\Sigma y)^2}} = \frac{10(67{,}937) - (414)(1512)}{\sqrt{10(20{,}314) - (414)^2}\sqrt{10(239{,}514) - (1512)^2}}$$

$$= \frac{53{,}402}{\sqrt{31{,}744}\sqrt{108{,}996}} \approx 0.908$$

(c) Strong positive linear correlation

17. (a)

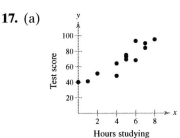

(b)

x	y	xy	x²	y²
0	40	0	0	1600
1	41	41	1	1681
2	51	102	4	2601
4	48	192	16	2304
4	64	256	16	4096
5	69	345	25	4761
5	73	365	25	5329
5	75	375	25	5625
6	68	408	36	4624
6	93	558	36	8649
7	84	588	49	7056
7	90	630	49	8100
8	95	760	64	9025
$\Sigma x = 60$	$\Sigma y = 891$	$\Sigma xy = 4620$	$\Sigma x^2 = 346$	$\Sigma y^2 = 65{,}451$

$$r = \frac{n\Sigma xy - (\Sigma x)(\Sigma y)}{\sqrt{n\Sigma x^2 - (\Sigma x)^2}\sqrt{n\Sigma y^2 - (\Sigma y)^2}} = \frac{13(4620) - (60)(891)}{\sqrt{13(346) - (60)^2}\sqrt{13(65{,}451) - (891)^2}}$$

$$= \frac{6600}{\sqrt{898}\sqrt{56{,}982}} \approx 0.923$$

(c) Strong positive linear correlation

19. (a)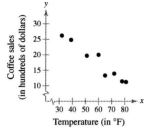

(b)

x	y	xy	x^2	y^2
32	26.2	838.4	1024	686.44
39	24.8	967.2	1521	615.04
51	19.7	1004.7	2601	388.09
60	20.0	1200.0	3600	400.00
65	13.3	864.5	4225	176.89
72	13.9	1000.8	5184	193.21
78	11.4	889.2	6084	129.96
81	11.2	907.2	6561	125.44
$\Sigma x = 478$	$\Sigma y = 140.5$	$\Sigma xy = 7672$	$\Sigma x^2 = 30{,}800$	$\Sigma y^2 = 2715.07$

$$r = \frac{n\Sigma xy - (\Sigma x)(\Sigma y)}{\sqrt{n\Sigma x^2 - (\Sigma x)^2}\sqrt{n\Sigma y^2 - (\Sigma y)^2}} = \frac{8(7672) - (478)(140.5)}{\sqrt{8(30800) - (478)^2}\sqrt{8(2715.07) - (140.5)^2}}$$

$$= \frac{-5783}{\sqrt{17{,}916}\sqrt{1980.31}} \approx -0.971$$

(c) Strong negative linear correlation

21. (a)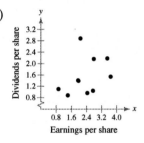

(b)

x	y	xy	x^2	y^2
2.78	2.16	6.005	7.728	4.666
1.41	0.88	1.241	1.988	0.774
2.74	1.04	2.850	7.508	1.082
0.92	1.1	1.012	0.846	1.210
2.44	0.96	2.342	5.954	0.922
3.5	2.18	7.630	12.250	4.752
3.68	1.54	5.667	13.542	2.372
1.97	1.39	2.738	3.881	1.932
1.95	1.41	2.750	3.803	1.988
2.07	2.88	5.962	4.285	8.294
$\Sigma x = 23.460$	$\Sigma y = 15.540$	$\Sigma xy = 38.196$	$\Sigma x^2 = 61.785$	$\Sigma y^2 = 27.992$

$$r = \frac{n\Sigma xy - (\Sigma x)(\Sigma y)}{\sqrt{n\Sigma x^2 - (\Sigma x)^2}\sqrt{n\Sigma y^2 - (\Sigma y)^2}} = \frac{10(38.196) - (23.460)(15.540)}{\sqrt{10(61.785) - (23.460)^2}\sqrt{10(27.992) - (15.540)^2}}$$

$$= \frac{17.392}{\sqrt{67.478}\sqrt{38.428}} = 0.342$$

(c) No linear correlation

23. $r \approx 0.623$

$n = 8$ and $\alpha = 0.01$

$cv = 0.834$

$|r| \approx 0.623 < 0.834 \Rightarrow$ The correlation is not significant.

or

$H_0: \rho = 0$; and $H_a: \rho \neq 0$

$\alpha = 0.01$

d.f. $= n - 2 = 6$

$cu = \pm 3.703$; Reject H_0 if $t < -3.707$ or $t > 3.707$.

$$t = \frac{r}{\sqrt{\frac{1-r^2}{n-2}}} = \frac{0.623}{\sqrt{\frac{1-(0.623)^2}{8-2}}} = \frac{0.623}{\sqrt{0.10198}} = 1.951$$

Fail to reject H_0. There is not a significant linear correlation between vehicle weight and the variability in braking distance.

25. $r \approx 0.923$

$n = 8$ and $\alpha = 0.01$

$cv = 0.834$

$|r| \approx 0.923 > 0.834 \Rightarrow$ The correlation is significant.

or

$H_0: \rho = 0$; $H_a: \rho \neq 0$

$\alpha = 0.01$

d.f. $= n - 2 = 11$

$cu = \pm 3.106$; Reject H_0 if $t < -3.106$ or $t > 3.106$.

$r \approx 0.923$

$$t = \frac{r}{\sqrt{\frac{1-r^2}{n-2}}} = \frac{0.923}{\sqrt{\frac{1-(0.923)^2}{13-2}}} = \frac{0.923}{\sqrt{0.01346}} \approx 7.955$$

Reject H_0. There is enough evidence to conclude that a significant linear correlation exists.

27. $r \approx 0.341$

$n = 10$ and $\alpha = 0.01$

$cv = 0.765$

$|r| \approx 0.341 < 0.765 \Rightarrow$ The correlation is not significant.

or

$H_0: \rho = 0$; $H_a: \rho \neq 0$

$\alpha = 0.01$

d.f. $= n - 2 = 8$

$cu = \pm 3.355$; Reject H_0 if $t < -3.355$ or $t > 3.355$.

$r \approx 0.341$

$$t = \frac{r}{\sqrt{\frac{1-r^2}{n-2}}} = \frac{0.341}{\sqrt{\frac{1-(0.341)^2}{10-2}}} = \frac{0.341}{\sqrt{0.110}} = 1.026$$

Fail to reject H_0. There is enough not evidence at the 1% level to conclude there is a significant linear correlation between earnings per share and dividends per share.

29. The correlation coefficient remains unchanged when the x-values and y-values are switched.

31. Answers will vary.

9.2 LINEAR REGRESSION

9.2 Try It Yourself Solutions

1a. $n = 6$

x	y	xy	x^2
42	9	378	1764
48	10	480	2304
50	8	400	2500
59	5	295	3481
65	6	390	4225
72	3	216	5184
$\Sigma x = 336$	$\Sigma y = 41$	$\Sigma xy = 2159$	$\Sigma x^2 = 19{,}458$

b. $m = \dfrac{n\Sigma xy - (\Sigma x)(\Sigma y)}{n\Sigma x^2 - (\Sigma x)^2} = \dfrac{6(2159) - (336)(41)}{6(19{,}458) - (336)^2} = \dfrac{-840}{3852} \approx -0.2134$

c. $b = \bar{y} - m\bar{x} = \left(\dfrac{41}{6}\right) - (-0.2134)\left(\dfrac{336}{6}\right) \approx 18.7837$

d. $\hat{y} = -0.213x + 18.784$

2a. Enter the data.

b. $m \approx 1.486;\ b \approx -3.819$

c. $\hat{y} = 1.486x - 3.819$

3a. (1) $\hat{y} = 11.824(2) + 35.301$ (2) $\hat{y} = 11.824(3.32) + 35.301$

b. (1) 58.949 (2) 74.557

c. (1) 58.949 minutes (2) 74.557 minutes

9.2 EXERCISE SOLUTIONS

1. c **3.** d **5.** g

7. h **9.** c **11.** a

13.

x	y	xy	x^2
17	110	1870	289
26	124	3224	676
37	146	5402	1369
48	140	6720	2304
50	200	10,000	2500
68	192	13,056	4624
72	200	14,400	5184
$\Sigma x = 318$	$\Sigma y = 1112$	$\Sigma xy = 54{,}672$	$\Sigma x^2 = 16{,}946$

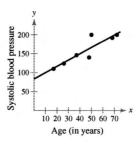

$m = \dfrac{n\Sigma xy - (\Sigma x)(\Sigma y)}{n\Sigma x^2 - (\Sigma x)^2} = \dfrac{7(54{,}672) - (318)(1112)}{7(16{,}946) - (318)^2} = \dfrac{29{,}088}{17{,}498} \approx 1.662$

$b = \bar{y} - m\bar{x} = \left(\dfrac{1112}{7}\right) - (1.662)\left(\dfrac{318}{7}\right) \approx 83.338$

$\hat{y} = 1.662x + 83.338$

(a) $\hat{y} = 1.662(18) + 83.338 \approx 113.254$

(b) $\hat{y} = 1.662(71) + 83.338 \approx 201.340$

(c) $\hat{y} = 1.662(29) + 83.338 \approx 131.536$

(d) $\hat{y} = 1.662(55) + 83.338 \approx 174.748$

15.

x	y	xy	x^2
0	40	0	0
1	41	41	1
2	51	102	4
4	48	192	16
4	64	256	16
5	69	345	25
5	73	365	25
5	75	375	25
6	68	408	36
6	93	558	36
7	84	588	49
7	90	630	49
8	95	760	64
$\Sigma x = 60$	$\Sigma y = 891$	$\Sigma xy = 4620$	$\Sigma x^2 = 346$

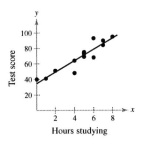

$$m = \frac{n\Sigma xy - (\Sigma x)(\Sigma y)}{n\Sigma x^2 - (\Sigma x)^2} = \frac{13(4620) - (60)(891)}{13(346) - (60)^2} = \frac{6600}{898} \approx 7.350$$

$$b = \bar{y} - m\bar{x} = \left(\frac{60}{13}\right) - (7.350)\left(\frac{891}{13}\right) \approx 34.617$$

$\hat{y} = 7.350x + 34.617$

(a) $\hat{y} = 7.350(3) + 34.617 \approx 56.7$

(b) $\hat{y} = 7.350(6.5) + 34.617 \approx 82.4$

(c) It is not meaningful to predict the value of y for $x = 13$ because $x = 13$ is outside the range of the original data.

(d) $\hat{y} = 7.350(4.5) + 34.617 \approx 67.7$

17.

x	y	xy	x^2
186	495	92,070	34,596
181	477	86,337	32,761
176	425	74,800	30,976
149	322	47,978	22,201
184	482	88,688	33,856
190	587	111,530	36,100
158	370	58,460	24,964
139	322	44,758	19,321
175	479	83,825	30,625
148	375	55,500	21,904
$\Sigma x = 1686$	$\Sigma y = 4334$	$\Sigma xy = 743,946$	$\Sigma x^2 = 287,304$

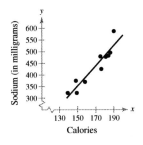

$$m = \frac{n\Sigma xy - (\Sigma x)(\Sigma y)}{n\Sigma x^2 - (\Sigma x)^2} = \frac{10(743,946) - (1686)(4334)}{10(287,304) - (1686)^2} = \frac{132,336}{30,444} \approx 4.347$$

$$b = \bar{y} - m\bar{x} = \left(\frac{4334}{10}\right) - 4.347\left(\frac{1686}{10}\right) = -299.482$$

$\hat{y} = 4.347x - 299.482$

(a) $\hat{y} = 4.347(170) - 299.482 = 439.508$

(b) It is not meaningful to predict the value of y for $x = 136$ because $x = 136$ is outside the range of the original data.

(c) $\hat{y} = 4.347(145) - 299.482 = 330.833$

(d) $\hat{y} = 4.347(183) - 299.482 = 496.019$

19.

x	y	xy	x^2
8.5	66.0	561.00	72.25
9.0	68.5	616.50	81.00
9.0	67.5	607.50	81.00
9.5	70.0	665.00	90.25
10.0	70.0	700.00	100.00
10.0	72.0	720.00	100.00
10.5	71.5	750.75	110.25
10.5	69.5	729.75	110.25
11.0	71.5	786.50	121.00
11.0	72.0	792.00	121.00
11.0	73.0	803.00	121.00
12.0	73.5	882.00	144.00
12.0	74.0	888.00	144.00
12.5	74.0	925.00	156.25
$\Sigma x = 146.5$	$\Sigma y = 993.0$	$\Sigma xy = 10,427.0$	$\Sigma x^2 = 1552.3$

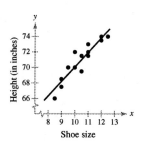

$m = \dfrac{n\Sigma xy - (\Sigma x)(\Sigma y)}{n\Sigma x^2 - (\Sigma x)^2} = \dfrac{14(10,427.0) - (146.5)(993.0)}{14(1552.3) - (146.5)^2} = \dfrac{503.5}{269.95} \approx 1.870$

$b = \bar{y} - m\bar{x} = \left(\dfrac{993.0}{14}\right) - (1.870)\left(\dfrac{146.5}{14}\right) \approx 51.360$

$\hat{y} = 1.870x + 51.360$

(a) $\hat{y} = 1.870(11.5) + 51.360 \approx 72.865$

(b) $\hat{y} = 1.870(8.0) + 51.360 \approx 66.320$

(c) It is not meaningful to predict the value of y for $x = 15.5$ because $x = 15.5$ is outside the range of the original data.

(d) $\hat{y} = 1.870(10.0) + 51.360 \approx 70.060$

21.

x	y	xy	x^2
5720	2.19	12,527	32,718,400
4050	1.36	5508	16,402,500
6130	2.58	15,815	37,576,900
5000	1.74	8700	25,000,000
5010	1.78	8917.8	25,100,100
4270	1.69	7216.3	18,232,900
5500	1.80	9900	30,250,000
5550	1.87	10,379	30,802,500
$\Sigma x = 41,230$	$\Sigma y = 15.01$	$\Sigma xy = 78,962.8$	$\Sigma x^2 = 216,083,300$

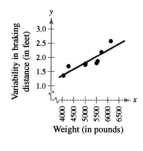

$m = \dfrac{n\Sigma xy - (\Sigma x)(\Sigma y)}{n\Sigma x^2 - (\Sigma x)^2} = \dfrac{8(78,962.8) - (41,230)(15.01)}{8(216,083,300) - (41,230)^2} = \dfrac{12,833.7}{28,753,500} = 0.000447$

$b = \bar{y} - m\bar{x} = \left(\dfrac{15.01}{8}\right) - (0.000447)\left(\dfrac{41,230}{8}\right) = -0.425$

$\hat{y} = 0.000447x - 0.425$

(a) $\hat{y} = 0.000447(4500) - 0.425 = 1.587$

(b) $\hat{y} = 0.000447(6000) - 0.425 = 2.257$

(c) It is not meaningful to predict the value of y for $x = 7500$ because $x = 7500$ is outside the range of the original data.

(d) $\hat{y} = 0.000447(5750) - 0.425 = 2.145$

23. Substitute a value x into the equation of a regression line and solve for y.

25. (a) $\hat{y} = 1.724x + 79.733$ (b) $\hat{y} = 0.453x - 26.448$ (c) The slope of the line keeps the same sign, but the values of m and b change.

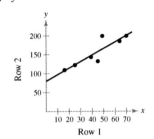
Row 1

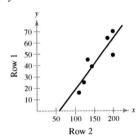
Row 2

27. Strong negative non-linear correlation; as the number of subscribers increases, the average monthly bill decreases.

29. $\hat{y} = -0.294(180) + 70.652 \approx 17.73$

This is not a valid prediction because $x = 180$ is outside the range of the original data.

31. As the number of cellular phone subscribers increases, the average monthly cellular phone bill decreases. (Answers will vary).

33. The exponential equation is a better fit to the data.

35. $\hat{y} = a + b \ln x = 13.667 - 0.471 \ln x$

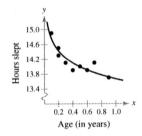

37. The logarithmic equation is a better fitting model to the data.

9.3 MEASURES OF REGRESSION AND PREDICTION INTERVALS

9.3 Try It Yourself Solutions

1a. $r = 0.970$

b. $r^2 = (0.970)^2 = 0.941$

c. 94.1% of the variation in the times is explained.
5.9% of the variation is unexplained.

2a.

x_i	y_i	$\hat{y}_i$	$(y_i - \hat{y}_i)^2$
15	26	28.386	5.693
20	32	35.411	11.635
20	38	35.411	6.703
30	56	49.461	42.759
40	54	63.511	90.459
45	78	70.536	55.711
50	80	77.561	5.949
60	88	91.611	13.039
			$\Sigma = 231.948$

b. $n = 8$

c. $s_e = \sqrt{\dfrac{\Sigma(y_i - \hat{y}_i)^2}{n - 2}} = \sqrt{\dfrac{231.948}{6}} \approx 6.218$

d. The standard deviation of the weekly sales for a specific radio ad time is about $621.80.

3a. $n = 8$, d.f. $= 6$, $t_c = 2.447$, $s_e \approx 10.290$

b. $\hat{y} = 50.729x + 104.061 = 50.729(2.5) + 104.061 \approx 230.884$

c. $E = t_c s_e \sqrt{1 + \dfrac{1}{n} + \dfrac{n(x - \bar{x})^2}{n(\Sigma x^2) - (\Sigma x)^2}} = (2.447)(10.290)\sqrt{1 + \dfrac{1}{8} + \dfrac{8(2.5 - 1.975)^2}{8(32.44) - (15.8)^2}}$

$= (2.447)(10.290)\sqrt{1.34818} \approx 29.236$

d. $\hat{y} \pm E \rightarrow (201.648, 260.120)$

e. You can be 95% confident that the company sales will be between $201,648 and $260,120 when advertising expenditures are $2500.

9.3 EXERCISE SOLUTIONS

1. Total variation $= \Sigma(y_i - \bar{y})^2$; the sum of the squares of the differences between the y-values of each ordered pair and the mean of the y-values of the ordered pairs.

3. Unexplained variation $= \Sigma(y_i - \hat{y}_i)^2$; the sum of the squares of the differences between the observed y-values and the predicted y-values.

5. $r^2 = (0.250)^2 \approx 0.063$

6.3% of the variation is explained. 93.7% of the variation is unexplained.

7. $r = (-0.891)^2 \approx 0.794$

79.4% of the variation is explained. 20.6% of the variation is unexplained.

9. (a) $r^2 = \dfrac{\Sigma(\hat{y}_i - \bar{y})^2}{\Sigma(y_i - \bar{y})^2} \approx 0.623$

62.3% of the variation in proceeds can be explained by the variation in the number of issues and 37.7% of the variation is unexplained.

(b) $s_e = \sqrt{\dfrac{\Sigma(y_i - \hat{y}_i)^2}{n - 2}} = \sqrt{\dfrac{770{,}153{,}382.32}{10}} \approx 8775.838$

The standard deviation of the proceeds for a specific number of issues is about $8,775,838,000.

11. (a) $r^2 = \dfrac{\Sigma(\hat{y}_i - \bar{y})^2}{\Sigma(y_i - \bar{y})^2} \approx 0.985$

98.5% of the variation in sales can be explained by the variation in the total square footage and 1.5% of the variation is unexplained.

(b) $s_e = \sqrt{\dfrac{\Sigma(y_i - \hat{y}_i)^2}{n-2}} = \sqrt{\dfrac{11{,}439.531}{9}} \approx 35.652$

The standard deviation of the sales for a specific total square footage is about 35,652,000,000.

13. (a) $r^2 = \dfrac{\Sigma(\hat{y}_i - \bar{y})^2}{\Sigma(y_i - \bar{y})^2} \approx 0.982$

98.2% of the variation in the median weekly earnings of female workers can be explained by the variation in the median weekly earnings of male workers and 1.8% of the variation is unexplained.

(b) $s_e = \sqrt{\dfrac{\Sigma(y_i - \hat{y}_i)^2}{n-2}} = \sqrt{\dfrac{203.004}{3}} \approx 8.226$

The standard deviation of the median weekly earnings of female workers for a specific median weekly earnings of male workers is about $8.226.

15. (a) $r^2 = \dfrac{\Sigma(\hat{y}_i - \bar{y})^2}{\Sigma(y_i - \bar{y})^2} \approx 0.992$

99.2% of the variation in the money spent can be explained by the variation in the money raised and 0.8% of the variation is unexplained.

(b) $s_e = \sqrt{\dfrac{\Sigma(y_i - \hat{y}_i)^2}{n-2}} = \sqrt{\dfrac{1{,}551.176}{6}} \approx 16.079$

The standard deviation of the money spent for a specified amount of money raised is about $16,079,000.

17. $n = 12$, d.f. $= 10$, $t_c = 2.228$, $s_e = 8775.838$

$\hat{y} = 51.185x - 35.462 = 51.185(712) - 35.462 \approx 36{,}408.258$

$E = t_c s_e \sqrt{1 + \dfrac{1}{n} + \dfrac{n(x - \bar{x})^2}{n(\Sigma x^2) - (\Sigma x)^2}} = (2.228)(8775.838)\sqrt{1 + \dfrac{1}{12} + \dfrac{12(712 - 5376/12)^2}{12(2{,}894{,}666) - (5376)^2}}$

$= (2.228)(8775.838)\sqrt{1.22667} \approx 21{,}655.529$

$\hat{y} \pm E \to (14{,}752.729, 58{,}063.787) \to (\$14{,}752{,}729{,}000, \$58{,}063{,}787{,}000)$

You can be 95% confident that the proceeds will be between 14,752,729,000 and $58,063,787,000 when the number of initial offerings is 712.

19. $n = 11$, d.f. $= 9$, $t_c = 1.833$, $s_e \approx 35.652$

$\hat{y} = 230.831x - 289.809 = 230.831(4.5) - 289.809 \approx 748.931$

$E = t_c s_e \sqrt{1 + \dfrac{1}{n} + \dfrac{n(x - \bar{x})^2}{n(\Sigma x^2) - (\Sigma x)^2}} = (1.833)(35.652)\sqrt{1 + \dfrac{1}{11} + \dfrac{11[4.5 - (43.3/11)]^2}{11(184.93) - (43.3)^2}}$

$= (1.833)(35.652)\sqrt{1.11284} \approx 68.939$

$\hat{y} \pm E \to (679.992, 817.870) \to (\$679{,}992{,}000, \$817{,}870{,}000)$

You can be 90% confident that the sales will be between $679,992,000,000 and $817,870,000,000 when the total square footage is 4.5 billion.

21. $n = 5$, d.f. $= 3$, $t_c = 5.841$, $s_e \approx 8.226$

$\hat{y} = 0.919x - 95.462 = 0.919(500) - 95.462 \approx 364.038$

$E = t_c s_e \sqrt{1 + \dfrac{1}{n} + \dfrac{n(x - \bar{x})^2}{n(\Sigma x^2) - (\Sigma x)^2}} = (5.841)(8.226)\sqrt{1 + \dfrac{1}{5} + \dfrac{5[500 - (3221/5)]^2}{5(2{,}087{,}823) - (3221)^2}}$

$= (5.841)(8.226)\sqrt{2.81758} \approx 80.652$

$\hat{y} \pm E \rightarrow (283.386, 444.690)$

You can be 99% confident that the median earnings of female workers will be between $283.386 and $444.690 when the median weekly earnings of male workers is $500.

23. $n = 8$, d.f. $= 6$, $t_c = 2.447$, $s_e \approx 16.079$

$\hat{y} = 1.020x - 25.854 = 1.020(775.8) - 25.854 \approx 765.462$

$E = t_c s_e \sqrt{1 + \dfrac{1}{n} + \dfrac{n(x - \bar{x})^2}{n(\Sigma x^2) - (\Sigma x)^2}} = (2.447)(16.079)\sqrt{1 + \dfrac{1}{8} + \dfrac{8[775.8 - (4363.5/8)]^2}{8(2{,}564{,}874) - (4363.5)^2}}$

$= (2.447)(16.079)\sqrt{1.41207} \approx 46.754$

$\hat{y} \pm E \rightarrow (\$718.708 \text{ million}, \$812.216 \text{ million})$

You can be 95% confident that the money spent in congressional campaigns will be between $729.046 million and $822.554 million when the money raised is $775.8 million.

25.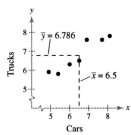

27.

x_i	y_i	$\hat{y}_i$	$\hat{y}_i - \bar{y}$	$y_i - \hat{y}_i$	$y_i - \bar{y}$
8.1	7.8	7.893	1.107	-0.093	1.014
7.7	7.6	7.616	0.830	-0.016	0.814
6.5	6.5	6.785	-0.001	-0.285	-0.286
6.9	7.6	7.062	0.276	0.538	0.814
6.0	6.3	6.438	-0.348	-0.138	-0.486
5.4	5.8	6.022	-0.764	-0.222	-0.986
4.9	5.9	5.676	-1.11	0.224	-0.886

29. $r^2 \approx 0.887$; About 88.7% of the variation in the median age of trucks can be explained by the variation in the median age of cars, and 11.3% of the variation is unexplained.

31. $\hat{y} = 0.693x + 2.280 = 0.693(7.3) + 2.280 \approx 7.339$

$E = t_c s_e \sqrt{1 + \dfrac{1}{n} + \dfrac{n(x - \bar{x})^2}{n(\Sigma x^2) - (\Sigma x)^2}} = (2.571)(0.316)\sqrt{1 + \dfrac{1}{7} + \dfrac{7[7.3 - (45.5/7)]^2}{7(303.930) - (45.5)^2}}$

$= (2.571)(0.316)\sqrt{1.22110} \approx 0.898$

$\hat{y} \pm E \rightarrow (6.441, 8.237)$

9.4 MULTIPLE REGRESSION

9.4 Try It Yourself Solutions

1a. Enter the data. **b.** $\hat{y} = 46.385 + 0.540x_1 - 4.897x_2$

2ab. (1) $\hat{y} = 46.385 + 0.540(89) - 4.897(1)$ (2) $\hat{y} = 46.385 + 0.540(78) - 4.897(3)$
(3) $\hat{y} = 46.385 + 0.540(83) - 4.897(2)$

c. (1) $\hat{y} = 89.548$ (2) $\hat{y} = 73.814$ (3) $\hat{y} = 81.411$

d. (1) 90   (2) 74   (3) 81

9.4 EXERCISE SOLUTIONS

1. $\hat{y} = 6503 - 14.8x_1 + 12.2x_2$

    (a) $\hat{y} = 6503 - 14.8(1458) + 12.2(1450) = 2614.6$

    (b) $\hat{y} = 6503 - 14.8(1500) + 12.2(1475) = 2298$

    (c) $\hat{y} = 6503 - 14.8(1400) + 12.2(1385) = 2680$

    (d) $\hat{y} = 6503 - 14.8(1525) + 12.2(1500) = 2233$

3. $\hat{y} = -52.2 + 0.3x_1 + 4.5x_2$

    (a) $\hat{y} = -52.2 + 0.3(70) + 4.5(8.6) = 7.5$

    (b) $\hat{y} = -52.2 + 0.3(65) + 4.5(11.0) = 16.8$

    (c) $\hat{y} = -52.2 + 0.3(83) + 4.5(17.6) = 51.9$

    (d) $\hat{y} = -52.2 + 0.3(87) + 4.5(19.6) = 62.1$

5. $\hat{y} = -256.293 + 103.502x_1 + 14.649x_2$

    (a) $s = 34.16$

    (b) $r^2 = 0.988$

    (c) The standard deviation of the predicted sales given a specific total square footage and number of shopping centers is $34.16 billion. The multiple regression model explains 98.8% of the variation in y.

7. $n = 11, k = 2, r = 0.988$

    $r^2_{adj} = 1 - \left[\frac{(1-r^2)(n-1)}{n-k-1}\right] = 0.985$

CHAPTER 9 REVIEW EXERCISE SOLUTIONS

1.

$r \approx -0.939$; negative linear correlation; milk production decreases with age.

3.

$r \approx 0.868$; positive linear correlation; calorie intake increases with weight.

5. $H_0: \rho = 0; H_a: \rho \neq 0$

    $\alpha = 0.01$, d.f. $= n - 2 = 24$

    $t_0 = 2.797$

    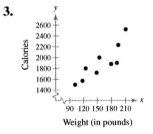

    $t = \dfrac{r}{\sqrt{\dfrac{1-r^2}{n-2}}} = \dfrac{0.24}{\sqrt{\dfrac{1-(0.24)^2}{26-2}}} = \dfrac{0.24}{\sqrt{0.03927}} = 1.211$

    Fail to reject H_0. There is not enough evidence to conclude that a significant linear correlation exists.

7. $H_0: \rho = 0; H_a: \rho \neq 0$

   $\alpha = 0.05$, d.f. $= n - 2 = 6$

   $t_0 = \pm 2.447$

   $t = \dfrac{r}{\sqrt{\dfrac{1-r^2}{n-2}}} = \dfrac{-0.939}{\sqrt{\dfrac{1-(-0.939)^2}{8-2}}} = \dfrac{-0.939}{\sqrt{0.01971}} = -6.688$

   Reject H_0. There is enough evidence to conclude that there is a significant linear correlation between the age of a cow and its milk production.

9. $H_0: \rho = 0; H_a: \rho \neq 0$

   $\alpha = 0.01$, d.f. $= n - 2 = 7$

   $t_0 = \pm 3.499$

   $t = \dfrac{r}{\sqrt{\dfrac{1-r^2}{n-2}}} = \dfrac{0.868}{\sqrt{\dfrac{1-(0.868)^2}{7}}} = 4.625$

   Reject H_0. There is enough evidence to conclude a significant linear correlation exists between weight and calories.

11. $\hat{y} = 0.757x + 21.5$

13. $\hat{y} = 10.450 - 0.086x$

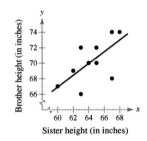

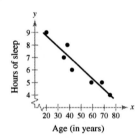

$r \approx 0.688$ (Moderate positive linear correlation)

$r \approx -0.949$ (Strong negative linear correlation)

15. (a) $\hat{y} = 0.757(61) + 21.5 = 67.677 \approx 68$ inches

    (b) $\hat{y} = 0.757(66) + 21.5 = 71.462 \approx 71$ inches

    (c) Not meaningful since $x = 71$ inches is outside range of data.

    (d) Not meaningful since $x = 50$ inches is outside range of data.

17. (a) Not meaningful since $x = 18$ years is outside range of data.

    (b) $\hat{y} = 10.450 = 0.086x = 10.450 - 0.086(25) = 8.3$

    (c) Not meaningful since $x = 80$ years is outside range of data.

    (d) $\hat{y} = 10.450 - 0.086x = 10.450 - 0.086(50) = 6.15$

19. $r^2 = (-0.553)^2 = 0.306$

    30.6% of the variation in y is explained by the model.
    69.4% of the variation in y is unexplained by the model.

21. $r^2 = (0.181)^2 = 0.033$

    3.3% of the variation in y is explained by the model.
    96.7% of the variation in y is unexplained by the model.

23. (a) $r^2 = 0.897$

    89.7% of the variation in y is explained by the model.
    10.3% of the variation in y is unexplained by the model.

    (b) $s_e = 568.011$

    The standard error of the cooling capacity for a specific living area is 568.0 Btu/hr.

25. $\hat{y} = 0.757(64) + 21.5 = 69.948$

    $$E = t_c s_e \sqrt{1 + \frac{1}{n} + \frac{n(x - \bar{x})^2}{n\Sigma x^2 - (\Sigma x)^2}} \approx (1.833)(2.206) \sqrt{1 + \frac{1}{11} + \frac{11(64 - 64.727)^2}{11(46,154) - 712^2}}$$

    $= (1.833)(2.206)\sqrt{1.09866} \approx 4.238$

    $\hat{y} - E < y < \hat{y} + E$
    $69.948 - 4.238 < y < 69.948 + 4.238$
    $65.71 < y < 74.19$

27. $\hat{y} = -0.086(45) + 10.450 = 6.580$

    $$E = t_c s_e \sqrt{1 + \frac{1}{n} + \frac{n(x - \bar{x})^2}{n\Sigma x^2 - (\Sigma x)^2}} \approx (2.571)(0.623) \sqrt{1 + \frac{1}{7} + \frac{7[45 - (337/7)]^2}{7(18,563) - (337)^2}}$$

    $= (2.571)(0.623)\sqrt{1.14708} = 1.715$

    $\hat{y} - E < y < \hat{y} + E$
    $6.580 - 1.715 < y < 6.58 + 1.715$
    $4.865 < y < 8.295$

29. $\hat{y} = 3002.991 + 9.468(720)$
    $= 9819.95$

    $$E = t_c s_e \sqrt{1 + \frac{1}{n} + \frac{n(x - \bar{x})^2}{n\Sigma x^2 - (\Sigma x)^2}} = (3.707)(568) \sqrt{1 + \frac{1}{8} + \frac{8(720 - 499.4)^2}{8(2,182,275) - 3995^2}}$$

    $= (3.707)(568)\sqrt{1.38486} \approx 2477.94$

    $\hat{y} - E < y < \hat{y} + E$
    $9819.96 - 2477.94 < y < 9819.96 + 2477.94$
    $7342.01 < y < 12,297.89$

31. $\hat{y} = 6.317 + 0.8227x_1 + 0.031x_2 - 0.004x_3$

33. (a) 21.705   (b) 25.210   (c) 30.100   (d) 25.860

CHAPTER 9 QUIZ SOLUTIONS

1. The data appear to have a positive correlation. The outlays increase as the incomes increase.

2. $r \approx 0.996 \rightarrow$ Strong positive linear correlation

3. $H_0: \rho = 0; H_a: \rho \neq 0$

$\alpha = 0.05$, d.f. $= n - 2 = 9$

$t_0 = \pm 2.262$

$t = \dfrac{r}{\sqrt{\dfrac{1-r^2}{n-2}}} = \dfrac{0.996}{\sqrt{\dfrac{1-(0.996)^2}{11-2}}} = \dfrac{0.996}{\sqrt{0.00088711}} \approx 33.44$

Reject H_0. There is enough evidence to conclude that a significant correlation exists.

4. $\Sigma y^2 = 441.140$
$\Sigma y = 68.600$
$\Sigma xy = 522.470$
$b = -0.359$
$m = 0.891$
$\hat{y} = 0.891x - 0.359$

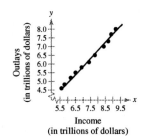

5. $\hat{y} = 0.891(5.3) - 0.359 = 4.363$

6. $r^2 = 0.992$

99.2% of the variation in y is explained by the regression model.
0.8% of the variation in y is unexplained by the regression model.

7. $s_e = 0.110$

The standard deviation of personal outlays for a specified personal income is $0.110 trillion.

8. $\hat{y} = 0.891(6.4) - 0.359$
$= 5.343$

$E = t_c s_e \sqrt{1 + \dfrac{1}{n} + \dfrac{n(x-\bar{x})^2}{n\Sigma x^2 - (\Sigma x)^2}}$

$= 2.262(0.110)\sqrt{1 + \dfrac{1}{11} + \dfrac{11[6.4 - (81.4/11)]^2}{11(619) - 81.4^2}} \approx (2.262)(0.110)\sqrt{1.15101}$

≈ 0.267

$\hat{y} - E < y < \hat{y} + E$
$5.343 - 0.267 < y < 5.343 + 0.267$
$5.076 < y < 5.610 \rightarrow (\$5.076$ trillion, $\$5.610$ trillion$)$

You can be 95% confident that the personal outlays will be between $5.076 trillion and $5.610 trillion when personal income is $6.4 trillion.

9. (a) 1311.150

(b) 961.110

(c) 1120.900

(d) 1386.740; x_2 has the greatest influence on y.

Chi-Square Tests and the *F*-Distribution

CHAPTER 10

10.1 GOODNESS OF FIT

10.1 Try It Yourself Solutions

1a.

Music	% of Listeners	Expected Frequency
Classical	4%	12
Country	36%	108
Gospel	11%	33
Oldies	2%	6
Pop	18%	54
Rock	29%	87

2a. The expected frequencies are 64, 80, 32, 56, 60, 48, 40, and 20, all of which are at least 5.

b. Claimed Distribution:

Ages	Distribution
0–9	16%
10–19	20%
20–29	8%
30–39	14%
40–49	15%
50–59	12%
60–69	10%
70+	5%

H_0: Distribution of ages is as shown in table above.

H_a: Distribution of ages differs from the claimed distribution.

c. $\alpha = 0.05$ **d.** d.f. $= n - 1 = 7$

e. $\chi_0^2 = 14.067$; Reject H_0 if $\chi^2 > 14.067$.

f.

Ages	Distribution	Observed	Expected	$\frac{(O-E)^2}{E}$
0–9	16%	76	64	2.250
10–19	20%	84	80	0.200
20–29	8%	30	32	0.125
30–39	14%	60	56	0.286
40–49	15%	54	60	0.600
50–59	12%	40	48	1.333
60–69	10%	42	40	0.100
70+	5%	14	20	1.800
				6.694

$\chi^2 \approx 6.694$

g. Fail to reject H_0.

h. There is not enough evidence to conclude that the distribution of ages differs from the claimed distribution.

3a. The expected frequencies are 88, 54, and 58, all of which are at least 5.

b. Claimed Distribution:

Response	Distribution
In favor of	44%
Against	27%
No Opinion	29%

H_0: Distribution of responses is as shown in table above.

H_a: Distribution of responses differs from the claimed distribution.

c. $\alpha = 0.01$ **d.** d.f. $= n - 1 = 2$ **e.** $\chi_0^2 = 9.210$; Reject H_0 if $\chi^2 > 9.210$.

f.

Response	Distribution	Observed	Expected	$\frac{(O - E)^2}{E}$
In favor of	44%	100	88	1.636
Against	27%	48	54	0.667
No opinion	29%	52	58	0.621
				2.924

$\chi^2 \approx 2.924$

g. Fail to reject H_0.

h. There is not enough evidence to conclude that the distribution of responses differs from the claimed distribution.

10.1 EXERCISE SOLUTIONS

1. A multinomial experiment is a probability experiment consisting of a fixed number of trials in which there are more than two possible outcomes for each independent trial.

3. $E_i = np_i = (150)(0.3) = 45$

5. $E_i = np_i = (230)(0.25) = 57.5$

7. (a) Claimed distribution:

Response	Distribution
Agree	78%
Disagree	17%
Neither	5%

H_0: Distribution of responses is as shown in the table above.

H_a: Distribution of responses differs from the claimed distribution.

(b) $\chi_0^2 = 5.991$; Reject H_0 if $\chi^2 > 5.991$.

(c)

Response	Distribution	Observed	Expected	$\frac{(O - E)^2}{E}$
Agree	78%	311	331.50	1.268
Disagree	17%	84	72.25	1.911
Neither	5%	30	21.25	3.603
				6.782

$\chi^2 = 6.782$

(d) Reject H_0. There is enough evidence at the 5% level to conclude that the distribution of responses differs from the claimed distribution.

9. (a) Claimed distribution:

Response	Distribution
Keeping Medicare, Social Security, and Medicaid	77%
Tax cuts	15%
Not sure	8%

H_0: Distribution of responses is as shown in the table above.

H_a: Distribution of responses differs from the claimed distribution.

(b) $\chi_0^2 = 9.210$; Reject H_0 if $\chi^2 > 9.210$.

(c)

Response	Distribution	Observed	Expected	$\frac{(O-E)^2}{E}$
Keeping Medicare, Social Security, and Medicaid	77%	431	462	2.080
Tax cuts	15%	107	90	3.211
Not sure	8%	62	48	4.083
				9.374

$\chi^2 = 9.374$

(d) Reject H_0. There is enough evidence at the 1% level to conclude that the distribution of responses differs from the claimed distribution.

11. (a) Claimed distribution:

Day	Distribution
Sunday	14.286%
Monday	14.286%
Tuesday	14.286%
Wednesday	14.286%
Thursday	14.286%
Friday	14.286%
Saturday	14.286%

H_0: The distribution of fatal bicycle accidents throughout the week is uniform as shown in the table above.

H_a: The distribution of fatal bicycle accidents throughout the week is not uniform.

(b) $\chi_0^2 = 10.645$; Reject H_0 if $\chi^2 > 10.645$.

(c)

Day	Distribution	Observed	Expected	$\frac{(O-E)^2}{E}$
Sunday	14.286%	118	130.15	1.133
Monday	14.286%	119	130.15	0.954
Tuesday	14.286%	127	130.15	0.076
Wednesday	14.286%	137	130.15	0.361
Thursday	14.286%	129	130.15	0.010
Friday	14.286%	146	130.15	1.931
Saturday	14.286%	135	130.15	0.181
				4.648

$\chi^2 \approx 4.648$

(d) Fail to reject H_0. There is not enough evidence at the 10% level to conclude that the distribution of fatal bicycle accidents throughout the week is not uniform.

13. (a) Claimed distribution:

Object struck	Distribution
Tree	28%
Embankment	10%
Utility pole	10%
Guardrail	9%
Ditch	7%
Curb	6%
Culvert	5%
Sign/Post/Fence	10%
Other	15%

H_0: Distribution of objects struck is as shown in table above.

H_a: Distribution of objects struck differs from the claimed distribution.

(b) $\chi_0^2 = 20.090$; Reject H_0 if $\chi^2 > 20.090$.

(c)

Object Struck	Distribution	Observed	Expected	$\frac{(O-E)^2}{E}$
Tree	28%	179	193.48	1.084
Embankment	10%	100	69.10	13.818
Utility pole	10%	107	69.10	20.787
Guardrail	9%	57	62.19	0.433
Ditch	7%	36	48.37	3.163
Curb	6%	43	41.46	0.057
Culvert	5%	28	34.55	1.242
Sign/Post/Fence	10%	68	69.10	0.018
Other	15%	73	103.65	9.063
			691	49.665

$\chi^2 \approx 49.665$

(d) Reject H_0. There is enough evidence at the 1% level to conclude that the distribution of the objects struck changed after warning signs were erected.

15. (a) Claimed distribution:

Response	Distribution
Not a HS grad	33.333%
HS graduate	33.333%
College (1yr+)	33.333%

H_0: Distribution of the responses is uniform as shown in table above.

H_a: Distribution of the responses is not uniform.

(b) $\chi_0^2 = 7.378$; Reject H_0 if $\chi^2 > 7.378$.

(c)

Response	Distribution	Observed	Expected	$\frac{(O-E)^2}{E}$
Not a HS grad	33.333%	37	33	0.485
HS graduate	33.333%	40	33	1.485
College (1yr+)	33.333%	22	33	3.667
			99	5.637

$\chi^2 \approx 5.637$

(d) Fail to reject H_0. There is not enough evidence at the 2.5% level to conclude that the distribution of the responses is not uniform.

17. (a) Claimed distribution:

Cause	Distribution
Trans. Accidents	41%
Assaults	20%
Objects/equipment	15%
Falls	10%
Exposure	10%
Other	4%

H_0: Distribution of the causes is as shown in table above.

H_a: Distribution of the causes differs from the claimed distribution.

(b) $\chi_0^2 = 11.071$; Reject H_0 if $\chi^2 > 11.071$.

(c)

Cause	Distribution	Observed	Expected	$\frac{(O-E)^2}{E}$
Trans. Accidents	41%	2500	2554.71	1.172
Assaults	20%	1300	1246.20	2.323
Objects/equipment	15%	985	934.65	2.712
Falls	10%	620	623.10	0.015
Exposure	10%	602	623.10	0.715
Other	4%	224	249.24	2.556
		6231		9.493

$\chi^2 \approx 9.493$

(d) Fail to reject H_0. There is not enough evidence at the 5% level to conclude that the distributions of the causes in the Western U.S. differs from the national distribution.

19. (a) Frequency distribution: $\mu = 69.435$; $\sigma \approx 8.337$

Lower Boundary	Upper Boundary	Lower z-score	Upper z-score	Area	Class Boundaries	Distribution	Frequency	Expected	$\frac{(O-E)^2}{E}$
49.5	58.5	−2.39	−1.31	0.0867	49.5–58.5	8.67%	19	17	0.235
58.5	67.5	−1.31	−0.23	0.3139	58.5–67.5	31.39%	61	63	0.063
67.5	76.5	−0.23	0.85	0.3933	67.5–76.5	39.33%	82	79	0.114
76.5	85.5	0.85	1.93	0.1709	76.5–85.5	17.09%	34	34	0
85.5	94.5	1.93	3.01	0.0255	85.5–94.5	2.55%	4	5	0.2
							200		0.612

H_0: Variable has a normal distribution

H_a: Variable does not have a normal distribution

(b) $\chi_0^2 = 13.277$; Reject H_0 if $\chi^2 > 13.277$.

(c) $\chi^2 = 0.612$

(d) Fail to reject H_0. There is not enough evidence at the 1% level to conclude that the distribution of test scores is not normal.

10.2 INDEPENDENCE

10.2 Try It Yourself Solutions

1ab.

	Hotel	Leg Room	Rental Size	Other	Total
Business	36	108	14	22	180
Leisure	38	54	14	14	120
Total	74	162	28	36	300

c. $n = 300$

d.

	Hotel	Leg Room	Rental Size	Other
Business	44.4	97.2	16.8	21.6
Leisure	29.6	64.8	11.2	14.4

2a. H_0: Travel concern is independent of travel purpose.

H_a: Travel concern is dependent on travel purpose. (claim)

b. $\alpha = 0.01$ **c.** $(r-1)(c-1) = 3$ **d.** $\chi_0^2 = 11.345$; Reject H_0 if $\chi^2 > 11.345$.

e. $\chi^2 = \sum \frac{(O-E)^2}{E} = \frac{(36-44.4)^2}{44.4} + \frac{(108-97.2)^2}{97.2} + \cdots + \frac{(14-14.4)^2}{14.4} \approx 8.158$

f. Fail to reject H_0.

g. There is not enough evidence to conclude that travel concern is dependent on travel purpose.

3a. $\chi_0^2 = 9.488$; Reject H_0 if $\chi^2 > 9.488$. **b.** Enter the data.

c. $\chi^2 \approx 65.619$ **d.** Reject H_0.

e. There is enough evidence to conclude that minutes spent online per day is dependent on gender.

10.2 EXERCISE SOLUTIONS

1. $E_{r,c} = \dfrac{(\text{Sum of row } r)(\text{Sum of column } c)}{\text{Sample size}}$

3. False. A contingency table with two rows and five columns will have $(2-1)(5-1) = 4$ degrees of freedom.

5.

	Athlete has	
Result	Stretched	Not stretched
Injury	20.818	19.182
No injury	208.182	191.818

7.

	Treatment		
Result	Brand-name	Generic	Placebo
Improvement	24	21	10
No change	12	13	45

9. (a) H_0: Skill level in a subject is independent of location. (claim)

H_a: Skill level in a subject is dependent on location.

(b) d.f. = $(r-1)(c-1) = 2$

$\chi_0^2 = 9.210$; Reject H_0 if $\chi^2 > 9.210$.

(c) $\chi^2 \approx 0.297$

(d) Fail to reject H_0. There is not enough evidence at the 1% level to conclude that skill level in a subject is dependent on location.

11. (a) H_0: Grades are independent of the institution.
    H_a: Grades are dependent on the institution. (claim)

    (b) d.f. = $(r-1)(c-1) = 8$

    $\chi_0^2 = 15.507$; Reject H_0 if $\chi^2 > 15.507$.

    (c) $\chi^2 \approx 48.488$

    (d) Reject H_0. There is enough evidence at the 5% level to conclude that grades are dependent on the institution.

13. (a) H_0: Results are independent of the type of treatment.
    H_a: Results are dependent on the type of treatment. (claim)

    (b) d.f. = $(r-1)(c-1) = 1$

    $\chi_0^2 = 2.706$; Reject H_0 if $\chi^2 > 2.706$.

    (c) $\chi^2 \approx 5.106$

    (d) Reject H_0. There is not enough evidence at the 10% level to conclude that results are dependent on the type of treatment. I would recommend using the drug.

15. (a) H_0: Reasons are independent of the type of worker.
    H_a: Reasons are dependent on the type of worker. (claim)

    (b) d.f. = $(r-1)(c-1) = 2$

    $\chi_0^2 = 9.210$; Reject H_0 if $\chi^2 > 9.210$.

    (c) $\chi^2 \approx 7.326$

    (d) Fail to reject H_0. There is not enough evidence at the 1% level to conclude that the reason(s) for continuing education are dependent on the type of worker. Based on these results, marketing strategies should not differ between technical and non-technical audiences in regard to reason(s) for continuing education.

17. (a) H_0: Type of crash is independent of the type of vehicle.
    H_a: Type of crash is dependent on the type of vehicle. (claim)

    (b) d.f. = $(r-1)(c-1) = 2$

    $\chi_0^2 = 5.991$; Reject H_0 if $\chi^2 > 5.991$.

    (c) $\chi^2 \approx 106.390$

    (d) Reject H_0. There is enough evidence at the 5% level to conclude that type of crash is dependent on the type of vehicle.

19. (a) H_0: Subject is independent of coauthorship.
    H_a: Subject and coauthorship are dependent. (claim)

(b) d.f. $= (r-1)(c-1) = 4$

$\chi_0^2 = 7.779$; Reject H_0 if $\chi^2 > 7.779$.

(c) $\chi^2 \approx 5.610$

(d) Fail to reject H_0. There is not enough evidence at the 10% level to conclude that the subject matter and coauthorship are related.

21. H_0: The proportions are equal. (claim)

H_a: At least one of the proportions is different from the others.

d.f. $= (r-1)(c-1) = 1$

Critical value is 2.706; Reject H_0 if $\chi^2 > 2.706$.

$\chi^2 \approx 5.106$

Reject H_0. There is enough evidence at the 10% level to conclude that at least one of the proportions is different from the others.

10.3 COMPARING TWO VARIANCES

10.3 Try It Yourself Solutions

1a. $\alpha = 0.01$ **b.** $F = 5.42$

2a. $\alpha = 0.01$ **b.** $F = 18.31$

3a. $H_0: \sigma_1^2 \le \sigma_2^2$; $H_a: \sigma_1^2 > \sigma_2^2$ (claim)

b. $\alpha = 0.01$

c. d.f.$_N = n_1 - 1 = 24$
d.f.$_D = n_2 - 1 = 19$

d. $F_0 = 2.92$; Reject H_0 if $F > 2.92$.

e. $F = \dfrac{s_1^2}{s_2^2} = \dfrac{180}{56} \approx 3.214$

f. Reject H_0.

g. There is enough evidence to support the claim.

4a. $H_0: \sigma_1 = \sigma_2$ (claim); $H_a: \sigma_1 \ne \sigma_2$

b. $\alpha = 0.01$

c. d.f.$_N = n_1 - 1 = 15$
d.f.$_D = n_2 - 1 = 21$

d. $F_0 = 3.43$; Reject H_0 if $F > 3.43$.

e. $F = \dfrac{s_1^2}{s_2^2} = \dfrac{(0.95)^2}{(0.78)^2} \approx 1.483$

f. Fail to reject H_0.

g. There is not enough evidence to reject the claim.

10.3 EXERCISE SOLUTIONS

1. Specify the level of significance α. Determine the degrees of freedom for the numerator and denominator. Use Table 7 to find the critical value F.

3. $F = 2.93$

5. $F = 5.32$

7. $F = 2.06$

9. $H_0: \sigma_1^2 \leq \sigma_2^2$; $H_a: \sigma_1^2 > \sigma_2^2$ (claim)

   d.f.$_N$ = 4
   d.f.$_D$ = 5

   $F_0 = 3.52$; Reject H_0 if $F > 3.52$.

   $F = \dfrac{s_1^2}{s_2^2} = \dfrac{773}{765} \approx 1.010$

   Fail to reject H_0. There is not enough evidence to support the claim.

11. $H_0: \sigma_1^2 \leq \sigma_2^2$ (claim); $H_a: \sigma_1^2 > \sigma_2^2$

    d.f.$_N$ = 10
    d.f.$_D$ = 9

    $F_0 = 5.26$; Reject H_0 if $F > 5.26$.

    $F = \dfrac{s_1^2}{s_2^2} = \dfrac{842}{836} \approx 1.007$

    Fail to reject H_0. There is not enough evidence to reject the claim.

13. $H_0: \sigma_1^2 = \sigma_2^2$ (claim); $H_a: \sigma_1^2 \neq \sigma_2^2$

    d.f.$_N$ = 6
    d.f.$_D$ = 8

    $F_0 = 4.65$; Reject H_0 if $F > 4.65$.

    $F = \dfrac{s_1^2}{s_2^2} = \dfrac{5.2}{4.8} \approx 1.083$

    Fail to reject H_0. There is not enough evidence to reject the claim.

15. Population 1: Company B
    Population 2: Company A

    (a) $H_0: \sigma_1^2 \leq \sigma_2^2$; $H_a: \sigma_1^2 > \sigma_2^2$ (claim)

    (b) d.f.$_N$ = 22
        d.f.$_D$ = 19

        $F_0 = 2.13$; Reject H_0 if $F > 2.13$.

    (c) $F = \dfrac{s_1^2}{s_2^2} = \dfrac{2.8}{2.6} \approx 1.077$

    (d) Fail to reject H_0. There is not enough evidence at the 5% level to support the Company A's claim that the variance of life of its appliances is less than the variance of life of Company B appliances.

17. Population 1: District 1
Population 2: District 2

(a) $H_0: \sigma_1^2 = \sigma_2^2$ (claim); $H_a: \sigma_1^2 \neq \sigma_2^2$

(b) $\text{d.f.}_N = 11$
$\text{d.f.}_D = 13$
$F_0 = 2.63$; Reject H_0 if $F > 2.63$.

(c) $F = \dfrac{s_1^2}{s_2^2} = \dfrac{(27.7)^2}{(26.1)^2} \approx 1.126$

(d) Fail to reject H_0. There is not enough evidence at the 10% level to reject the claim that the standard deviation of physical science assessment test scores for eighth-grade students is the same in Districts 1 and 2.

19. Population 1: Before new admissions procedure
Population 2: After new admissions procedure

(a) $H_0: \sigma_1^2 \leq \sigma_2^2$; $H_a: \sigma_1^2 > \sigma_2^2$ (claim)

(b) $\text{d.f.}_N = 24$
$\text{d.f.}_D = 20$
$F_0 = 1.77$; Reject H_0 if $F > 1.77$.

(c) $F = \dfrac{s_1^2}{s_2^2} = \dfrac{(0.7)^2}{(0.5)^2} \approx 1.96$

(d) Reject H_0. There is enough evidence at the 10% level to support the claim that the standard deviation of waiting times has decreased.

21. (a) Population 1: California
Population 2: New York

$H_0: \sigma_1^2 \leq \sigma_2^2$; $H_a: \sigma_1^2 > \sigma_2^2$ (claim)

(b) $\text{d.f.}_N = 15$
$\text{d.f.}_D = 16$
$F_0 = 2.35$; Reject H_0 if $F > 2.35$.

(c) $F = \dfrac{s_1^2}{s_2^2} = \dfrac{(13{,}900)^2}{(8{,}800)^2} \approx 2.495$

(d) Reject H_0. There is enough evidence at the 5% level to conclude the standard deviation of annual salaries for actuaries is greater in California than in New York.

23. Right-tailed: $F_R = 14.73$

Left-tailed:

(1) $\text{d.f.}_N = 3$ and $\text{d.f.}_D = 6$

(2) $F = 6.60$

(3) Critical value is $\dfrac{1}{F} = \dfrac{1}{6.60} \approx 0.152$.

25. $\dfrac{s_1^2}{s_2^2} F_L < \dfrac{\sigma_1^2}{\sigma_2^2} < \dfrac{s_1^2}{s_2^2} F_R \rightarrow \dfrac{9.61}{8.41}\, 0.32 < \dfrac{\sigma_1^2}{\sigma_2^2} < \dfrac{9.61}{8.41}\, 3.36 \rightarrow 0.366 < \dfrac{\sigma_1^2}{\sigma_2^2} < 3.839$

10.4 ANALYSIS OF VARIANCE

10.4 Try It Yourself Solutions

1a. $H_0: \mu_1 = \mu_2 = \mu_3 = \mu_4$
$H_a:$ At least one mean is different from the others.

b. $\alpha = 0.05$

c. $\text{d.f.}_N = 3$
$\text{d.f.}_D = 14$

d. $F_0 = 3.34$; Reject H_0 if $F > 3.34$.

e.

Variation	Sum of Squares	Degrees of Freedom	Mean Squares	F
Between	549.8	3	183.3	4.22
Within	608.0	14	43.4	

$F \approx 4.22$

f. Reject H_0.

g. There is enough evidence to conclude that at least one mean is different from the others.

2a. Enter the data.

b. $H_0: \mu_1 = \mu_2 = \mu_3 = \mu_4$
$H_a:$ At least one mean is different from the others.

Variation	Sum of Squares	Degrees of Freedom	Mean Squares	F
Between	0.584	3	0.195	1.34
Within	4.360	30	0.145	

$F = 1.34 \rightarrow P\text{-value} = 0.280$

c. Fail to reject H_0.

d. There is not enough evidence to conclude that at least one mean is different from the others.

10.4 EXERCISE SOLUTIONS

1. $H_0: \mu_1 = \mu_2 = \ldots = \mu_k$
$H_a:$ At least one of the means is different from the others.

3. MS_B measures the differences related to the treatment given to each sample.
MS_W measures the differences related to entries within the same sample.

5. (a) $H_0: \mu_1 = \mu_2 = \mu_3$
$H_a:$ At least one mean is different from the others. (claim)

(b) $\text{d.f.}_N = k - 1 = 2$
$\text{d.f.}_D = N - k = 21$

$F_0 = 3.47$; Reject H_0 if $F > 3.47$.

(c)

Variation	Sum of Squares	Degrees of Freedom	Mean Squares	F
Between	1.253	2	0.627	0.771
Within	17.067	21	0.813	

$F \approx 0.771$

(d) Fail to reject H_0. There is not enough evidence at the 5% level to conclude that the mean costs per month are different.

7. (a) $H_0: \mu_1 = \mu_2 = \mu_3$ (claim)

H_a: At least one mean is different from the others.

(b) $d.f._N = k - 1 = 2$

$d.f._D = N - k = 12$

$F_0 = 2.81$; Reject H_0 if $F > 2.81$.

(c)

Variation	Sum of Squares	Degrees of Freedom	Mean Squares	F
Between	302.6	2	151.3	1.77
Within	1024.2	12	85.4	

$F \approx 1.77$

(d) Fail to reject H_0. There is not enough evidence at the 10% level to reject the claim that the mean prices are all the same for the three types of treatment.

9. (a) $H_0: \mu_1 = \mu_2 = \mu_3 = \mu_4$ (claim)

H_a: At least one mean is different from the others.

(b) $d.f._N = k - 1 = 3$

$d.f._D = N - k = 29$

$F_0 = 4.54$; Reject H_0 if $F > 4.54$.

(c)

Variation	Sum of Squares	Degrees of Freedom	Mean Squares	F
Between	37.703	3	12.568	4.979
Within	73.206	29	2.524	

$F \approx 4.979$

(d) Reject H_0. There is enough evidence at the 1% level to reject the claim that the mean number of days patients spend in the hospital is the same for all four regions.

11. (a) $H_0: \mu_1 = \mu_2 = \mu_3 = \mu_4$ (claim)

H_a: At least one mean is different from the others.

(b) $d.f._N = k - 1 = 3$

$d.f._D = N - k = 43$

$F_0 = 2.22$; Reject H_0 if $F > 2.22$.

(c)

Variation	Sum of Squares	Degrees of Freedom	Mean Squares	F
Between	15,532	3	5177	3.04
Within	73,350	43	1706	

$F \approx 3.04$

(d) Reject H_0. There is not enough evidence at the 10% level to reject the claim that the mean price is the same for all four cities.

13. (a) H_0: $\mu_1 = \mu_2 = \mu_3 = \mu_4$

H_a: At least one mean is different from the others. (claim)

(b) d.f.$_N$ = $k - 1 = 3$

d.f.$_D$ = $N - k = 26$

$F_0 = 2.31$; Reject H_0 if $F > 2.31$

(c)

Variation	Sum of Squares	Degrees of Freedom	Mean Squares	F
Between	877,803,059	3	292,601,020	6.36
Within	1,195,691,981	26	45,988,153	

$F \approx 6.36$

(d) Reject H_0. There is enough evidence at the 10% level to conclude that at least one of the mean prices is different.

15. (a) H_0: $\mu_1 = \mu_2 = \mu_3$ (claim)

H_a: At least one mean is different from the others.

(b) d.f.$_N$ = $k - 1 = 3$

d.f.$_D$ = $N - k = 19$

$F_0 = 5.01$; Reject H_0 if $F > 5.01$.

(c)

Variation	Sum of Squares	Degrees of Freedom	Mean Squares	F
Between	1,711.791	3	570.597	0.19
Within	57,303.65	19	3015.982	

$F \approx 0.19$

(d) Fail to reject H_0. There is not enough evidence at the 1% level to reject the claim that the mean numbers of female students are equal for all grades.

17. H_0: Advertising medium has no effect on mean rating.

H_a: Advertising medium has an effect on mean ratings.

H_0: Length of ad has no effect on mean ratings.

H_a: Length of ad has an effect on mean ratings.

H_0: There is no interaction effect between advertising medium and length of ad on mean ratings.

H_a: There is an interaction effect between advertising medium and length of ad on mean ratings.

Source	d.f.	SS	MS	F	P
Ad medium	1	1.25	1.25	0.57	0.459
Length of ad	1	0.45	0.45	0.21	0.655
Interaction	1	0.45	0.45	0.21	0.655
Error	16	34.80	2.17		
Total	19	36.95			

None of the null hypotheses can be rejected at the 10% level.

19. H_0: Age has no effect on mean GPA.
H_a: Age has an effect on mean GPA.

H_0: Gender has no effect on mean GPA.
H_a: Gender has an effect on mean GPA.

H_0: There is no interaction effect between age and gender on mean GPA.
H_a: There is an interaction effect between age and gender on mean GPA.

Source	d.f.	SS	MS	F	P
Age	3	0.41	0.14	0.12	0.948
Gender	1	0.18	0.18	0.16	0.697
Interaction	3	0.29	0.10	0.08	0.968
Error	16	18.66	1.17		
Total	23	19.55			

None of the null hypotheses can be rejected at the 10% level.

21.

	Mean	Size
Pop 1	15.62	13
Pop 2	11.60	13
Pop 3	14.16	13
Pop 4	11.81	13

$SS_w = 1024.5$

$\Sigma(n_i - 1) = N - k = 48$

$F_0 = 2.20 \rightarrow CV_{\text{Scheffé}} = 2.20(4 - 1) = 6.60$

$$\dfrac{(\bar{x}_1 - \bar{x}_2)^2}{\dfrac{SS_w}{\Sigma(n_i - 1)}\left[\dfrac{1}{n_1} + \dfrac{1}{n_2}\right]} \approx 4.927 \rightarrow \text{No difference}$$

$$\dfrac{(\bar{x}_1 - \bar{x}_3)^2}{\dfrac{SS_w}{\Sigma(n_i - 1)}\left[\dfrac{1}{n_1} + \dfrac{1}{n_3}\right]} \approx 0.650 \rightarrow \text{No difference}$$

$$\dfrac{(\bar{x}_1 - \bar{x}_4)^2}{\dfrac{SS_w}{\Sigma(n_i - 1)}\left[\dfrac{1}{n_1} + \dfrac{1}{n_4}\right]} \approx 4.426 \rightarrow \text{No difference}$$

$$\dfrac{(\bar{x}_2 - \bar{x}_3)^2}{\dfrac{SS_w}{\Sigma(n_i - 1)}\left[\dfrac{1}{n_2} + \dfrac{1}{n_3}\right]} \approx 1.998 \rightarrow \text{No difference}$$

$$\dfrac{(\bar{x}_2 - \bar{x}_4)^2}{\dfrac{SS_w}{\Sigma(n_i - 1)}\left[\dfrac{1}{n_2} + \dfrac{1}{n_4}\right]} \approx 0.013 \rightarrow \text{No difference}$$

$$\dfrac{(\bar{x}_3 - \bar{x}_4)^2}{\dfrac{SS_w}{\Sigma(n_i - 1)}\left[\dfrac{1}{n_3} + \dfrac{1}{n_4}\right]} \approx 1.684 \rightarrow \text{No difference}$$

23.

	Mean	Size
Pop 1	135.3	10
Pop 2	156.373	11
Pop 3	65.478	9
Pop 4	73.62	10

$SS_w \approx 86{,}713.313$

$\Sigma(n_i - 1) = N - k = 36$

$F_0 = 4.38 \rightarrow CV_{\text{Scheffé}} = 4.38(4 - 1) = 13.14$

$$\frac{(\bar{x}_1 - \bar{x}_2)^2}{\frac{SS_w}{\Sigma(n_i - 1)}\left[\frac{1}{n_1} + \frac{1}{n_2}\right]} \approx 0.966 \rightarrow \text{No difference}$$

$$\frac{(\bar{x}_1 - \bar{x}_3)^2}{\frac{SS_w}{\Sigma(n_i - 1)}\left[\frac{1}{n_1} + \frac{1}{n_3}\right]} \approx 9.587 \rightarrow \text{No difference}$$

$$\frac{(\bar{x}_1 - \bar{x}_4)^2}{\frac{SS_w}{\Sigma(n_i - 1)}\left[\frac{1}{n_1} + \frac{1}{n_4}\right]} \approx 7.897 \rightarrow \text{No difference}$$

$$\frac{(\bar{x}_2 - \bar{x}_3)^2}{\frac{SS_w}{\Sigma(n_i - 1)}\left[\frac{1}{n_2} + \frac{1}{n_3}\right]} \approx 16.979 \rightarrow \text{Significant difference}$$

$$\frac{(\bar{x}_2 - \bar{x}_4)^2}{\frac{SS_w}{\Sigma(n_i - 1)}\left[\frac{1}{n_2} + \frac{1}{n_4}\right]} \approx 14.892 \rightarrow \text{Significant difference}$$

$$\frac{(\bar{x}_3 - \bar{x}_4)^2}{\frac{SS_w}{\Sigma(n_i - 1)}\left[\frac{1}{n_3} + \frac{1}{n_4}\right]} \approx 0.130 \rightarrow \text{No difference}$$

CHAPTER 10 REVIEW EXERCISE SOLUTIONS

1. Claimed distribution:

Category	Distribution
New parents	25%
Old/New	25%
Old/Recurring	50%

H_0: Distribution of office visits is as shown in table above.

H_a: Distribution of office visits differs from the claimed distribution.

$\chi_0^2 = 5.991$

Category	Distribution	Observed	Expected	$\frac{(O - E)^2}{E}$
New parents	25%	97	174	34.075
Old/New	25%	142	174	5.885
Old/Recurring	50%	457	348	34.141
		696		74.101

3. Claimed distribution:

Price	Distribution
Less than $1.50	6%
$1.50–$1.99	22%
$2.00	29%
$2.01–$2.49	5%
$2.50 or more	38%

H_0: Distribution of prices is as shown in the table above.

H_a: Distribution of prices differs from the claimed distribution.

$\chi_0^2 = 7.779$

Price	Distribution	Observed	Expected	$\frac{(O-E)^2}{E}$
Less than $1.50	6%	54	48	0.750
$1.50–$1.99	22%	172	176	0.091
$2.00	29%	244	232	0.621
$2.01–$2.49	5%	28	40	3.600
$2.50 or more	38%	302	304	0.013
		800		5.075

$\chi^2 \approx 5.075$

Fail to reject H_0. There is not enough evidence at the 10% level to reject the claimed distribution.

5. (a) Expected frequencies:

	HS–did not complete	HS complete	College 1–3 years	College 4+ years	Total
25–44	13,836.62	27,733.60	20,686.43	20,962.35	83,219
45+	15,047.38	30,160.40	22,496.57	22,796.65	90,501
Total	28,884	57,894	43,183	43,759	173,720

(b) H_0: Education is independent of age.

H_a: Education is dependent on age.

d.f. = 3

$\chi_0^2 = 6.251$

$\chi^2 = \sum \frac{(O-E)^2}{E} \approx 2{,}904.408$

Reject H_0.

(c) There is enough evidence at the 10% level of significance to conclude that education level of people in the U.S. and their age are dependent.

7. (a) Expected frequencies:

	Age Group						
Gender	16–20	21–30	31–40	41–50	51–60	61+	Total
Male	143.76	284.17	274.14	237.37	147.10	43.46	1130
Female	71.24	140.83	135.86	117.63	72.90	21.54	560
Total	215	425	410	355	220	65	1690

(b) H_0: Gender is independent of age.

H_a: Gender and age are dependent.

d.f. = 5

$\chi_0^2 = 11.071$

$\chi^2 = \sum \dfrac{(O-E)^2}{E} = 9.951$

Fail to reject H_0.

(c) There is not enough evidence at the 10% level to conclude that gender and age are dependent.

9. $F_0 \approx 2.295$

11. $F_0 = 2.39$

13. $H_0: \sigma_1^2 \leq \sigma_2^2$ (claim); $H_a: \sigma_1^2 > \sigma_2^2$

d.f.$_N$ = 15

d.f.$_D$ = 20

$F_0 = 3.09$; Reject H_0 if $F > 3.09$.

$F = \dfrac{s_1^2}{s_2^2} = \dfrac{653}{270} \approx 2.419$

Fail to reject H_0. There is not enough evidence to reject the claim.

15. Population 1: Garfield County

Population 2: Kay County

$H_0: \sigma_1^2 \leq \sigma_2^2$; $H_a: \sigma_1^2 > \sigma_2^2$ (claim)

d.f.$_N$ = 20

d.f.$_D$ = 15

$F_0 = 1.92$; Reject H_0 if $F > 1.92$.

$F = \dfrac{s_1^2}{s_2^2} = \dfrac{(0.76)^2}{(0.58)^2} \approx 1.717$

Fail to reject H_0. There is not enough evidence at the 10% level to support the claim that the variation in wheat production is greater in Garfield County than in Kay County.

17. Population 1: Male $\rightarrow s_1^2 = 18{,}486.26$

Population 2: Female $\rightarrow s_2^2 = 12{,}102.78$

$H_0: \sigma_1^2 = \sigma_2^2$; $H_a: \sigma_1^2 \neq \sigma_2^2$ (claim)

d.f.$_N$ = 13

d.f.$_D$ = 8

$F_0 = 6.94$; Reject H_0 if $F > 6.94$.

$F = \dfrac{s_1^2}{s_2^2} = \dfrac{18{,}486.26}{12{,}102.78} \approx 1.527$

Fail to reject H_0. There is not enough evidence at the 1% level to support the claim that the test score variance for females is different than that for males.

19. H_0: $\mu_1 = \mu_2 = \mu_3 = \mu_4$

    H_a: At least one mean is different from the others. (claim)

    d.f.$_N$ = $k - 1 = 3$
    d.f.$_D$ = $N - k = 28$

    $F_0 = 2.29$; Reject H_0 if $F > 2.29$.

Variation	Sum of Squares	Degrees of Freedom	Mean Squares	F
Between	2,318,032	3	772,677	6.60
Within	3,277,311	28	117,047	

    $F \approx 6.60$

    Reject H_0. There is enough evidence at the 10% level to conclude that the mean residential energy expenditures are not the same for all four regions.

CHAPTER 10 QUIZ SOLUTIONS

1. (a) Population 1: San Jose → $s_1^2 \approx 429.984$

    Population 2: Dallas → $s_2^2 \approx 112.779$

    H_0: $\sigma_1^2 = \sigma_2^2$; H_a: $\sigma_1^2 \neq \sigma_2^2$ (claim)

   (b) $\alpha = 0.01$

   (cd) d.f.$_N$ = 12
    d.f.$_D$ = 17

    $F_0 = 3.97$; Reject H_0 if $F > 3.97$.

   (e) $F = \dfrac{s_1^2}{s_2^2} = \dfrac{429.984}{112.779} \approx 3.813$

   (f) Fail to reject H_0.

   (g) There is not enough evidence at the 1% level to conclude that the variances in annual wages for San Jose, CA and Dallas, TX are different.

2. (a) H_0: $\mu_1 = \mu_2 = \mu_3$ (claim)

    H_a: At least one mean is different from the others.

   (b) $\alpha = 0.10$

   (cd) d.f.$_N$ = $k - 1 = 2$
    d.f.$_D$ = $N - k = 44$

    $F_0 = 2.43$; Reject H_0 if $F > 2.43$.

Variation	Sum of Squares	Degrees of Freedom	Mean Squares	F
Between	2676	2	1338	7.39
Within	7962	44	181	

   (e) $F \approx 7.39$

   (f) Reject H_0.

   (g) There is enough evidence at the 10% level to conclude that the mean annual wages for the three cities are not all equal.

3. (a) Claimed distribution:

Education	25 & Over
Not a HS graduate	16.6%
HS graduate	33.3%
Some college, no degree	17.3%
Associate degree	7.5%
Bachelor's degree	17.0%
Advanced degree	8.2%

H_0: Distribution of educational achievement for people in the United States aged 35–44 is as shown in table above.

H_a: Distribution of educational achievement for people in the United States aged 35–44 differs from the claimed distribution.

(b) $\alpha = 0.01$

(cd) $\chi_0^2 = 15.086$; Reject H_0 if $\chi^2 > 15.086$.

(e)

Education	25 & Over	Observed	Expected	$\frac{(O-E)^2}{E}$
Not a HS graduate	16.6%	37	49.966	3.365
HS graduate	33.3%	102	100.233	0.0312
Some college, no degree	17.3%	59	52.073	0.921
Associate degree	7.5%	27	22.575	0.867
Bachelor's degree	17.0%	51	51.17	0.001
Advanced degree	8.2%	25	24.682	0.004
		301		5.189

$\chi^2 = 5.189$

(f) Fail to reject H_0.

(g) There is not enough evidence at the 1% level to conclude that the distribution of educational achievement for people in the United States aged 35–44 differs from the claimed distribution.

4. (a) Claimed distribution:

Education	25 & Over
Not a HS graduate	16.6%
HS graduate	33.3%
Some college, no degree	17.3%
Associate degree	7.5%
Bachelor's degree	17.0%
Advanced degree	8.2%

H_0: Distribution of educational achievement for people in the United States aged 65–74 is as shown in table above.

H_a: Distribution of educational achievement for people in the United States aged 65–74 differs from the claimed distribution.

(b) $\alpha = 0.05$

(c) $\chi_0^2 = 11.071$

(d) Reject H_0 if $\chi^2 > 11.071$.

(e)

Education	25 & Over	Observed	Expected	$\frac{(O-E)^2}{E}$
Not a HS graduate	16.6%	124	67.562	47.146
HS graduate	33.3%	148	135.531	1.147
Some college, no degree	17.3%	61	70.411	1.258
Associate degree	7.5%	15	30.525	7.896
Bachelor's degree	17.0%	37	69.19	14.976
Advanced degree	8.3%	22	33.781	4.109
		407		76.532

$\chi^2 = 76.532$

(f) Reject H_0.

(g) There is enough evidence at the 5% level to conclude that the distribution of educational achievement for people in the United States aged 65–74 differs from the claimed distribution.

Nonparametric Tests

CHAPTER 11

11.1 THE SIGN TEST

11.1 Try It Yourself Solutions

1a. H_0: median $\leq$ 2500; H_a: median > 2500 (claim)

b. $\alpha = 0.025$

c. $n = 22$

d. The critical value is 5.

e. $x = 10$

f. Fail to reject H_0.

g. There is not enough evidence to support the claim.

2a. H_0: median = 134,500 (claim); H_a: median $\neq$ 134,500

b. $\alpha = 0.10$

c. $n = 81$

d. The critical value is $z_0 = -1.645$.

e. $x = 30$

$$z = \frac{(x + 0.5) - 0.5(n)}{\frac{\sqrt{n}}{2}} = \frac{(30 + 0.5) - 0.5(81)}{\frac{\sqrt{81}}{2}} = \frac{-10}{4.5} = -2.22$$

f. Reject H_0.

g. There is enough evidence to reject the claim.

3a. H_0: The number of colds will not decrease.
H_a: The number of colds will decrease. (claim)

b. $\alpha = 0.05$

c. $n = 11$

d. The critical value is 2.

e. $x = 2$

f. Reject H_0.

g. There is enough evidence to support the claim.

11.1 EXERCISE SOLUTIONS

1. A nonparametric test is a hypothesis test that does not require any specific conditions concerning the shape of populations or the value of any population parameters.

A nonparametric test is usually easier to perform than its corresponding parametric test, but the nonparametric test is usually less efficient.

3. (a) H_0: median ≤ 200; H_a: median > 200 (claim)

   (b) Critical value is 1.

   (c) $x = 5$

   (d) Fail to reject H_0.

   (e) There is not enough evidence at the 1% level to support the claim that the median amount of new credit card charges for the previous month was more than $200.

5. (a) H_0: median $\leq 148{,}000$ (claim); H_a: median $> 148{,}000$

   (b) Critical value is 1.

   (c) $x = 3$

   (d) Fail to reject H_0.

   (e) There is not enough evidence at the 5% level to reject the claim that the median sales price of new privately owned one-family homes sold in the past year is $148,000 or less.

7. (a) H_0: median ≥ 1900 (claim); H_a: median < 1900

   (b) Critical value is $z_0 = -2.055$.

   (c) $x = 36$

   $$z = \frac{(x + 0.5) - 0.5(n)}{\frac{\sqrt{n}}{2}} = \frac{(36 + 0.5) - 0.5(104)}{\frac{\sqrt{104}}{2}} = \frac{-15.5}{5.099} \approx -3.040$$

   (d) Reject H_0.

   (e) There is enough evidence at the 2% level to reject the claim that the median amount of credit card debt for families holding such debts is at least $1900.

9. (a) H_0: median ≤ 34; H_a: median > 34 (claim)

   (b) Critical value is 3.

   (c) $x = 8$

   (d) Fail to reject H_0.

   (e) There is not enough evidence at the 1% level to support the claim that the median age of recipients of social science doctorates is greater than 34.

11. (a) H_0: median $= 4$ (claim); H_a: median $\neq 4$

    (b) Critical value is $z_0 = -1.96$.

    (c) $x = 15$

    $$z = \frac{(x + 0.5) - 0.5(n)}{\frac{\sqrt{n}}{2}} = \frac{(15 + 0.5) - 0.5(47)}{\frac{\sqrt{47}}{2}} = \frac{-8}{3.429} \approx -2.334$$

    (d) Reject H_0.

    (e) There is enough evidence at the 5% level to reject the claim that the median number of rooms in renter-occupied units is 4.

13. (a) H_0: median $= \$11.63$ (claim); H_a: median $\neq \$11.63$

    (b) Critical value is $z_0 = -2.575$.

(c) $x = 16$

$$z = \frac{(x + 0.5) - 0.5(n)}{\frac{\sqrt{n}}{2}} = \frac{(16 + 0.5) - 0.5(39)}{\frac{\sqrt{39}}{2}} = \frac{-3}{3.1225} \approx -0.961$$

(d) Fail to reject H_0.

(e) There is not enough evidence at the 1% level to reject the claim that the median hourly earnings of male workers paid hourly rates is $11.63.

15. (a) H_0: The headache hours have not decreased.
    H_a: The headache hours have decreased. (claim)

    (b) Critical value is 1.

    (c) $x = 3$

    (d) Fail to reject H_0.

    (e) There is not enough evidence at the 5% level to support the claim that the daily headache hours were reduced after the soft tissue therapy and spinal manipulation.

17. (a) H_0: The SAT scores have not improved.
    H_a: The SAT scores have improved. (claim)

    (b) Critical value is 2.

    (c) $x = 4$

    (d) Fail to reject H_0.

    (e) There is not enough evidence at the 5% level to support the claim that the verbal SAT scores improved.

19. (a) H_0: The proportion of companies offering transportation is equal to the proportion of companies that do not. (claim)
    H_a: The proportion of companies offering transportation is not equal to the proportion of companies that do not.
    Critical value is 5.
    $\alpha = 0.05$
    $x = 8$
    Fail to reject H_0.

    (b) There is not enough evidence at the 5% level to reject the claim that the proportion of companies providing transportation home is equal to the proportions of companies that do not.

21. (a) H_0: median ≤ 530 (claim); H_a: median > 530

    (b) Critical value is $z_0 = 2.33$.

    (c) $x = 29$

    $$z = \frac{(x + 0.5) - 0.5(n)}{\frac{\sqrt{n}}{2}} = \frac{(29 + 0.5) - 0.5(47)}{\frac{\sqrt{47}}{2}} = \frac{6}{3.428} \approx 1.459$$

    (d) Fail to reject H_0. There is not enough evidence at the 1% level to reject the claim that the median weekly earnings of female workers is less than or equal to $530.

176 CHAPTER 11 | NONPARAMETRIC TESTS

23. (a) H_0: median ≤ 25.3; H_a: median > 25.3 (claim)

(b) Critical value is $z_0 = 1.645$.

(c) $x = 38$

$$z = \frac{(x + 0.5) - 0.5(n)}{\frac{\sqrt{n}}{2}} = \frac{(38 + 0.5) - 0.5(60)}{\frac{\sqrt{60}}{2}} = \frac{8.5}{3.873} \approx 1.936$$

(d) Reject H_0. There is enough evidence at the 5% level to support the claim that the median age of first-time brides is greater than 25.3 years.

11.2 THE WILCOXON TESTS

11.2 Try It Yourself Solutions

1a. H_0: The water repellent is not effective.

H_a: The water repellent is effective. (claim)

b. $\alpha = 0.01$

c. $n = 11$

d. Critical value is 7.

e.

No repellent	Repellent applied	Difference	Absolute value	Rank	Signed rank
8	15	−7	7	11	−11
7	12	−5	5	9	−9
7	11	−4	4	7.5	−7.5
4	6	−2	2	3.5	−3.5
6	6	0	0	—	—
10	8	2	2	3.5	3.5
9	8	1	1	1.5	1.5
5	6	−1	1	1.5	−1.5
9	12	−3	3	5.5	−5.5
11	8	3	3	5.5	5.5
8	14	−6	6	10	−10
4	8	−4	4	7.5	−7.5

Sum of negative ranks = −55.5

Sum of positive ranks = 10.5

$w_s = 10.5$

f. Reject H_0.

g. There is enough evidence at the 1% level to support the claim.

2a. H_0: There is no difference in the claims paid by the companies.

H_a: There is a difference in the claims paid by the companies. (claim)

b. $\alpha = 0.05$

c. The critical values are $z_0 = \pm 1.96$.

d. $n_1 = 12$ and $n_2 = 12$

e.

Ordered data	Sample	Rank
1.7	B	1
1.8	B	2
2.2	B	3
2.5	A	4
3.0	A	5.5
3.0	B	5.5
3.4	B	7
3.9	A	8
4.1	B	9
4.4	B	10
4.5	A	11
4.7	B	12

Ordered data	Sample	Rank
5.3	B	13
5.6	B	14
5.8	A	15
6.0	A	16
6.2	A	17
6.3	A	18
6.5	A	19
7.3	B	20
7.4	A	21
9.9	A	22
10.6	A	23
10.8	B	24

R = sum ranks of company B = 120.5

f. $\mu_R = \dfrac{n_1(n_1 + n_2 + 1)}{2} = \dfrac{12(12 + 12 + 1)}{2} = 150$

$\sigma_R = \sqrt{\dfrac{n_1 n_2 (n_1 + n_2 + 1)}{12}} = \sqrt{\dfrac{(12)(12)(12 + 12 + 1)}{12}} \approx 17.321$

$z = \dfrac{R - \mu_R}{\sigma_R} = \dfrac{120.5 - 150}{17.321} \approx -1.703$

g. Fail to reject H_0.

h. There is not enough evidence to conclude that there is a difference in the claims paid by both companies.

11.2 EXERCISE SOLUTIONS

1. The Wilcoxon signed-rank test is used to determine whether two dependent samples were selected from populations having the same distribution. The Wilcoxon rank sum test is used to determine whether two independent samples were selected from populations having the same distribution.

3. (a) H_0: There is no reduction in diastolic blood pressure. (claim)

   H_a: There is a reduction in diastolic blood pressure.

   (b) Wilcoxon signed-rank test

   (c) Critical value is 10.

   (d) $w_s = 17$

   (e) Fail to reject H_0.

   (f) There is not enough evidence at the 1% level to reject the claim that there was no reduction in diastolic blood pressure.

5. (a) H_0: There is no difference in the earnings.

   H_a: There is a difference in the earnings. (claim)

   (b) Wilcoxon rank sum test

   (c) The critical values are $z_0 = \pm 1.96$.

(d) $R = 55$

$$\mu_R = \frac{n_1(n_1 + n_2 + 1)}{2} = \frac{11(11 + 10 + 1)}{2} = 110$$

$$\sigma_R = \sqrt{\frac{n_1 n_2(n_1 + n_2 + 1)}{12}} = \sqrt{\frac{(11)(10)(11 + 10 + 1)}{12}} \approx 14.201$$

$$z = \frac{R - \mu_R}{\sigma_R} = \frac{55 - 110}{14.201} \approx -3.873$$

(e) Reject H_0.

(f) There is enough evidence at the 5% level to support the claim that there is a difference in the earnings.

7. (a) H_0: There is not a difference in salaries.
   H_a: There is a difference in salaries. (claim)

   (b) Wilcoxon rank sum test

   (c) The critical values are $z_0 = \pm 1.96$.

   (d) $R = 118.5$

   $$\mu_R = \frac{n_1(n_1 + n_2 + 1)}{2} = \frac{12(12 + 12 + 1)}{2} = 150$$

   $$\sigma_R = \sqrt{\frac{n_1 n_2(n_1 + n_2 + 1)}{12}} = \sqrt{\frac{(12)(12)(12 + 12 + 1)}{12}} \approx 17.321$$

   $$z = \frac{R - \mu_R}{\sigma_R} = \frac{118.5 - 150}{17.321} \approx -1.819$$

   (e) Fail to reject H_0.

   (f) There is not enough evidence at the 5% level to support the claim that there is a difference in salaries.

9. H_0: The fuel additive does not improve gas mileage.
   H_a: The fuel additive does improve gas mileage. (claim)
   Critical value is $z_0 = 1.282$.
   $$w_s = 43.5$$

   $$z = \frac{w_s - \frac{n(n + 1)}{4}}{\sqrt{\frac{n(n + 1)(2n + 1)}{24}}} = \frac{43.5 - \frac{32(32 + 1)}{4}}{\sqrt{\frac{32(32 + 1)[(2)32 + 1]}{24}}} = \frac{-220.5}{\sqrt{2860}} \approx -4.123$$

   Note: $n = 32$ because one of the differences is zero and should be discarded.
   Reject H_0. There is enough evidence at the 10% level to conclude that the gas mileage is improved.

11.3 THE KRUSKAL-WALLIS TEST

11.3 Try It Yourself Solutions

1a. H_0: There is no difference in the salaries in the three states.
   H_a: There is a difference in the salaries in the three states. (claim)

b. $\alpha = 0.10$

c. d.f. $= k - 1 = 2$

d. Critical value is $\chi_0^2 = 4.605$; Reject H_0 if $\chi^2 > 4.605$.

e.

Ordered data	State	Rank	Ordered data	State	Rank
25.57	CT	1	40.31	CA	16
26.42	CA	2	41.33	CA	17
29.73	NJ	3	43.26	NJ	18
29.86	CT	4	43.89	NJ	19
30.00	CT	5	44.57	NJ	20
33.68	CA	6.5	45.04	CT	21
33.68	CT	6.5	45.29	CT	22
33.91	NJ	8	46.55	CA	23
34.29	NJ	9	46.72	NJ	24
36.35	NJ	10	47.17	CA	25
36.55	CA	11	48.46	CA	26
37.18	CA	12	50.16	NJ	27
37.24	NJ	13	51.03	CT	28
37.39	CT	14	57.07	CT	29
38.36	CA	15	61.46	CT	30

$R_1 = 153.5 \quad R_2 = 160.5 \quad R_3 = 151$

f. $H = \dfrac{12}{N(N+1)} \left(\dfrac{R_1^2}{n_1} + \dfrac{R_2^2}{n_2} + \dfrac{R_3^2}{n_3} \right) - 3(N+1)$

$= \dfrac{12}{30(30+1)} \left(\dfrac{(153.5)^2}{10} + \dfrac{(160.5)^2}{10} + \dfrac{(151)^2}{10} \right) - 3(30+1) = 0.063$

g. Fail to reject H_0.

h. There is not enough evidence to support the claim.

11.3 EXERCISE SOLUTIONS

1. Each sample must be randomly selected and the size of each sample must be at least 5.

3. (a) H_0: There is no difference in the premiums.

   H_a: There is a difference in the premiums. (claim)

   (b) Critical value is 5.991.

   (c) $H \approx 14.05$

   (d) Reject H_0.

   (e) There is enough evidence at the 5% level to support the claim that the distributions of the annual premiums in California, Florida, and Illinois are different.

5. (a) H_0: There is no difference in the salaries.

   H_a: There is a difference in the salaries. (claim)

   (b) Critical value is 6.251.

   (c) $H \approx 1.45$

   (d) Fail to reject H_0.

   (e) There is enough evidence at the 10% level to support the claim that the distributions of the annual salaries in the four states are different.

180 CHAPTER 11 | NONPARAMETRIC TESTS

7. (a) H_0: There is no difference in the number of days spent in the hospital.

H_a: There is a difference in the number of days spent in the hospital. (claim)

The critical value is 11.345.

$H = 1.51$;

Fail to reject H_0.

(b)

Variation	Sum of squares	Degrees of freedom	Mean Squares	F
Between	9.17	3	3.06	0.52
Within	194.72	33	5.90	

For $\alpha = 0.01$, the critical value is about 4.45. Because $F = 0.52$ is less than the critical value, the decision is to fail to reject H_0. There is not enough evidence to support the claim.

(c) Both tests come to the same decision, which is, that there is not enough evidence to support the claim that there is a difference in the number of days spent in the hospital.

11.4 RANK CORRELATION

11.4 Try It Yourself Solutions

1a. $H_0: \rho_s = 0; H_a: \rho_s \neq 0$

b. $\alpha = 0.05$

c. Critical value is 0.700.

d.

Oat	Rank	Wheat	Rank	d	d^2
1.67	7	4.55	9	−2	4
1.96	9	4.30	8	1	1
1.60	6	3.38	6	0	0
1.10	1.5	2.65	3	−1.5	2.25
1.12	3	2.48	1	2	4
1.10	1.5	2.62	2	−0.5	0.25
1.59	5	2.78	4	1	1
1.81	8	3.56	7	1	1
1.45	4	3.35	5	−1	1
					14.5

$\Sigma d^2 = 14.5$

e. $r_s = 1 - \dfrac{6\Sigma d^2}{n(n^2 - 1)} = 0.879$

f. Reject H_0.

g. There is enough evidence to conclude that a significant correction exists.

11.4 EXERCISE SOLUTIONS

1. The Spearman rank correlation coefficient can (1) be used to describe the relationship between linear and nonlinear data, (2) be used for data at the ordinal level, and (3) is easier to calculate by hand than the Pearson coefficient.

3. (a) $H_0: \rho_s = 0; H_a: \rho_s \neq 0$ (claim)

   (b) Critical value is 0.929.

   (c) $\Sigma d^2 = 14$
   $$r_s = 1 - \frac{6\Sigma d^2}{n(n^2-1)} \approx 0.750$$

   (d) Fail to reject H_0.

   (e) There is not enough evidence at the 1% level to support the claim that there is a correlation between debt and income in the farming business.

5. (a) $H_0: \rho_s = 0; H_a: \rho_s \neq 0$ (claim)

   (b) The critical value is 0.881.

   (c) $\Sigma d^2 = 69.5$
   $$r_s = 1 - \frac{6\Sigma d^2}{n(n^2-1)} \approx 0.173$$

   (d) Fail to reject H_0.

   (e) There is not enough evidence at the 1% level to support the claim that there is a correlation between the overall score and price.

7. $H_0: \rho_s = 0; H_a: \rho_s \neq 0$ (claim)

   Critical value is 0.700.

   $\Sigma d^2 = 124$
   $$r_s = 1 - \frac{6\Sigma d^2}{n(n^2-1)} \approx -0.033$$

   Fail to reject H_0. There is not enough evidence at the 5% level to conclude that there is a correlation between science achievement scores and GNP.

9. $H_0: \rho_s = 0; H_a: \rho_s \neq 0$ (claim)

   Critical value is 0.700.

   $\Sigma d^2 = 42$
   $$r_s = 1 - \frac{6\Sigma d^2}{n(n^2-1)} \approx 0.650$$

   Fail to reject H_0. There is not enough evidence at the 5% level to conclude that there is a correlation between science and mathematics achievement scores.

11. $H_0: \rho_s = 0; H_a: \rho_s \neq 0$ (claim)

    The critical value is $= \dfrac{\pm z}{\sqrt{n-1}} = \dfrac{\pm 1.96}{\sqrt{33-1}} \approx \pm 0.346$.

    $\Sigma d^2 = 6673$
    $$r_s = 1 - \frac{6\Sigma d^2}{n(n^2-1)} \approx -0.115$$

    Fail to reject H_0. There is not enough evidence to support the claim.

11.5 THE RUNS TEST

11.5 Try It Yourself Solutions

1a. PPP F P F $PPPP$ FF P F PP FFF PPP F PPP

b. 13 groups $\Rightarrow$ 13 runs

c. 3, 1, 1, 1, 4, 2, 1, 1, 2, 3, 3, 1, 3

2a. H_0: The sequence of genders is random.
H_a: The sequence of genders is not random. (claim)

b. $\alpha = 0.05$

c. FFF MM FF M F MM FFF
n_1 = number of Fs = 9
n_2 = number of Ms = 5
G = number of runs = 7

d. $cv = 3$ and 12

e. $G = 7$

f. Fail to reject H_0.

g. At the 5% level, there is not enough evidence to support the claim that the sequence of genders is not random.

3a. H_0: The sequence of weather conditions is random.
H_a: The sequence of weather conditions is not random. (claim)

b. $\alpha = 0.05$

c. n_1 = number of Ns = 21
n_2 = number of Ss = 10
G = number of runs = 17

d. $cv = \pm 1.96$

e. $\mu_G = \dfrac{2n_1 n_2}{n_1 + n_2} + 1 = \dfrac{2(21)(10)}{21 + 10} + 1 \approx 14.5$

$\sigma_G = \sqrt{\dfrac{2n_1 n_2 (2n_1 n_2 - n_1 - n_2)}{(n_1 + n_2)^2 (n_1 + n_2 - 1)}} = \sqrt{\dfrac{2(21)(10)(2(21)(10) - 21 - 10)}{(21 + 10)^2 (21 + 10 - 1)}} \approx 2.4$

$z = \dfrac{G - \mu_G}{\sigma_G} = \dfrac{17 - 14.5}{2.4} = 1.04$

f. Fail to reject H_0.

g. At the 5% level, there is not enough evidence to support the claim that the sequence of weather conditions each day is not random.

11.5 EXERCISE SOLUTIONS

1. Number of runs = 8
Run lengths = 1, 1, 1, 1, 3, 3, 1, 1

3. Number of runs = 9
Run lengths = 1, 1, 1, 1, 1, 6, 3, 2, 4

5. $n_1 = $ number of Ts $= 6$
   $n_2 = $ number of Fs $= 6$

7. $n_1 = $ number of Ms $= 10$
   $n_2 = $ number of Fs $= 10$

9. $n_1 = $ number of Ts $= 6$
   $n_1 = $ number of Fs $= 6$
   $cv = 3$ and 11

11. $n_1 = $ number of Ns $= 11$
    $n_1 = $ number of Ss $= 7$
    $cv = 5$ and 14

13. (a) H_0: The coin tosses were random.
    H_a: The coin tosses were not random. (claim)

    (b) $n_1 = $ number of Hs $= 7$
    $n_2 = $ number of Ts $= 9$

    $cv = 4$ and 14

    (c) $G = 9$ runs

    (d) Fail to reject H_0.

    (e) At the 5% level, there is not enough evidence to support the claim that the coin tosses were not random.

15. (a) H_0: The sequence of digits were randomly generated.
    H_a: The sequence of digits were not randomly generated. (claim)

    (b) $n_1 = $ number of Os $= 16$
    $n_2 = $ number of Es $= 16$

    $cv = 11$ and 23

    (c) $G = 9$

    (d) Reject H_0.

    (e) At the 5% level, there is enough evidence to support the claim that the sequence of digits were not randomly generated.

17. (a) H_0: The production of defective parts is random.
    H_a: The production of defective parts is not random. (claim)

    (b) $n_1 = $ number of Ns $= 40$
    $n_2 = $ number of Ds $= 9$

    $cv = \pm 1.96$

    (c) $G = 14$ runs

    $$\mu_G = \frac{2n_1 n_2}{n_1 + n_2} - 1 = \frac{2(40)(9)}{40 + 9} - 1 = 13.7$$

    $$\sigma_G = \sqrt{\frac{2n_1 n_2(2n_1 n_2 - n_1 - n_2)}{(n_1 + n_2)^2(n_1 + n_2 - 1)}} = \sqrt{\frac{2(40)(9)(2(40)(9) - 40 - 9)}{(40 + 9)^2(40 + 9 - 1)}} = 2.0$$

    $$z = \frac{G - \mu_G}{\sigma_G} = \frac{14 - 13.7}{2.0} = 0.15$$

    (d) Fail to reject H_0.

    (e) At the 5% level, there is not enough evidence to support the claim that the production of defective parts is not random.

19. H_0: Daily high temperatures occur randomly.

H_a: Daily high temperatures do not occur randomly. (claim)

median = 87

n_1 = number above median = 14

n_2 = number below median = 13

cv = 9 and 20

G = 11 runs

Fail to reject H_0.

At the 5% level, there is not enough evidence to support the claim that the daily high temperatures do not occur randomly.

21. Answers will vary.

CHAPTER 11 REVIEW EXERCISE SOLUTIONS

1. (a) H_0: median = $34,300 (claim); H_a: median ≠ $34,300

(b) Critical value is 2.

(c) $x = 7$

(d) Fail to reject H_0.

(e) There is not enough evidence at the 1% level to reject the claim that the median value of stock among families that own stock is $34,300.

3. (a) H_0: median ≤ 6 (claim); H_a: median > 6

(b) Critical value is $z_0 \approx -1.28$.

(c) $x = 44$

$$z = \frac{(x - 0.5) - 0.5(n)}{\frac{\sqrt{n}}{2}} = \frac{(44 - 0.5) - 0.5(70)}{\frac{\sqrt{70}}{2}} = \frac{8.5}{4.1833} \approx 2.032$$

(d) Reject H_0.

(e) There is enough evidence at the 10% level to reject the claim that the median turnover time is no more than 6 hours.

5. (a) H_0: There is no reduction in diastolic blood pressure. (claim)

H_a: There is a reduction in diastolic blood pressure.

(b) Critical value is 2.

(c) $x = 3$

(d) Fail to reject H_0.

(e) There is not enough evidence at the 5% level to reject the claim that there was no reduction diastolic blood pressure.

7. (a) Independent; Wilcoxon Rank Sum Test

(b) H_0: There is no difference in the amount of time that it takes to earn a doctorate.

H_a: There is a difference in the amount of time that it takes to earn a doctorate. (claim)

(c) Critical values are $z_0 = \pm 2.575$.

(d) $R = 95$

$$\mu_R = \frac{n_1(n_1 + n_2 + 1)}{2} = \frac{12(12 + 12 + 1)}{2} = 150$$

$$\sigma_R = \sqrt{\frac{n_1 n_2(n_1 + n_2 + 1)}{12}} \sqrt{\frac{(12)(12)(12 + 12 + 1)}{12}} \approx 17.321$$

$$z = \frac{R - \mu_R}{\sigma_R} = \frac{95 - 150}{17.321} \approx -3.175$$

(e) Reject H_0.

(f) There is enough evidence at the 1% level to support the claim that there is a difference in the amount of time that it takes to earn a doctorate.

9. (a) H_0: There is no difference in salaries between the fields of study.
   H_a: There is a difference in salaries between the fields of study. (claim)

   (b) Critical value is 5.991.

   (c) $H \approx 22.98$

   (d) Reject H_0.

   (e) There is enough evidence at the 5% level to conclude that there is a difference in salaries between the fields of study.

11. (a) $H_0: \rho_s = 0$; $H_a: \rho_s \neq 0$ (claim)

    (b) Critical value is 0.881.

    (c) $\Sigma d^2 = 120$

    $$r_s = 1 - \frac{6\Sigma d^2}{n(n^2 - 1)} \approx -0.429$$

    (d) Fail to reject H_0.

    (e) There is not enough evidence at the 1% level to conclude that there is a correlation between overall score and price.

13. (a) H_0: The traffic stops are random by gender.
    H_a: The traffic stops are not random by gender. (claim)

    (b) $n_1 =$ number of Fs $= 12$
    $n_2 =$ number of Ms $= 13$

    $cv = 8$ and 19

    (c) $G = 14$ runs

    (d) Fail to reject H_0.

    (e) There is not enough evidence at the 5% level to conclude that the traffic stops are not random.

CHAPTER 11 QUIZ SOLUTIONS

1. (a) H_0: There is no difference in the salaries between genders.
   H_a: There is a difference in the salaries between genders. (claim)

   (b) Wilcoxon Rank Sum Test

(c) Critical values are $z_0 = \pm 1.645$.

(d) $R = 66$

$$\mu_R = \frac{n_1(n_1 + n_2 + 1)}{2} = \frac{9(9 + 9 + 1)}{2} = 85.500$$

$$\sigma_R = \sqrt{\frac{n_1 n_2(n_1 + n_2 + 1)}{12}} = \sqrt{\frac{(9)(9)(9 + 9 + 1)}{12}} \approx 11.325$$

$$z = \frac{R - \mu_R}{\sigma_R} = \frac{66 - 85.5}{11.325} \approx -1.722$$

(e) Reject H_0.

(f) There is enough evidence at the 10% level to support the claim that there is a difference in the salaries between genders.

2. (a) H_0: median = 28 (claim); H_a: median ≠ 28

(b) Sign Test

(c) Critical value is 6.

(d) $x = 10$

(e) Fail to reject H_0.

(f) There is not enough evidence at the 5% level to reject the claim that the median age in Puerto Rico is 28 years.

3. (a) H_0: There is no difference in the annual premiums between the states.
H_a: There is a difference in the annual premiums between the states. (claim)

(b) Kruskal-Wallis Test

(c) Critical value is 5.991.

(d) $H \approx 1.43$

(e) Fail to reject H_0.

(f) There is not enough evidence at the 5% level to conclude that there is a difference in the annual premiums between the states.

4. (a) H_0: The days with rain are random.
H_a: The days with rain are not random. (claim)

(b) The Runs test

(c) n_1 = number of Ns = 15
n_2 = number of Rs = 15

cv = 10 and 22

(d) $G = 16$ runs

(e) Fail to reject H_0.

(f) There is not enough evidence at the 5% level to conclude that days with rain are not random.

Alternative Presentation of the Standard Normal Distribution

APPENDIX A

Try It Yourself Solutions

1a. (1) 0.4857

b. (2) $z = \pm 2.17$

2a.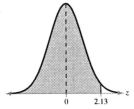

b. 0.4834

c. Area $= 0.5 + 0.4834 = 0.9834$

3a.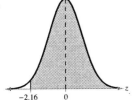

b. 0.4846

c. Area $= 0.5 + 0.4846 = 0.9846$

4a.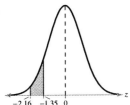

b. $z = -2.16$: Area $= 0.4846$
$z = -1.35$: Area $= 0.4115$

CONTENTS

CONTENTS

Solutions to Even-Numbered Problems

Chapter 1	Introduction to Statistics	191
Chapter 2	Descriptive Statistics	195
Chapter 3	Probability	219
Chapter 4	Discrete Probability Distributions	229
Chapter 5	Normal Probability Distributions	237
Chapter 6	Confidence Intervals	251
Chapter 7	Hypothesis Testing with One Sample	259
Chapter 8	Hypothesis Testing with Two Samples	275
Chapter 9	Correlation and Regression	291
Chapter 10	Chi-Square Tests and the F-Distribution	305
Chapter 11	Nonparametric Tests	321
Case Studies		331
Uses and Abuses		341
Real Statistics–Real Decisions		345
Technologies		353

Introduction to Statistics

1.1 AN OVERVIEW OF STATISTICS

1.1 EXERCISE SOLUTIONS

2. It is usually impractical (too expensive and time consuming) to obtain all the population data.

4. Descriptive statistics and inferential statistics.

6. True

8. False. Inferential statistics involves using a sample to draw conclusions about a population.

10. True

12. The data set is a sample since only the speed of every fifth car is recorded.

14. The data set is a population since it is a collection of the annual salaries of all employees at the company.

16. The data set is a population since it is a collection of the number of pets in all U.S. households.

18. Population: Major of college students at Central College.
    Sample: Major of college students at Central College who take statistics.

20. Population: Income of all homeowners in Ohio.
    Sample: Income of homeowners in Ohio with mortgages.

22. Population: Collection of all U.S. households.
    Sample: Collection of the 2104 U.S. households surveyed.

24. Population: Collection of all U.S. vacationers.
    Sample: Collection of the 791 U.S. vacationers surveyed.

26. Statistic. 43% is a numerical description of a sample of high school students.

28. Parameter. 12% is a numerical description of all new magazines.

30. Parameter. 20.7 is a numerical description of ACT scores for all graduates.

32. The statement "spending at least $2000 for their next vacation" is an application of descriptive statistics.

    An inference drawn from the sample is that U.S. vacationers are associated with spending more than $2000 for their next vacation.

34. (a) The volunteers in the study represent the sample.

    (b) The population is the collection of all individuals who completed the math test.

    (c) The statement "were three times more likely to answer correctly" is an application of descriptive statistics.

    (d) An inference drawn from the sample is that individuals who are not sleep deprived will be three times more likely to answer math questions correctly than individuals who are sleep deprived.

192 CHAPTER 1 | INTRODUCTION TO STATISTICS

36. (a) An inference drawn from the sample is that the number of people who have strokes has increased every year for the past 15 years.

(b) It implies the same trend will continue for the next 15 years.

1.2 DATA CLASSIFICATION

1.2 EXERCISE SOLUTIONS

2. Ordinal, Interval, and Ratio

4. False. For data at the interval level, you can calculate meaningful differences between data entries. You cannot calculate meaningful differences at the nominal or ordinal level.

6. False. Data at the ratio level can be placed in a meaningful order.

8. Quantitative, because the daily low temperature is a numerical measure.

10. Qualitative, because the player numbers are merely labels.

12. Nominal. No mathematical computations can be made and data are categorized using names.

14. Ratio. A ratio of two data values can be formed so one data value can be expressed as a multiple of another.

16. Ratio. The ratio of two data values can be formed so one data value can be expressed as a multiple of another.

18. Ratio

20. Interval

22. Answers will vary.

1.3 EXPERIMENTAL DESIGN

1.3 EXERCISE SOLUTIONS

2. False. A census is a count of an entire population.

4. True

6. It would be nearly impossible to ask every consumer whether he or she would still buy a product with a warning label. So, you should use sampling to collect these data.

8. Because the U.S. Congress keeps accurate financial records of all members, you could take a census.

10. Because the persons are divided into strata (rural and urban), and a sample is selected from each stratum, this is a stratified sample.

12. Because the disaster area was divided into grids and thirty grids were then entirely selected, this is a cluster sample. Certain grids may have been much more severely damaged than others, so this is a possible source of bias.

14. Because every twentieth engine part is sampled from an assembly line, this is a systematic sample. It is possible for bias to enter into the sample if, for some reason, the assembly line performs differently on a consistent basis.

16. Because a sample is taken from members of a population that are readily available, this is a convenience sample. The sample may be biased if the teachers sampled are not representative of the population of teachers. For example, some teachers may frequent the lounge more often than others.

18. Each telephone has an equal chance of being dialed and all samples of 1012 phone numbers have an equal chance of being selected, so this is a simple random sample. Telephone sampling only samples those individuals who have telephones, are available, and are willing to respond, so this is a possible source of bias.

20. Sampling, since the population of cars is too large to easily record their color. Cluster sampling would be advised since it would be easy to randomly select car dealerships then record color for every car sold at the selected dealerships.

22. Question is biased since it already suggests that drivers who change lanes several times are dangerous. The question might be rewritten as "Are drivers who change lanes several times dangerous?"

24. Question is biased since it already suggests that the media has a negative effect on teen girls' dieting habits. The question might be rewritten as "Do you think the media has an effect on teen girls' dieting habits?"

26. Stratified sampling ensures that each segment of the population is represented.

28. (a) Advantage: Usually results in a savings in the survey cost.

    (b) Disadvantage: There tends to be a lower response rate and this can introduce a bias into the sample.
Sampling Technique: Convenience sampling

CHAPTER 1 REVIEW EXERCISE SOLUTIONS

2. Population: Collection of all nurses in SF area.
Sample: Collection of 38 nurses in SF area that were sampled.

4. Population: Collection of all small-business managers
Sample: Collection of 787 small-business managers

6. Since 62% is describing a characteristic of the sample, this is a statistic.

8. Since 19% is describing a characteristic of a sample of Indiana ninth graders, this is a statistic.

10. 72% of the small-business managers surveyed are optimistic about the future of their company is representative of the descriptive branch of statistics. An inference drawn from the sample is that 72% of all small-business managers are optimistic about the future of their company.

12. Qualitative because social security numbers are merely labels for employees.

14. Qualitative because zip codes are merely labels for the nursing homes.

16. Ordinal. The data is categorical but could be arranged in order of car size.

18. Ratio. The data is numerical, and it makes sense saying that one player is twice as tall as another player.

20. Because it is impractical to create this situation, you would want to perform a simulation.

22. Since it would be nearly impossible to ask every college student about his/her opinion on space exploration, you should use sampling to collect the data.

24. Since the student sampled a convenient group of friends, this is a convenience sample.

26. Because every tenth car is measured, this is a systematic sample.

28. Because of the convenience of surveying people leaving one restaurant, this is a convenience sample.

30. Due to the convenience sample taken, the study may be biased toward the opinions of the student's friends.

32. In heavy interstate traffic, it may be difficult to identify every tenth car that passes the law enforcement official.

CHAPTER 1 QUIZ SOLUTIONS

1. Population: Collection of all individuals with sleep disorders.
   Sample: Collection of 254 patients in study.

2. (a) Statistic. 53% is a characteristic of a sample of parents.

   (b) Parameter. 67% is a characteristic of the entire union (population).

3. (a) Qualitative, since student identification numbers are merely labels.

   (b) Quantitative, since a test score is a numerical measure.

4. (a) Nominal. Players may be ordered numerically, but there is no meaning in this order and no mathematical computations can be made.

   (b) Ratio. It makes sense to say that the number of products sold during the 1st quarter was twice as many as sold in the 2nd quarter.

   (c) Interval because meaningful differences between entries can be calculated, but a zero entry is not an inherent zero.

5. (a) In this study, you want to measure the effect of a treatment (low dietary intake of vitamin C and iron) on lead levels in adults. You want to perform an experiment.

   (b) Since it would be difficult to survey every individual within 500 miles of your home, sampling should be used.

6. (a) Because people were chosen due to their convenience of location (on the beach), this is a convenience sample.

   (b) Since every fifth part is selected from an assembly line, this is a systematic sample.

   (c) Stratified sample because the population is first stratified and then a sample is collected from each strata.

7. Convenience

8. (a) False. A statistic is a numerical measure that describes a sample characteristic.

   (b) False. Ratio data represents the highest level of measurement.

Descriptive Statistics

CHAPTER 2

2.1 FREQUENCY DISTRIBUTIONS AND THEIR GRAPHS

2.1 EXERCISE SOLUTIONS

2. Sometimes it is easier to identify patterns of a data set by looking at a graph of the frequency distribution.

4. Frequency is the number of entries in each class. Relative frequency is the proportion of entries in each class.

6. False. The relative frequency of a class is the frequency of the class divided by the sample size.

8. False. Class boundaries are used so that consecutive bars of a histogram do touch.

10a. Class width = 21 − 16 = 5

bc.

Class	Frequency	Midpoint	Class boundaries
16–20	100	18	15.5 – 20.5
21–25	122	23	20.5 – 25.5
26–30	900	28	25.5 – 30.5
31–35	207	33	30.5 – 35.5
36–40	795	38	35.5 – 40.5
41–45	568	43	40.5 – 45.5
46–50	322	48	45.5 – 50.5
	$\Sigma f = 3014$		

12.

Class	Frequency	Midpoint	Relative frequency	Cumulative frequency
16–20	100	18	0.03	100
21–25	122	23	0.04	222
26–30	900	28	0.30	1122
31–35	207	33	0.07	1329
36–40	795	38	0.26	2124
41–45	568	43	0.19	2692
46–50	322	48	0.11	3014
	$\Sigma f = 3014$		$\Sigma \dfrac{f}{n} = 1$	

14. (a) Number of classes = 7

(b) Least frequency ≈ 100

(c) Greatest frequency ≈ 900

(d) Class width = 5

16. (a) 50   (b) 68–70 inches

18. (a) 44   (b) 70 inches

20. (a) Class with greatest relative frequency: 19–20 min.
Class with least relative frequency: 21–22 min.

(b) Greatest relative frequency ≈ 40%
Least relative frequency ≈ 2%

(c) Approximately 33%

22. Class with greatest frequency: 7.75–8.25
Class with least frequency: 6.25–6.75

24. Class width $= \dfrac{\text{Max} - \text{Min}}{\text{Number of classes}} = \dfrac{530 - 30}{6} = 83.3 \Rightarrow 84$

Class	Frequency	Midpoint	Relative frequency	Cumulative frequency
30–113	5	71.5	0.1724	5
114–197	7	155.5	0.2414	12
198–281	8	239.5	0.2759	20
282–365	2	323.5	0.0690	22
366–449	3	407.5	0.1034	25
450–533	4	491.5	0.1379	29
	$\Sigma f = 29$		$\Sigma \dfrac{f}{n} = 1$	

26. Class width $= \dfrac{\text{Max} - \text{Min}}{\text{Number of classes}} = \dfrac{51 - 32}{5} = 3.8 \Rightarrow 4$

Class	Frequency	Midpoint	Relative frequency	Cumulative frequency
32–35	3	33.5	0.1250	3
36–39	9	37.5	0.3750	12
40–43	8	41.5	0.3333	20
44–47	3	45.5	0.1250	23
48–51	1	49.5	0.0417	24
	$\Sigma f = 24$		$\Sigma \dfrac{f}{n} = 1$	

Class with greatest frequency: 36–39
Class with least frequency: 48–51

28. Class width $= \dfrac{\text{Max} - \text{Min}}{\text{Number of classes}} = \dfrac{2888 - 2456}{5} = 86.4 \Rightarrow 87$

Class	Frequency	Midpoint	Relative frequency	Cumulative frequency
2456–2542	7	2499	0.28	7
2543–2629	3	2586	0.12	10
2630–2716	2	2673	0.08	12
2717–2803	4	2760	0.16	16
2804–2890	9	2847	0.36	25
	$\Sigma f = 25$		$\Sigma \dfrac{f}{n} = 1$	

Class with greatest frequency: 2804–2890
Class with least frequency: 2630–2716

30. Class width = $\dfrac{\text{Max} - \text{Min}}{\text{Number of classes}} = \dfrac{80 - 10}{5} = 14$

Class	Frequency	Midpoint	Relative frequency	Cumulative frequency
10–23	11	16.5	0.3438	11
24–37	9	30.5	0.2813	20
38–51	6	44.5	0.1875	26
52–65	2	58.5	0.0625	28
66–80	4	72.5	0.1250	32
	$\Sigma f = 32$		$\Sigma \dfrac{f}{n} = 1$	

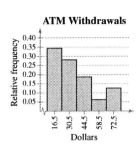

ATM Withdrawals

Class with greatest relative frequency: 10–23
Class with least relative frequency: 52–65

32. Class width = $\dfrac{\text{Max} - \text{Min}}{\text{Number of classes}} = \dfrac{16 - 7}{5} = 1.8 \Rightarrow 2$

Class	Frequency	Midpoint	Relative frequency	Cumulative frequency
7–8	7	7.5	0.28	7
9–10	8	9.5	0.32	15
11–12	6	11.5	0.24	21
13–14	3	13.5	0.12	24
15–16	1	15.5	0.04	25
	$\Sigma f = 25$		$\Sigma \dfrac{f}{n} = 1$	

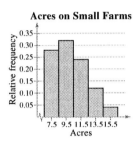

Acres on Small Farms

Class with greatest relative frequency: 9–10
Class with least relative frequency: 15–16

34. Class width = $\dfrac{\text{Max} - \text{Min}}{\text{Number of classes}} = \dfrac{57 - 16}{6} = 6.83 \Rightarrow 7$

Class	Frequency	Relative frequency	Cumulative frequency
16–22	2	0.10	2
23–29	3	0.15	5
30–36	8	0.40	13
37–43	5	0.25	18
44–50	0	0.00	18
51–57	2	0.10	20
	$\Sigma f = 20$	$\Sigma \dfrac{f}{n} = 1$	

Daily Saturated Fat Intake

Location of the greatest increase in frequency: 30–36

36. Class width = $\dfrac{\text{Max} - \text{Min}}{\text{Number of classes}} = \dfrac{29 - 1}{6} = 4.67 \Rightarrow 5$

Class	Frequency	Relative frequency	Cumulative frequency
1–5	5	0.2083	5
6–10	9	0.3750	14
11–15	3	0.1250	17
16–20	4	0.1667	21
21–25	2	0.0833	23
26–30	1	0.0417	24
	$\Sigma f = 24$	$\Sigma \dfrac{f}{n} = 1$	

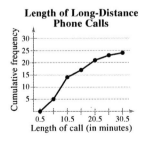

Length of Long-Distance Phone Calls

Location of the greatest increase in frequency: 6–10

198 CHAPTER 2 | DESCRIPTIVE STATISTICS

38.

Class	Frequency	Midpoint	Relative frequency	Cumulative frequency
0–2	17	1	0.4048	17
3–5	16	4	0.3810	33
6–8	7	7	0.1667	40
9–11	1	10	0.0238	41
12–14	0	13	0.0000	41
15–17	1	16	0.0238	42
	$\Sigma f = 42$		$\Sigma \frac{f}{N} = 1$	

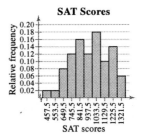
Number of Children of First 42 Presidents

Classes with greatest frequency: 0–2
Classes with least frequency: 15–17

40. (a) Class width $= \dfrac{\text{Max} - \text{Min}}{\text{Number of classes}} = \dfrac{1359 - 410}{10} = 94.9 \Rightarrow 95$

Class	Frequency	Midpoint	Relative frequency
410–505	1	457.5	0.02
506–601	1	553.5	0.02
602–697	4	649.5	0.08
698–793	6	745.5	0.12
794–889	8	841.5	0.16
890–985	6	937.5	0.12
986–1081	9	1033.5	0.18
1082–1177	5	1129.5	0.10
1178–1273	7	1225.5	0.14
1274–1369	3	1321.5	0.06
	$\Sigma f = 50$		$\Sigma \frac{f}{N} = 1$

(b) 48%, because the sum of the relative frequencies for the last four classes is 0.48.

(c) 698, because the sum of the relative frequencies for the last seven classes is 0.88.

2.2 MORE GRAPHS AND DISPLAYS

2.2 EXERCISE SOLUTIONS

2. Unlike the histogram, the stem-and-leaf plot still contains the original data values. However, some data are difficult to organize in a stem-and-leaf plot.

4. d

6. c

8. 129, 133, 136, 137, 137, 141, 141, 141, 141, 143, 144, 144, 146, 149, 149, 150, 150, 150, 151, 152, 154, 156, 157, 158, 158, 158, 159, 161, 166, 167

Max: 167   Min: 129

10. 214, 214, 214, 216, 216, 217, 218, 218, 220, 221, 223, 224, 225, 225, 227, 228, 228, 228, 228, 230, 230, 231, 235, 237, 239

Max: 239   Min: 214

12. The value of the stock portfolio has increased fairly steadily over the past five years with the greatest increase happening between 2000 and 2003. (Answers will vary.)

14. The most frequent incident occurring while driving and using a cell phone is swerving. Twice as many people "sped up" than "cut off a car." (Answers will vary.)

16. Key: 31|9 = 319   It appears that the majority of the elephants eat between 390 and 480 lbs. of hay each day. (Answers will vary.)

```
29 | 8
30 | 5
31 | 9
32 | 7
33 |
34 | 5
35 | 1
36 |
37 |
38 |
39 | 03
40 | 39
41 | 059
42 |
43 |
44 | 689
45 | 05
46 | 05
47 | 99
48 |
49 | 1
50 | 3
```

18.

It appears that most of the 30 people from the U.S. see or hear between 450 and 750 advertisements per week. (Answers will vary.)

20.

Category	Frequency	Relative frequency	Angle
United States	270	0.40	(0.40)(360°) ≈ 144°
United Kingdom	100	0.15	(0.15)(360°) ≈ 54°
France	49	0.07	(0.07)(360°) ≈ 25°
Sweden	30	0.04	(0.04)(360°) ≈ 14°
Germany	77	0.11	(0.11)(360°) ≈ 40°
Other	157	0.23	(0.23)(360°) ≈ 83°
	$\sum f = 683$	$\sum \frac{f}{n} = 1$	

It appears that 55% of the Nobel Prize laureates come from either the U.S. or the United Kingdom. The greatest number of Nobel Prize laureates came from the U.S. during the years 1901–2002. (Answers will vary.)

22. 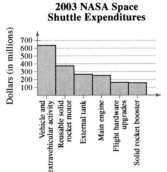 The greatest NASA space shuttle operations expenditures in 2003 were for vehicle and extravehicle activity while the least were for solid rocket booster. (Answers will vary.)

24.  It appears that hourly wage increases as the number of hours worked increases. (Answers will vary.)

26. 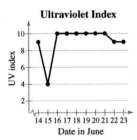 Of the period from June 14–23, 2001, the ultra voliet index was highest from June 16–21, 2001 in Memphis, TN. (Answers will vary.)

28. 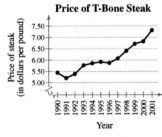 It appears that the price of a T-bone steak rose from 1991–2001. (Answers will vary.)

30. (a) The pie chart should be displaying all four quarters, not just the first three.

(b)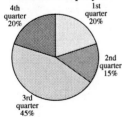

2.3 MEASURES OF CENTRAL TENDENCY

2.3 EXERCISE SOLUTIONS

2. False. Not all data sets must have a mode.

4. False. The mode is the only measure of central tendency that can be used for data at the nominal level of measurement.

6. Any data set that is symmetric has the same median and mode.

8. Symmetric since the left and right halves of the distribution are approximately mirror images.

10. Skewed left since the tail of the distribution extends to the left.

12. (7), since the distribution has values in the thousands of dollars and is skewed right due to the few executives that make a much higher salary than the majority of the employees.

14. (8), since the distribution is rather symmetric due to the nature of the weights of seventh grade boys.

16. (a) $\bar{x} = \dfrac{\Sigma x}{n} = \dfrac{196}{10} = 19.6$

    14 17 18 19 19 20 20 21 22 26

    two middle values $\Rightarrow$ median $= \dfrac{19 + 20}{2} = 19.5$

    mode $= 19, 20$ (occurs 2 times each)

    (b) Mean appears to be the best measure of central tendency since there are no outliers.

18. (a) $\bar{x} = \dfrac{\Sigma x}{n} = \dfrac{1846}{10} = 184.6$

    147 154 171 173 181 184 188 203 216 229

    two middle values $\Rightarrow$ median $= \dfrac{181 + 184}{2} = 182.5$

    mode $=$ none

    (b) Mean appears to be the best measure of central tendency since there are no outliers.

20. (a) $\bar{x} = \dfrac{\Sigma x}{n} = \dfrac{1223}{20} = 61.2$

    12 18 26 28 31 33 40 44 45 49 61 63 75 80 80 89 96 103 125 125

    two middle values $\Rightarrow$ median $= \dfrac{49 + 61}{2} = 55$

    mode $= 80, 125$

    (b) Median appears to be the best measure of central tendency since the distribution is skewed.

202 CHAPTER 2 | DESCRIPTIVE STATISTICS

22. (a) $\bar{x}$ = not possible (nominal data)

median = not possible (nominal data)

mode = "Watchful"

(b) Mode appears to be the best measure of central tendency since the data is at the nominal level of measurement.

24. (a) $\bar{x}$ = not possible (nominal data)

median = not possible (nominal data)

mode = "Domestic"

(b) Mode appears to be the best measure of central tendency since the data is at the nominal level of measurement.

26. (a) $\bar{x} = \frac{\Sigma x}{n} = \frac{83}{5} = 16.6$

1, 10, ⑮, 25.5, 31.5

↑── middle value ⇒ median = 15

mode = none

(b) Mean appears to be the best measure of central tendency since there are no outliers.

28. (a) $\bar{x} = \frac{\Sigma x}{n} = \frac{3734}{11} \approx 339.5$

26, 37, 102, 145, 280, ㉞, 375, 444, 567, 573, 819

↑── middle value ⇒ median = 366

mode = none

(b) Median appears to be the best measure of central tendency since the distribution is skewed.

30. (a) $\bar{x} = \frac{\Sigma x}{n} = \frac{299}{12} \approx 2.5$

0.8, 1.5, 1.6, 1.8, 2.1, 2.3, 2.4, 2.5, 3.0, 3.9, 4.0, 4.0

two middle values ⇒ median = $\frac{2.3 + 2.4}{2} = 2.35$

mode = 4.0 (occurs 2 times)

(b) Mean appears to be the best measure of central tendency since there are no outliers.

32. (a) $\bar{x} = \frac{\Sigma x}{n} = \frac{2987}{14} \approx 213.4$

205, 208, 210, 212, 212, 214, 214, 214, 215, 215, 217, 217, 217, 217

two middle values ⇒ median = $\frac{214 + 214}{2} = 214$

mode = 217 (occurs 4 times)

(b) Median appears to be the best measure of central tendency since the data is skewed.

34. A = mean (left of median in skewed left dist.)

B = median (right of mean in skewed left dist.)

C = mode (data entry that occurred most often)

36. Median since the data is skewed.

38. Median since the data is skewed.

40.

Source	Score, x	Weight, w	$x \cdot w$
MBAs	$42,500	8	$(42,500)(8) = 340,000$
Bas	$28,000	17	$(28,000)(17) = 476,000$
		$\Sigma w = 25$	$\Sigma(x \cdot w) = 816,000$

$\bar{x} = \dfrac{\Sigma(x \cdot w)}{\Sigma w} = \dfrac{816,000}{25} = \$32,640$

42.

Source	Score, x	Weight, w	$x \cdot w$
Engineering	83	8	$(83)(8) = 664$
Business	79	11	$(79)(11) = 869$
Math	87	5	$(87)(5) = 435$
		$\Sigma(x \cdot w) = 24$	$\Sigma w = 1968$

$\bar{x} = \dfrac{\Sigma(x \cdot w)}{\Sigma w} = \dfrac{1968}{24} = 82$

44.

Midpoint, x	Frequency, f	$x \cdot f$
64	2	$(64)(2) = 128$
67	4	$(67)(4) = 268$
70	8	$(70)(8) = 560$
73	5	$(73)(5) = 365$
76	2	$(76)(2) = 152$
	$\Sigma(x \cdot f) = 21$	$\Sigma(x \cdot f) = 1473$

$\bar{x} = \dfrac{\Sigma(x \cdot f)}{n} = \dfrac{1473}{21} \approx 70.1$

46.

Midpoint, x	Frequency, f	$x \cdot f$
3	12	$(3)(12) = 36$
8	26	$(8)(26) = 208$
13	20	$(13)(20) = 260$
18	7	$(18)(7) = 126$
23	11	$(23)(11) = 253$
28	7	$(28)(7) = 196$
33	4	$(33)(4) = 132$
38	4	$(38)(4) = 152$
43	1	$(43)(1) = 43$
	$n = 92$	$\Sigma(x \cdot f) = 1406$

$\bar{x} = \dfrac{\Sigma(x \cdot f)}{n} = \dfrac{1406}{92} \approx 15.3$

48. Class width $= \dfrac{\text{Max} - \text{Min}}{\text{Number of classes}} = \dfrac{297 - 127}{5} = 34$

Class	Midpoint, x	Frequency, f
127–161	144	9
162–196	179	8
197–231	214	3
232–266	249	3
267–301	284	1
		$\Sigma f = 24$

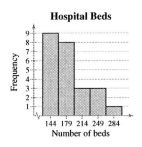

Shape: Positively skewed

50. Class width = $\dfrac{\text{Max} - \text{Min}}{\text{Number of classes}} = \dfrac{6-1}{6} = 0.8333 \Rightarrow 1$

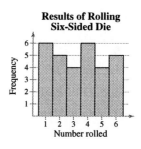

Class	Frequency
1	6
2	5
3	4
4	6
5	4
6	5
	$\Sigma f = 30$

Shape: Uniform

52. (a) $\bar{x} = \dfrac{\Sigma x}{n} = \dfrac{562.4}{19} = 29.63$

4.8, 4.9, 7.8, 10.1, 10.3, 12.4, 13.1, 13.3, 16.2, ⟨18.3⟩, 18.4, 19, 22.1, 22.6, 26.6, 33.3, 51.4, 97.5, 160.8

middle value ⇒ median = 18.3

(b) $\bar{x} = \dfrac{\Sigma x}{n} = \dfrac{402.1}{18} \approx 22.34$

4.8, 4.9, 7.8, 10.1, 10.3, 12.4, 13.1, 13.3, 16.2, 18.3, 18.4, 19, 22.1, 22.6, 26.6, 33.3, 51.4, 97.5

two middle values ⇒ median = $\dfrac{16.2 + 18.3}{2} = 17.25$

(c) mean

54. Car A because it has the highest midrange of the three.

Car A: Midrange = $\dfrac{34 + 28}{2} = 32$

Car B: Midrange = $\dfrac{31 + 29}{2} = 30$

Car C: Midrange = $\dfrac{32 + 28}{2} = 30$

56. (a) Order the data values.

11 13 22 28 36 36 36 37 37 37 38 41 43 44 46
47 51 51 51 53 61 62 63 64 72 72 74 76 85 90

Delete the lowest 10%, smallest 3 observations (11, 13, 22).
Delete the highest 10%, largest 3 observations (76, 85, 90).

Find the 10% trimmed mean using the remaining 24 observations.

10% trimmed mean = $49.1\overline{6}$

(b) $\bar{x} = 49.2\overline{3}$

median = 46.5

mode = 36, 37, 51

(c) Using a trimmed mean eliminates potential outliers that may affect the mean of all the observations.

58. A distribution with one data entry in each class would be an example of a rectangular distribution whose mean and median would be equal but the mode does not exist.

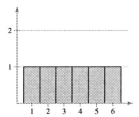

2.4 MEASURES OF VARIATION

2.4 EXERCISE SOLUTIONS

2. Range = max − min = 23 − 13 = 10

$$\mu = \frac{\Sigma x}{N} = \frac{232}{14} \approx 16.6$$

x	$x - \mu$	$(x - \mu)^2$
13	13 − 16.6 = −3.6	$(-3.6)^2 = 12.96$
23	23 − 16.6 = 6.4	$(6.4)^2 = 40.96$
15	15 − 16.6 = −1.6	$(-1.6)^2 = 2.56$
13	13 − 16.6 = −3.6	$(-3.6)^2 = 12.96$
18	18 − 16.6 = 1.4	$(1.4)^2 = 1.96$
13	13 − 16.6 = −3.6	$(-3.6)^2 = 12.96$
15	15 − 16.6 = −1.6	$(-1.6)^2 = 2.56$
14	14 − 16.6 = −2.6	$(-2.6)^2 = 6.76$
20	20 − 16.6 = 3.4	$(3.4)^2 = 11.56$
20	20 − 16.6 = 3.4	$(3.4)^2 = 11.56$
18	18 − 16.6 = 1.4	$(1.4)^2 = 1.96$
17	17 − 16.6 = 0.4	$(0.4)^2 = 0.16$
20	20 − 16.6 = 3.4	$(3.4)^2 = 11.56$
13	13 − 16.6 = −3.6	$(-3.6)^2 = 12.96$
$\Sigma x = 232$	$\Sigma(x - \mu) = 0$	$\Sigma(x - \mu)^2 \approx 143.44$

$$\sigma^2 = \frac{\Sigma(x - \mu)^2}{N} = \frac{143.44}{14} \approx 10.2$$

$$\sigma = \sqrt{\frac{\Sigma(x - \mu)^2}{N}} = \sqrt{\frac{143.44}{14}} \approx \sqrt{10.2} \approx 3.2$$

4. Range = max − min = 27 − 8 = 19

$$\bar{x} = \frac{\Sigma x}{n} = \frac{233}{13} \approx 17.9$$

x	$x - \bar{x}$	$(x - \bar{x})^2$
24	24 − 17.9 = 6.1	$(6.1)^2 = 37.21$
26	26 − 17.9 = 8.1	$(8.1)^2 = 65.61$
27	27 − 17.9 = 9.1	$(9.1)^2 = 82.81$
23	23 − 17.9 = 5.1	$(5.1)^2 = 26.81$
9	9 − 17.9 = −8.9	$(-8.9)^2 = 79.21$
14	14 − 17.9 = −3.9	$(-3.9)^2 = 15.21$
8	8 − 17.9 = −9.9	$(-9.9)^2 = 98.01$
8	8 − 17.9 = −9.9	$(-9.9)^2 = 98.01$
26	26 − 17.9 = 8.1	$(8.1)^2 = 65.61$
15	15 − 17.9 = −2.9	$(-2.9)^2 = 8.41$
15	15 − 17.9 = −2.9	$(-2.9)^2 = 8.41$
27	27 − 17.9 = 9.1	$(9.1)^2 = 82.81$
11	11 − 17.9 = −6.9	$(-6.9)^2 = 47.61$
$\Sigma x = 233$	$\Sigma(x - \bar{x}) = 0$	$\Sigma(x - \bar{x})^2 \approx 715.73$

$$s^2 = \frac{\Sigma(x - \bar{x})^2}{n - 1} = \frac{715.73}{12} \approx 59.6$$

$$s = \sqrt{\frac{\Sigma(x - \bar{x})^2}{n - 1}} = \sqrt{\frac{715.73}{12}} \approx \sqrt{59.6} \approx 7.7$$

6. Range = max − min = 34 − 24 = 10

8. The deviation, $(x - \mu)$, is the difference between an observation, x, and the mean of the data, μ. The sum of the deviations is always zero.

10. The standard deviation is the positive square root of the variance.

Since squared deviations can never be negative, the standard deviation and variance can never be negative.

$$\{7, 7, 7, 7, 7\} \rightarrow n = 5$$
$$\bar{x} = 7$$
$$s = 0$$

12. $\{3, 3, 3, 7, 7, 7\} \rightarrow n = 6$
$$\mu = 5$$
$$s = 2$$

14. Graph (a) has a standard deviation of 2.4 and graph (b) has a standard deviation of 5. Graph (b) has more variability.

16. When given a data set, one would have to determine if it represented the population or was a sample taken from a population. If the data are a population, then σ is calculated. If the data are a sample, then s is calculated.

18. Player B. Due to the smaller standard deviation in number of strokes, player B would be the more consistent player.

20. (a) Dal: range = max − min = 34.9 − 16.8 = 18.1

x	$x - \bar{x}$	$(x - \bar{x})^2$
34.9	8.92	79.61
25.7	−0.28	0.08
17.3	−8.68	75.30
16.8	−9.18	84.23
26.8	0.82	0.68
24.7	−1.28	1.63
29.4	3.42	11.71
32.7	6.72	45.19
25.5	−0.48	0.23
$\Sigma x = 233.8$		$\Sigma(x - \bar{x})^2 = 298.66$

$$\bar{x} = \frac{\Sigma x}{n} = \frac{233.8}{9} = 25.98$$

$$s^2 = \frac{\Sigma(x - \bar{x})^2}{(n - 1)} = \frac{298.66}{8} \approx 37.33$$

$$s = \sqrt{s^2} \approx 6.11$$

Hou: range = max − min = 31.3 − 18.3 = 13

x	$x - \bar{x}$	$(x - \bar{x})^2$
25.6	−0.03	0.00
23.2	−2.43	5.92
26.7	1.07	1.14
27.7	2.07	4.27
25.4	−0.23	0.05
26.4	0.77	0.59
18.3	−7.33	53.78
26.1	0.47	0.22
31.3	5.67	32.11
$\Sigma x = 230.7$		$\Sigma(x - \bar{x})^2 = 98.08$

$$\bar{x} = \frac{\Sigma x}{n} = \frac{230.7}{9} = 25.63$$

$$s^2 = \frac{\Sigma(x - \bar{x})^2}{(n - 1)} = \frac{98.08}{8} = 12.26$$

$$s = \sqrt{s^2} \approx 3.50$$

(b) It appears from the data that the annual salaries in Dallas are more variable than the salaries in Houston.

22. (a) Public: range = max − min = 39.9 − 34.8 = 5.1

x	$x - \bar{x}$	$(x - \bar{x})^2$
38.6	1.23	1.50
38.1	0.73	0.53
38.7	1.33	1.76
36.8	−0.58	0.33
34.8	−2.58	6.63
35.9	−1.48	2.18
39.9	2.53	6.38
36.2	−1.18	1.38
$\Sigma x = 299$		$\Sigma(x - \bar{x})^2 = 20.68$

$$\bar{x} = \frac{\Sigma x}{n} = \frac{299}{8} = 37.38$$

$$s^2 = \frac{\Sigma(x - \bar{x})^2}{(n-1)} = \frac{20.68}{7} \approx 2.95$$

$$s = \sqrt{s^2} \approx 1.72$$

Private: range = max − min = 21.8 − 17.6 = 4.2

x	$x - \bar{x}$	$(x - \bar{x})^2$
21.8	2.26	5.12
18.4	−1.14	1.29
20.3	0.76	0.58
17.6	−1.94	3.75
19.7	0.16	0.03
18.3	−1.24	1.53
19.4	−0.14	0.02
20.8	1.26	1.59
$\Sigma x = 156.3$		$\Sigma(x - \bar{x})^2 = 13.92$

$$\bar{x} = \frac{\Sigma x}{n} = \frac{156.3}{8} = 19.54$$

$$s^2 = \frac{\Sigma(x - \bar{x})^2}{(n-1)} = \frac{13.92}{7} \approx 1.99$$

$$s = \sqrt{s^2} \approx 1.41$$

(b) It appears from the data, that the annual salaries for public teachers are more variable than the salaries for private teachers.

24. (a) Greatest sample standard deviation: (i)

Data set (i) has more entries that are farther away from the mean.

Least same standard deviation: (iii)

Data set (iii) has more entries that are close to the mean.

(b) The three data sets have the same mean, median, and mode, but have different standard deviations.

26. (a) Greatest sample standard deviation: (iii)

Data set (iii) has more entries that are farther away from the mean.

Least same standard deviation: (i)

Data set (i) has more entries that are close to the mean.

(b) The three data sets have the same mean and median, but have different standard deviations.

28. You must know the distribution is bell-shaped.

30. 95% of the data falls between $\bar{x} - 2s$ and $\bar{x} + 2s$.

$\bar{x} - 2s = 1200 - 2(350) = 500$

$\bar{x} + 2s = 1200 + 2(350) = 1900$

95% of the farm values lie between $500 and $1900 per acre.

32. (a) $n = 40$

95% of the data lie within 2 standard deviations of the mean.

$(95\%)(40) = (0.95)(40) = 38$ farm values lie between $500 and $1900 per acre.

(b) $n = 60$

$(95\%)(20) = (0.95)(20) = 19$ of these farm values lie between $500 and $1900 per acre.

34. $\bar{x} = 1200$ $\quad$ {1950, 475, 2050} are outliers. They are more than 2 standard deviations from the mean (500, 1900).
$s = 350$

36. $1 - \dfrac{1}{k^2} = 1 - \dfrac{1}{(2)^2} = 1 - \dfrac{1}{4} = .75 \rightarrow$ At least 75% of the 400-meter dash times lie within 2 standard deviations of mean.

$(\bar{x} - 2s, \bar{x} + 2s) \rightarrow (48.07, 56.67) \rightarrow$ At least 75% of the 400-meter dash times lie between 48.07 and 56.67 seconds.

38.

x	f	xf	$x - \bar{x}$	$(x - \bar{x})^2$	$(x - \bar{x})^2 f$
0	3	0	−1.74	3.03	9.08
1	15	15	−0.74	0.55	8.21
2	24	48	0.26	0.07	1.62
3	8	24	1.26	1.59	12.70
	$n = 50$	$\sum xf = 87$			$\sum(x - \bar{x})^2 f = 31.62$

$\bar{x} = \dfrac{\sum xf}{n} = \dfrac{87}{50} \approx 1.7$

$s = \sqrt{\dfrac{\sum(x - \bar{x})^2 f}{n - 1}} \approx \sqrt{\dfrac{31.62}{49}} \approx \sqrt{0.645} \approx 0.8$

210 CHAPTER 2 | DESCRIPTIVE STATISTICS

40. Class width $= \dfrac{\text{Max} - \text{Min}}{5} = \dfrac{244 - 145}{5} = 19.8 \to 20$

Class	Midpoint, x	f	xf
145–164	154.5	8	1236
165–184	174.5	7	1221.5
185–204	194.5	3	583.5
205–224	214.5	1	214.5
225–244	234.5	1	234.5
		$N = 20$	$\Sigma xf = 3490$

$x - \mu$	$(x - \mu)^2$	$(x - \mu)^2 f$
−20	400	3200
0	0	0
20	400	1200
40	1600	1600
60	3600	3600
		$\Sigma(x - \mu)^2 f = 9600$

$\mu = \dfrac{\Sigma xf}{N} = \dfrac{3490}{20} = 174.5$

$\sigma = \sqrt{\dfrac{\Sigma(x - \mu)^2 f}{N}} = \sqrt{\dfrac{9600}{20}} = \sqrt{480} \approx 21.9$

42.

x	f	xf
0	1	0
1	9	9
2	13	26
3	5	15
4	2	8
	$n = 30$	$\Sigma xf = 58$

$\bar{x} = \dfrac{\Sigma xf}{n} = \dfrac{58}{30} \approx 1.9$

$x - \bar{x}$	$(x - \bar{x})^2$	$(x - \bar{x})^2 f$
−1.93	3.72	3.72
−0.93	0.86	7.74
0.07	0.00	0.00
1.07	1.14	5.70
2.07	4.28	8.56
		$\Sigma(x - \bar{x})^2 f = 25.72$

$s = \sqrt{\dfrac{\Sigma(x - \bar{x})^2 f}{n - 1}} = \sqrt{\dfrac{25.72}{29}} \approx \sqrt{0.89} \approx 0.9$

44.

Class	Midpoint, x	f	xf
0.5–9.5	5	11.9	59.5
10.5–19.5	15	12.1	181.5
20.5–29.5	25	14.0	350
30.5–39.5	35	18.5	647.5
40.5–49.5	45	16.6	747
50.5–59.5	55	16.3	896.5
60.5–69.5	65	17.8	1157
70.5–79.5	75	12.4	930
80.5–89.5	85	6.3	535.5
90.5–99.5	95	1.3	123.5
		$n = 127.2$	$\Sigma xf = 5628$

$\bar{x} = \dfrac{\Sigma xf}{n} = \dfrac{5628}{127.2} \approx 44.25$

$(x - \bar{x})$	$(x - \bar{x})^2$	$(x - \bar{x})^2 f$
−39.25	1540.5625	18,332.69
−29.25	855.5625	10,352.31
−19.25	370.5625	5,187.88
−9.25	85.5625	1,582.91
0.75	0.5625	9.34
10.75	115.5625	1,883.67
20.75	430.5625	7,664.01
30.75	945.5625	11,724.98
40.75	1660.5625	10,461.54
50.75	2575.5625	3,348.23
		$\sum(x - \bar{x})^2 f = 70{,}547.56$

$$s = \sqrt{\frac{\sum(x - \bar{x})^2 f}{n - 1}} = \sqrt{\frac{70{,}547.56}{126.2}} \approx \sqrt{559.01} \approx 23.64$$

46. (a)

x	x^2
1059	1,121,481
1328	1,763,584
1175	1,380,625
1123	1,261,129
923	851,929
1017	1,034,289
1214	1,473,796
1042	1,085,764
$\sum x = 8881$	$\sum x^2 = 9{,}972{,}597$

Male: $s = \sqrt{\dfrac{\sum x^2 - [(\sum x)^2/n]}{n - 1}} = \sqrt{\dfrac{9{,}972{,}597 - [(8881)^2/8]}{7}}$

$= \sqrt{\dfrac{113{,}576.875}{7}} = \sqrt{16{,}225.268} \approx 127.4$

x	x^2
1226	1,503,076
965	931,225
841	707,281
1053	1,108,809
1056	1,115,136
1393	1,940,449
1312	1,721,344
1222	1,493,284
$\sum x = 9068$	$\sum x^2 = 10{,}520{,}604$

Female: $s = \sqrt{\dfrac{\sum x^2 - [(\sum x)^2/n]}{n - 1}} = \sqrt{\dfrac{10{,}520{,}604 - [(9068)^2/8]}{7}}$

$= \sqrt{\dfrac{242{,}026}{7}} = \sqrt{34{,}575.143} \approx 185.9$

(b) The answers from exercise 40(a) are the same as from Exercise 19.

48. (a) $\bar{x} = 550$, $s = 302.8$

(b) $\bar{x} = 560$, $s = 302.8$

(c) $\bar{x} = 540$, $s = 302.8$

(d) By adding or subtracting a constant k to each entry, the new sample mean will be $\bar{x} + k$ with the sample standard deviation being unaffected.

50. (a) $P = \dfrac{3(\bar{x} - \text{median})}{s} = \dfrac{3(17 - 19)}{2.3} \approx -2.61$; skewed left

(b) $P = \dfrac{3(\bar{x} - \text{median})}{s} = \dfrac{3(32 - 25)}{5.1} \approx 4.12$; skewed right

2.5 MEASURES OF POSITION

2.5 EXERCISE SOLUTIONS

2. (a)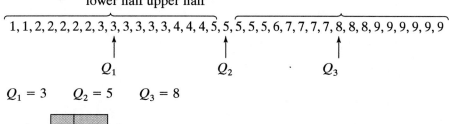

$Q_1 = 3 \quad Q_2 = 5 \quad Q_3 = 8$

(b)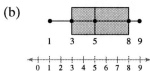

4. The salesperson sold more hardware equipment than 80% of the other sales people.

6. The child is taller than 87% of the other children in the same age group.

8. False. The five numbers you need to graph a box-and-whisker plot are the minimum, the maximum, Q_1, Q_3, and the median (Q_2).

10. False. Any score below the mean will have a corresponding negative z-score.

12. (a) Min = 100

(b) Max = 320

(c) $Q_1 = 130$

(d) $Q_2 = 205$

(e) $Q_3 = 270$

(f) IQR = $Q_3 - Q_1 = 270 - 130 = 140$

14. (a) Min = 25

(b) Max = 85

(c) $Q_1 = 50$

(d) $Q_2 = 65$

(e) $Q_3 = 70$

(f) IQR = $Q_3 - Q_1 = 70 - 50 = 20$

16. (a) Min $= -1.3$

(b) Max $= 2.1$

(c) $Q_1 = -0.3$

(d) $Q_2 = 0.2$

(e) $Q_3 = 0.4$

(f) IQR $= Q_3 - Q_1 = 0.4 - (-0.3) = 0.7$

18. $P_{10} = T$, $P_{50} = R$, $P_{80} = S$

10% of the entries are below T, 50% are below R, and 80% are below S.

20. (a) $Q_1 = 2$, $Q_2 = 4.5$, $Q_3 = 6.5$

(b)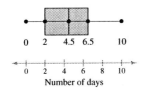

22. (a) $Q_1 = 15.125$, $Q_2 = 15.8$, $Q_3 = 17.650$

(b)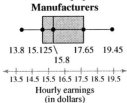

24. (a) $17.65

(b) 50%

(c) 50%

26. A $\to z = -1.54$

B $\to z = 0.77$

C $\to z = 1.54$

None of the z-scores are unusual.

28. Statistics: $x = 60 \Rightarrow z = \dfrac{x - \mu}{\sigma} = \dfrac{60 - 63}{7} \approx -0.43$

Biology: $x = 20 \Rightarrow z = \dfrac{x - \mu}{\sigma} = \dfrac{20 - 23}{3.9} \approx -0.77$

The student had a better score on the statistics test.

30. Statistics: $x = 63 \Rightarrow z = \dfrac{x - \mu}{\sigma} = \dfrac{63 - 63}{7} = 0$

Biology: $x = 23 \Rightarrow z = \dfrac{x - \mu}{\sigma} = \dfrac{23 - 23}{3.9} = 0$

The student performed equally well on the two tests.

32. (a) $x = 34 \Rightarrow z = \dfrac{x - \mu}{\sigma} = \dfrac{34 - 33}{4} = 0.25$

$x = 30 \Rightarrow z = \dfrac{x - \mu}{\sigma} = \dfrac{30 - 33}{4} = -0.75$

$x = 42 \Rightarrow z = \dfrac{x - \mu}{\sigma} = \dfrac{42 - 33}{4} = 2.25$

The life span of 42 days is unusual due to a rather large z-score.

(b) $x = 29 \Rightarrow z = \dfrac{x - \mu}{\sigma} = \dfrac{29 - 33}{4} = -1 \Rightarrow$ 16th percentile

$x = 41 \Rightarrow z = \dfrac{x - \mu}{\sigma} = \dfrac{41 - 33}{4} = 2 \Rightarrow$ 97.5th percentile

$x = 25 \Rightarrow z = \dfrac{x - \mu}{\sigma} = \dfrac{25 - 33}{4} = -2 \Rightarrow$ 2.5th percentile

34. 99th percentile.
99% of the heights are below 76 inches.

36. $x = 70$: $z = \dfrac{x - \mu}{\sigma} = \dfrac{70 - 69.2}{2.9} \approx 0.28$

$x = 66$: $z = \dfrac{x - \mu}{\sigma} = \dfrac{66 - 69.2}{2.9} \approx -1.10$

$x = 68$: $z = \dfrac{x - \mu}{\sigma} = \dfrac{68 - 69.2}{2.9} \approx -0.41$

None of the heights are unusual.

38. $x = 66.3$: $z = \dfrac{x - \mu}{\sigma} = \dfrac{66.3 - 69.2}{2.9} = -1$

Approximately the 11th percentile.

40. 1, 2, 3, 3, 5, 5, 7, 7, 8, 10
       ↑   ↑   ↑
      Q_1 Q_2 Q_3

Midquartile $= \dfrac{Q_1 + Q_3}{2} = \dfrac{3 + 7}{2} = 5$

42. 7.9, 8, 8.1, 9.7, 10.3, 11.2, 11.8, 12.2, 12.3, 12.7, 13.4, 15.4, 16.1

$Q_1 = 8.9 \quad\quad Q_2 \quad\quad Q_3 = 13.05$

Midquartile $= \dfrac{Q_1 + Q_3}{2} = \dfrac{8.9 + 13.05}{2} = 10.975$

CHAPTER 2 REVIEW EXERCISE SOLUTIONS

2.

Greatest relative frequency: 32–35

Least relative frequency: 20–23

4.

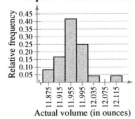

6.

8.

10.

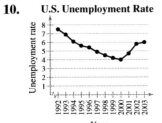

12.

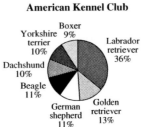

14. $\bar{x} = 30.8$

median $= 30$

mode $= 9$

216 CHAPTER 2 | DESCRIPTIVE STATISTICS

16.

x	f	xf
0	13	0
1	9	9
2	19	38
3	8	24
4	5	20
5	2	10
6	4	24
	$n = 60$	$\Sigma xf = 125$

$\bar{x} = \dfrac{\Sigma xf}{n} = \dfrac{125}{60} = 2.08$

18. $\bar{x} = \dfrac{\Sigma xw}{\Sigma w} = \dfrac{(81)(0.20) + (95)(0.20) + (89)(0.20) + (87)(0.40)}{0.20 + 0.20 + 0.20 + 0.40} = \dfrac{8780}{1} = 87.8$

20. Skewed   **22.** Skewed right   **24.** Mean

26. Range = max − min = 19.73 − 15.89 = 3.84

28. $\mu = \dfrac{\Sigma x}{N} = \dfrac{621}{9} \approx 69$

$\sigma = \sqrt{\dfrac{\Sigma(x - \mu)^2}{N}}$

$= \sqrt{\dfrac{(78 - 69)^2 + (83 - 69)^2 + \cdots + (70 - 69)^2 + (65 - 69)^2}{9}}$

$\approx \sqrt{\dfrac{550}{9}} \approx \sqrt{61.11} \approx 7.8$

30. $\bar{x} = \dfrac{\Sigma x}{n} = \dfrac{309{,}228}{8} = 38{,}653.5$

$s = \sqrt{\dfrac{\Sigma(x - \bar{x})^2}{n - 1}} = \sqrt{\dfrac{(46{,}098 - 38{,}653.5)^2 + \cdots + (37{,}109 - 38{,}653.5)^2}{7}}$

$= \sqrt{\dfrac{320{,}129{,}278}{7}} = \sqrt{45{,}732{,}754} \approx 6762.6$

32. $(26.75, 32.25) \rightarrow (29.50 - (1)(2.75), 29.50 + (1)(2.75)) \rightarrow (\mu - \sigma, \mu + \sigma)$

68% of the cable rates lie between \$26.75 and \$32.25.

34. $n = 20 \quad \mu = 7 \quad \sigma = 2$

$(3, 11) \rightarrow (7 - 2(2), 7 + 2(2)) \rightarrow (\mu - 2\sigma, \mu + 2\sigma) \rightarrow k = 2$

$1 - \dfrac{1}{k^2} = 1 - \dfrac{1}{(2)^2} = 1 - \dfrac{1}{4} = 0.75$

At least $(20)(0.75) = 15$ shuttle flights lasted between 3 days and 11 days.

36. $\bar{x} = \dfrac{\Sigma xf}{n} = \dfrac{61}{25} \approx 2.4$

$s = \sqrt{\dfrac{\Sigma(x - \bar{x})^2 f}{n - 1}}$

$= \sqrt{\dfrac{(0 - 2.44)^2(4) + (1 - 2.44)^2(5) + \cdots + (6 - 2.44)^2(1)}{24}}$

$= \sqrt{\dfrac{72.16}{24}} \approx 1.7$

38. $Q_3 = 70$

40. Height of Students

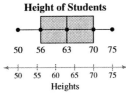

42. Weight of Football Players

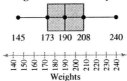

44. $\frac{84}{728} \approx 0.115 \rightarrow 12\%$ have larger audiences.

The station would represent the 88th percentile, P_{88}.

46. $x = 156 \Rightarrow z = \frac{x - \mu}{\sigma} = \frac{156 - 192}{24} = -1.5$

This is not an unusual player.

48. $x = 141 \Rightarrow z = \frac{x - \mu}{\sigma} = \frac{141 - 192}{24} \approx -2.13$

This is an unusually heavy player.

CHAPTER 2 QUIZ SOLUTIONS

1. (a)

Class limits	Midpoint	Class boundaries	Frequency	Relative frequency	Cumulative frequency
101–112	106.5	100.5–112.5	3	0.12	3
113–124	118.5	112.5–124.5	11	0.44	14
125–136	130.5	124.5–136.5	7	0.28	21
137–148	142.5	136.5–148.5	2	0.08	23
149–160	154.5	148.5–160.5	2	0.08	25

(b) Frequency Histogram and Polygon

(c) Relative Frequency Histogram

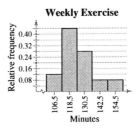

(d) Skewed

(e) Weekly Exercise

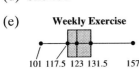

(f) Weekly Exercise

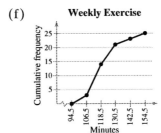

2. $\bar{x} = \dfrac{\Sigma xf}{n} = \dfrac{3130.5}{25} \approx 125.22$

$s = \sqrt{\dfrac{\Sigma(x - \bar{x})^2 f}{n - 1}} = \sqrt{\dfrac{4055.04}{24}} \approx 13.00$

3. (a)  (b)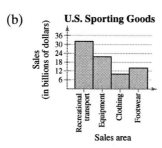

4. (a) $\bar{x} = \dfrac{\Sigma x}{n} = 751.6$

   median = 784.5

   mode = (none)

   The mean best describes a typical salary because there are no outliers.

   (b) range = max − min = 575

   $s^2 = \dfrac{\Sigma(x - \bar{x})^2}{n - 1} = 48{,}135.1$

   $s = \sqrt{\dfrac{\Sigma(x - \bar{x})^2}{n - 1}} = 219.4$

5. $\bar{x} - 2s = 155{,}000 - 2 \cdot 15{,}000 = \$125{,}000$

   $\bar{x} + 2s = 155{,}000 + 2 \cdot 15{,}000 = \$185{,}000$

   95% of the new home prices fall between \$125,000 and \$185,000.

6. (a) $x = 200{,}000 \quad z = \dfrac{x - \mu}{\sigma} = \dfrac{200{,}000 - 155{,}000}{15{,}000} = 3 \Rightarrow$ unusual price

   (b) $x = 55{,}000 \quad z = \dfrac{x - \mu}{\sigma} = \dfrac{55{,}000 - 155{,}000}{15{,}000} \approx -6.67 \Rightarrow$ very unusual price

   (c) $x = 175{,}000 \quad z = \dfrac{x - \mu}{\sigma} = \dfrac{175{,}000 - 155{,}000}{15{,}000} \approx 1.33 \Rightarrow$ not unusual

   (d) $x = 122{,}000 \quad z = \dfrac{x - \mu}{\sigma} = \dfrac{122{,}000 - 155{,}000}{15{,}000} = -2.2 \Rightarrow$ unusual price

7. (a) $Q_1 = 71 \quad Q_2 = 84.5 \quad Q_3 = 90$

   (b) IQR = $Q_3 - Q_1 = 90 - 71 = 19$

   (c) **Wins for Each Team**

   43   71 84.5 90   101

   40 50 60 70 80 90 100
   Number of wins

Probability

CHAPTER 3

3.1 BASIC CONCEPTS OF PROBABILITY

3.1 EXERCISE SOLUTIONS

2. It is impossible to have more than a 100% chance of rain.

4. {HHH, HHT, HTH, HTT, THH, THT, TTH, TTT}

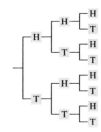

6. {(1, 1), (1, 2), (1, 3), (1, 4), (1, 5), (1, 6), (2, 1), (2, 2), (2, 3), (2, 4), (2, 5), (2, 6), (3, 1), (3, 2), (3, 3), (3, 4), (3, 5), (3, 6), (4, 1), (4, 2), (4, 3), (4, 4), (4, 5), (4, 6), (5, 1), (5, 2), (5, 3), (5, 4), (5, 5), (5, 6), (6, 1), (6, 2), (6, 3), (6, 4), (6, 5), (6, 6)}

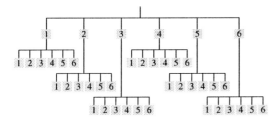

8. Not a simple event because it is an event that consists of more than a single outcome.

   Number less than 200 = {1, 2, 3 ... 199}

10. Simple event because it is an event that consists of a single outcome.

11. Classical probability since it is possible to generate the sample space and count the total number of outcomes.

14. $P(\text{greater than 1000}) = \frac{5296}{6296} \approx 0.841$

16. $P(\text{number not divisible by 1000}) = \frac{6290}{6296} \approx 0.999$

18. $P(\text{did not vote Democratic}) = \frac{39{,}979{,}372}{73{,}844{,}526} \approx 0.541$

20. $P(\text{between 35 and 44}) = \frac{28.4}{129.5} \approx 0.219$

22. $P(\text{not between 25 and 34}) = 1 - \frac{20.4}{129.5} \approx 1 - 0.158 \approx 0.842$

24. $P(\text{Associate}) = \frac{18}{89} \approx 0.202$

220 CHAPTER 3 | PROBABILITY

26. $P(\text{Bachelor's}) = \frac{33}{89} \approx 0.371$

28. $P(\text{same coloring as one of its parents}) = \frac{8}{16} = 0.5$

30. $P(\text{manufacturing industry}) = \frac{18,970}{135,073} \approx 0.140$

32. $P(\text{not in agriculture, forestry, or fishing industry})$

$$= 1 - P(\text{agriculture, forestry, or fishing industry})$$

$$= 1 - \left(\frac{3144}{135,073}\right) \approx 0.977$$

34. The probability of choosing a smoker whose mother did not smoke.

36. (a) $P(\text{less than } \$21) = 0.25$

(b) $P(\text{between } \$21 \text{ and } \$50) = 0.50$

(c) $P(\$30 \text{ or more}) = 0.50$

38. (a) {(SSS), (SSR), (SRS), (SRR), (RSS), (RSR), (RRS), (RRR)}

(b) {(RRR)}

(c) {(SSR), (SRS), (RSS)}

(d) {(SSR), (SRS), (SRR), (RSS), (RSR), (RRS), (RRR)}

40.

42. (a) $P(\text{event will occur}) = \frac{4}{9} \approx 0.444$

(b) $P(\text{event will not occur}) = \frac{5}{9} \approx 0.556$

44. $39:13 = 3:1$

3.2 CONDITIONAL PROBABILITY AND THE MULTIPLICATION RULE

3.2 EXERCISE SOLUTIONS

2. (a) Roll a die twice. The outcome of the 2nd toss is independent of the outcome of the 1st toss.

(b) Draw two cards (without replacement) from a standard 52 card deck. The outcome of the 2nd card is dependent on the outcome of the 1st card.

4. False. If events A and B are independent, then $P(A \text{ and } B) = P(A) \cdot P(B)$.

6. These events are dependent since parking beside a fire hydrant affects getting a parking ticket.

8. These events are independent since the outcome of the 1st ball drawn does not affect the outcome of the 2nd ball drawn.

10. $A = \{\text{drives pickup truck}\}$ and $B = \{\text{drives a Ford}\}$

    $P(A) = \frac{1}{6}$, $P(B) = \frac{3}{10}$, and $P(A|B) = \frac{2}{9}$

    (a) $P(A|B) = \frac{2}{9}$

    (b) $P(A \text{ and } B) = P(P)P(A|B) = \left(\frac{3}{10}\right)\left(\frac{2}{9}\right) = 0.067$

    (c) Dependent since $P(A|B) \neq P(A)$.

12. Let $A = \{\text{male}\}$ and $B = \{\text{Nursing major}\}$

    (a) $P(B|A) = \frac{65}{1036} = 0.063$

    (b) $P(A \text{ and } B) = P(A)P(B|A) = \left(\frac{1036}{3000}\right)\left(\frac{65}{1036}\right) = 0.022$

    (c) Dependent since $P(B|A) \neq P(B)$.

14. Let $A = \{\text{has the opinion that race relations have improved}\}$ and $B = \{\text{has the opinion that the rate of civil rights progress is too slow}\}$. Thus, $P(A) = 0.6$ and $P(B|A) = 0.4$.

    (a) $P(A \text{ and } B) = P(A) \cdot P(B|A) = (0.6) \cdot (0.4) = 0.24$

    (b) $P(B'|A) = 1 - P(B|A) = 1 - 0.4 = 0.6$

    (c) It is not unusual since the probability of an adult who thinks race relations have improved and says civil rights progress is too slow is 0.24.

16. Let $A = \{\text{survives bypass surgery}\}$ and $B = \{\text{heart damage will heal}\}$

    $P(A \text{ and } B) = P(A)P(B|A) = (0.60)(0.50) = 0.30$

18. Let $A = \{\text{1st bulb fails}\}$ and $B = \{\text{2nd bulb fails}\}$

    (a) $P(A \text{ and } B) = P(A) \cdot P(B|A) = \left(\frac{3}{12}\right)\left(\frac{2}{11}\right) \approx 0.045$

    (b) $P(A' \text{ and } B') = P(A') \cdot P(B'|A') = \left(\frac{9}{12}\right)\left(\frac{8}{11}\right) \approx 0.545$

    (c) $P(\text{at least one bulb failed}) = 1 - P(\text{none failed})$
    $= 1 - P(A' \text{ and } B')$
    $= 1 - 0.545$
    $= 0.455$

222 CHAPTER 3 | PROBABILITY

20. Let $A = \{\$100 \text{ or more}\}$ and $B = \{\text{purebred}\}$

(a) $P(A) = \dfrac{50}{90} \approx 0.556$

(b) $P(B'|A') = \dfrac{21}{40} = 0.525$

(c) $P(A \text{ and } B') = P(A) \cdot P(B'|A) = \left(\dfrac{50}{90}\right) \cdot \left(\dfrac{15}{50}\right) \approx 0.167$

(d) Dependent since $P(A) \approx 0.556 \neq 0.417 \approx P(A|B')$ The probability that the owner spent $100 or more on health care depends on whether or not the dog was a mixed breed.

22. (a) $P(\text{all three have O+}) = (0.38) \cdot (0.38) \cdot (0.38) \approx 0.055$

(b) $P(\text{none have O+}) = (0.62) \cdot (0.62) \cdot (0.62) \approx 0.238$

(c) $P(\text{at least one has O+}) = 1 - P(\text{none have O+}) = 1 - 0.238 = 0.762$

24. (a) $P(\text{none are defective}) = (0.995)^3 \approx 0.985$

(b) $P(\text{at least one defective}) = 1 - P(\text{none are defective}) = 1 - 0.985 = 0.015$

(c) $P(\text{all are defective}) = (0.005)^3 = 0.000000125$

26. (a) $P(\text{all share same birthday}) = \left(\dfrac{365}{365}\right)\left(\dfrac{1}{365}\right)\left(\dfrac{1}{365}\right)$

≈ 0.00000751

(b) $P(\text{none share same birthday}) = \left(\dfrac{365}{365}\right)\left(\dfrac{364}{365}\right)\left(\dfrac{363}{365}\right) = 0.992$

28. $P(A|B) = \dfrac{P(A) \cdot P(B|A)}{P(A) \cdot P(B|A) + P(A') \cdot P(B|A')} = \dfrac{\left(\frac{3}{8}\right) \cdot \left(\frac{2}{3}\right)}{\left(\frac{3}{8}\right) \cdot \left(\frac{2}{3}\right) + \left(\frac{5}{8}\right) \cdot \left(\frac{3}{5}\right)} = \dfrac{0.25}{0.25 + 0.375} = 0.4$

30. $P(A|B) = \dfrac{P(A)P(A|B)}{P(A)P(A|B) + P(A')P(B|A')}$

$= \dfrac{(0.62)(0.41)}{(0.62)(0.41) + (0.38)(0.17)} = \dfrac{0.254}{0.254 + 0.065} = 0.797$

32. (a) $P(\text{different birthdays}) = \dfrac{364}{365} \cdot \dfrac{363}{365} \cdot \ldots \cdot \dfrac{342}{365} \approx 0.462$

(b) $P(\text{at least two have same birthday}) = 1 - P(\text{different birthdays}) = 1 - 0.462 = 0.538$

(c) Yes, there were 2 birthdays on the 118th day.

(d) Answers will vary.

34. Let $A = \{\text{flight departs on time}\}$ and $B = \{\text{flight arrives on time}\}$

$P(B|A) = \dfrac{P(A \text{ and } B)}{P(A)} = \dfrac{(0.83)}{(0.89)} \approx 0.933$

3.3 THE ADDITION RULE

3.3 EXERCISE SOLUTIONS

2. (a) Toss coin once: $A = \{\text{head}\}$ and $B = \{\text{tail}\}$

   (b) Draw one card: $A = \{\text{ace}\}$ and $B = \{\text{spade}\}$

4. False, two events that are independent does not imply they are mutually exclusive. Example: Toss a coin then roll a 6-sided die. Let $A = \{\text{head}\}$ and $B = \{6 \text{ on die}\}$. $P(B|A) = \frac{1}{6} = P(B)$ implies A and B are independent events. However, $P(A \text{ and } B) = \frac{1}{12}$ implies A and B are not mutually exclusive.

6. True

8. Mutually exclusive since the two events cannot occur at the same time.

10. Not mutually exclusive since the two events can occur at the same time.

12. Not mutually exclusive since the two events can occur at the same time.

14. (a) Not mutually exclusive since the two events can occur at the same time.

    (b) $P(W \text{ or business}) = P(W) + P(\text{business}) - P(W \text{ and business})$
    $$= \frac{1800}{3500} + \frac{860}{3500} - \frac{425}{3500} \approx 0.639$$

16. (a) Not mutually exclusive since the two events can occur at the same time.

    (b) $P(\text{does not have puncture or does not have smashed edge})$
    $$= P(\text{does not have puncture}) + P(\text{does not have smashed edge})$$
    $$- P(\text{does not have puncture and does not have smashed edge})$$
    $$= 0.96 + 0.93 - 0.8928 = 0.9972$$

18. (a) $P(6 \text{ or greater than } 4) = P(6) + P(\text{greater than } 4) - P(6 \text{ and greater than } 4)$
    $$= \frac{1}{6} + \frac{2}{6} - \frac{1}{6} = 0.333$$

    (b) $P(\text{less than } 5 \text{ or odd}) = P(\text{less than } 5) + P(\text{odd}) - P(\text{less than } 5 \text{ and odd})$
    $$= \frac{4}{6} + \frac{3}{6} - \frac{2}{6} = 0.833$$

    (c) $P(3 \text{ or even}) = P(3) + P(\text{even}) - P(3 \text{ and even})$
    $$= \frac{1}{6} + \frac{3}{6} - 0 = 0.667$$

20. (a) $P(2) = 0.298$

    (b) $P(2 \text{ or more}) = 1 - P(1) = 1 - 0.555 = 0.445$

    (c) $P(\text{between 2 and 5}) = 0.298 + 0.076 + 0.047 + 0.014 = 0.435$

22. (a) $P(\text{not as good as used to be}) = \dfrac{171}{1005} \approx 0.17$

(b) $P(\text{too much violence or tickets too costly}) = P(\text{too much violence}) + P(\text{tickets too costly})$
$$= \dfrac{322}{1005} + \dfrac{302}{1005} = \dfrac{624}{1005} \approx 0.62$$

24. (a) $P(\text{Both are LH men}) = \left(\dfrac{63}{1000}\right) \cdot \left(\dfrac{62}{999}\right) \approx 0.0039$

(b) $P(\text{Both are LH women}) = \left(\dfrac{50}{1000}\right) \cdot \left(\dfrac{49}{999}\right) \approx 0.0025$

(c) $P(\text{at least one is LH}) = 1 - P(\text{neither are LH}) = 1 - \left(\dfrac{887}{1000}\right) \cdot \left(\dfrac{886}{999}\right) \approx 0.213$

(d) $P(\text{1st LH man and 2nd LH woman}) = \left(\dfrac{63}{1000}\right) \cdot \left(\dfrac{50}{999}\right) \approx 0.0032$

(e) $P(\text{LH man and LH woman})$
$= P(\{\text{1st LH man and 2nd LH woman}\} \text{ or } \{\text{1st LH woman and 2nd LH man}\})$
$= P(\{\text{1st LH man and 2nd LH woman}\}) + P(\{\text{1st LH woman and 2nd LH man}\})$
$= \left(\dfrac{63}{1000}\right) \cdot \left(\dfrac{50}{999}\right) + \left(\dfrac{50}{1000}\right) \cdot \left(\dfrac{62}{999}\right) \approx 0.0063$

26. $P(A \text{ or } B \text{ or } C) = P(A) + P(B) + P(C) - P(A \text{ and } B) - P(A \text{ and } C) - P(B \text{ and } C)$
$\qquad + P(A \text{ and } B \text{ and } C)$
$= 0.40 + 0.10 + 0.50 - 0.05 - 0.25 - 0.10 + 0.03$
$= 0.63$

3.4 COUNTING PRINCIPLES

3.4 EXERCISE SOLUTIONS

2. Permutation: Order matters
Combination: Order does not matter

4. True

6. $_{14}P_2 = \dfrac{14!}{(14-2)!} = \dfrac{14!}{12!} = \dfrac{(14)(13)\ldots(3)(2)(1)}{(12)(11)\ldots(3)(2)(1)} = 14 \cdot 13 = 182$

8. $\dfrac{_8C_3}{_{12}C_3} = \dfrac{\dfrac{8!}{3!(8-3)!}}{\dfrac{12!}{3!(12-3)!}} = \dfrac{\dfrac{8!}{3!\,5!}}{\dfrac{12!}{3!\,9!}} = \dfrac{8!}{3!\,5!} \cdot \dfrac{3!\,9!}{12!} = \dfrac{(8!)(9!)}{(5!)(12!)}$

$= \dfrac{[(18)(7)(6)(5)\ldots(2)(1)][(9)(8)(7)\ldots(2)(1)]}{[(5)(4)\ldots(2)(1)][(12)(11)\ldots(1)]} = \dfrac{(8)(7)(6)}{(12)(11)(10)} = 0.255$

10. Combination, since order of the committee members does not matter.

12. $3 \cdot 6 \cdot 4 = 72$

14. $2 \cdot 2 \cdot 2 \cdot 2 \cdot 2 \cdot 2 = 64$

16. $6! = 720$

18. $4! = 24$

20. $_{25}P_9 = 741{,}354{,}768{,}000$

22. $13! = 6{,}227{,}020{,}800$

24. 3-S's, 3-T's, 1-A, 2-I's, 1-C

$$\frac{10!}{3! \cdot 3! \cdot 1! \cdot 2! \cdot 1!} = 50{,}400$$

26. $_{20}C_4 = 4845$

28. $_{52}C_6 = 20{,}358{,}520$

30. (a) $26 \cdot 26 \cdot 10 \cdot 10 \cdot 10 \cdot 10 = 6{,}760{,}000$

(b) $24 \cdot 24 \cdot 10 \cdot 10 \cdot 10 \cdot 10 = 5{,}760{,}000$

(c) 0.50

32. (a) $(10)(10)(10) = 1000$

(b) $(8)(10)(10) = 800$

(c) $\dfrac{(8)(10)(5)}{(8)(10)(10)} = \dfrac{1}{2}$

34. (a) $(_8C_3) \cdot (_2C_0) = (56) \cdot (1) = 56$

(b) $(_8C_2) \cdot (_2C_1) = (28) \cdot (2) = 56$

(c) At least two good units = one or fewer defective units.
$56 + 56 = 112$

(d) $P(\text{at least 2 defective units}) = \dfrac{(_8C_1) \cdot (_2C_2)}{_{10}C_3} = \dfrac{(8) \cdot (1)}{120} \approx 0.067$

36. $(16\%)(1200) = (0.16)(1200) = 192$ of 1200 rate financial shape as poor.

$P(\text{all 10 rate poor}) = \dfrac{_{192}C_{10}}{_{1200}C_{10}} \approx 9.000 \times 10^{-9}$

38. $(38\%)(500) = (0.38)(500) = 190$ of 500 rate financial shape

as good $\Rightarrow 500 - 190 = 310$ rate as not good.

$P(\text{none of 55 selected rate good}) = \dfrac{_{310}C_{55}}{_{500}C_{55}} \approx 5.09 \times 10^{-13}$

40. (a) $_{200}C_{15} = 1.4629 \times 10^{22}$

(b) $_{144}C_{15} = 8.5323 \times 10^{19}$

(c) $P(\text{no minorities}) = \dfrac{(_{144}C_{15})}{(_{200}C_{15})} \approx 0.00583$

(d) Yes, there is a very low probability of randomly selecting 15 non-minorities.

42. $_{14}P_4 = 24{,}024$

226 CHAPTER 3 | PROBABILITY

44. Let $A = \{$team with the worst record wins second pick$\}$ and

$B = \{$team with the best record, ranked 13th, wins first pick$\}$.

$$P(A|B) \frac{250}{(1000-5)} \cdot \frac{250}{995} \approx 0.251$$

46. Let $A = \{$neither the first- nor the second-worst teams will get the first pick$\}$ and

$B = \{$the first- or second-worst team will get the first pick$\}$.

$$P(A) = 1 - P(B) = 1 - \left(\frac{200}{1000} + \frac{250}{100}\right) = 1 - \left(\frac{450}{1000}\right) = 1 - 0.45 = 0.55$$

CHAPTER 3 REVIEW EXERCISE SOLUTIONS

2. Sample space:

$\{(1, 1), (1, 2), (1, 3), (1, 4), (1, 5), (1, 6), (2, 1), (2, 2), (2, 3), (2, 4), (2, 5), (2, 6),$
$(3, 1), (3, 2), (3, 3), (3, 4), (3, 5), (3, 6), (4, 1), (4, 2), (4, 3), (4, 4), (4, 5), (4, 6),$
$(5, 1), (5, 2), (5, 3), (5, 4), (5, 5), (5, 6), (6, 1), (6, 2), (6, 3), (6, 4), (6, 5), (6, 6)\}$

Event: sum of 4 or 5
$\{(1, 3), (1, 4), (2, 2), (2, 3), (3, 1), (3, 2), (4, 1)\}$

4. Sample space: $\{GGG, GGB, GBG, GBB, BGG, BGB, BBG, BBB\}$

Event: The family has two boys
$\{BGB, BBG, GBB\}$

6. Classical probability

8. Empirical probability

10. Empirical probability

12. $P(\text{less than } 60) = 0.28 + 0.20 + 0.35 = 0.83$

14. $P(\text{graduate}|-) = 0.07$

16. Dependent, the first event does affect the outcome of the second event.

18. $P(\text{1st pair black and 2nd pair blue or white})$

$= P(\text{1st pair black}) \cdot P(\text{2nd pair blue or white}|\text{1st black})$

$= \left(\frac{4}{12}\right) \cdot \left(\frac{8}{11}\right) = 0.242$

20. Not mutually exclusive since both events can occur at the same time.

22. $P(\text{home or work}) = P(\text{home}) + P(\text{work}) - P(\text{home and work})$

$= 0.56 + 0.21 - 0.20$

$= 0.57$

24. $P(\text{red or queen}) = P(\text{red}) + P(\text{queen}) - P(\text{red and queen})$

$= \frac{26}{52} + \frac{4}{52} - \frac{2}{52} = 0.538$

26. $P(\text{even or greater than 6}) = P(\text{even}) + P(\text{greater than 6}) - P(\text{even and greater than 6})$

$$= \frac{4}{8} + \frac{2}{8} - \frac{1}{8} = \frac{5}{8} = 0.625$$

28. $P(300 - 999) = P(300 - 599) + P(600 - 999)$

$$= 0.226 + 0.192 = 0.418$$

30. $P(\text{no time to cook or confused about nutrition})$

$= P(\text{no time to cook}) + P(\text{confused about nutrition})$

$$= \frac{175}{500} + \frac{30}{500} = \frac{205}{500} = 0.41$$

32. $36 \cdot 36 \cdot 36 \cdot 36 = 1{,}679{,}616$

34. $5! = 120$

36. $_{13}C_2 = 78$

38. $\dfrac{1}{(23)(26)(26)(10)} = \dfrac{1}{155{,}480}$

40. (a) $\dfrac{347}{350} \cdot \dfrac{346}{349} \cdot \dfrac{345}{348} \cdot \dfrac{344}{347} \cdot \dfrac{343}{346} \approx 0.958$

(b) $1 - 0.958 = 0.042$

(c) $P(\text{at least one winner}) = 1 - P(\text{no winners}) = 1 - 0.974 = 0.026$

(d) $P(\text{at least one non-winner}) = 1 - P(\text{all winners})$

$$= 1 - \left(\frac{3}{350}\right)\left(\frac{2}{349}\right)\left(\frac{1}{348}\right) = 0.9999998589$$

CHAPTER 3 QUIZ SOLUTIONS

1. (a) $P(\text{bachelor}) = \dfrac{1322}{2466} \approx 0.536$

(b) $P(\text{bachelor}|F) = \dfrac{769}{1460} \approx 0.527$

(c) $P(\text{bachelor}|M) = \dfrac{553}{1006} \approx 0.550$

(d) $P(\text{associate or bachelor}) = P(\text{associate}) + P(\text{bachelor})$

$$= \frac{632}{2466} + \frac{1322}{2466} \approx 0.792$$

(e) $P(\text{doctorate}|M) = \dfrac{25}{1006} \approx 0.025$

(f) $P(\text{master or female}) = P(\text{master}) + P(\text{female}) - P(\text{master and female})$

$$= \frac{467}{2466} + \frac{1460}{2466} - \frac{270}{2466} = 0.672$$

(g) $P(\text{associate and male}) = P(\text{associate}) \cdot P(\text{male}|\text{associate})$

$$= \frac{632}{2466} \cdot \frac{231}{632} = \frac{145{,}992}{1{,}558{,}512} \approx 0.094$$

(h) $P(\text{F}|\text{bachelor}) = \frac{769}{1322} \approx 0.582$

2. Not mutually exclusive since both events can occur at the same time.

Dependent since one event can affect the occurrence of the second event.

3. (a) $_{147}C_3 = 518{,}665$ (b) $_3C_3 = 1$ (c) $_{150}C_3 - {_3C_3} = 551{,}300 - 1 = 551{,}299$

4. (a) $\dfrac{_{147}C_3}{_{150}C_3} = \dfrac{518{,}665}{551{,}300} \approx 0.94$

(b) $\dfrac{_3C_3}{_{150}C_3} = \dfrac{1}{551{,}300} \approx 0.00000181$

(c) $\dfrac{_{150}C_3 - {_3C_3}}{_{150}C_3} = \dfrac{551{,}299}{551{,}300} \approx 0.999998$

5. $9 \cdot 10 \cdot 10 \cdot 5 = 4500$

6. $_{25}P_4 = 303{,}600$

Discrete Probability Distributions

CHAPTER 4

4.1 PROBABILITY DISTRIBUTIONS

4.1 EXERCISE SOLUTIONS

2. A random variable is discrete if it has a finite or countable number of possible outcomes that can be listed.

   Condition 1: $0 \leq P(x) \leq 1$
   Condition 2: $\Sigma P(x) = 1$

4. True

6. False. The expected value of a discrete random variable is equal to the mean of the random variable.

8. Continuous, because length of time is a random variable that has an infinite number of possible outcomes and cannot be counted.

10. Discrete, because the number of fatalities is a random variable that is countable.

12. Continuous, because the random variable has an infinite number of possible outcomes and cannot be counted.

14. Discrete, because the random variable is countable.

16. Continuous, because the random variable has an infinite number of possible outcomes and cannot be counted.

18. Discrete, because the random variable is countable.

20. (a) $P(x > 1) = 1 - P(x < 2) = 1 - (0.30 + 0.25) = 0.45$

    (b) $P(x < 3) = 0.30 + 0.25 + 0.25 = 0.80$

22. $\Sigma P(x) = 1 \rightarrow P(1) = 0.15$

24. No, $\Sigma P(x) \approx 1.201$

26. Yes

28. (a)

x	f	$P(x)$	$xP(x)$	$(x - \mu)$	$(x - \mu)^2$	$(x - \mu)^2 P(x)$
0	1941	0.7275	0	−0.5321	0.2822	(0.2822)(0.7275) = 0.2053
1	349	0.1308	0.1308	0.4688	0.2198	(0.2198)(0.1308) = 0.0287
2	203	0.0761	0.1522	1.4688	2.1574	(2.1574)(0.0761) = 0.1642
3	78	0.0292	0.0876	2.4688	6.0950	(6.0950)(0.0292) = 0.1780
4	57	0.0214	0.0856	3.4688	12.0326	(12.0326)(0.0214) = 0.2575
5	40	0.0150	0.075	4.4688	19.9702	(19.9702)(0.0150) = 0.2996
	$n = 2668$	$\Sigma P(x) = 1$	$\Sigma x P(x) = 0.531$			$\Sigma (x - \mu)^2 P(x) = 1.1333$

(b) $\mu = \Sigma x P(x) \approx 0.5$

(c) $\sigma^2 = \Sigma (x - \mu)^2 P(x) \approx 1.1$

(d) $\sigma = \sqrt{\sigma^2} \approx 1.0$

(e) A household has on average 0.5 cats with a standard deviation of 1.0.

30. (a)

x	f	P(x)	xP(x)	(x − μ)	(x − μ)²	(x − μ)²P(x)
0	260	0.152	0.000	−1.862	3.467	(3.467)(0.152) = 0.527
1	500	0.292	0.292	−0.862	0.743	(0.743)(0.292) = 0.217
2	425	0.249	0.498	0.138	0.019	(0.019)(0.294) = 0.005
3	305	0.178	0.534	1.138	1.295	(1.295)(0.178) = 0.231
4	175	0.102	0.408	2.138	4.571	(4.571)(0.102) = 0.466
5	45	0.026	0.130	3.138	9.847	(9.847)(0.026) = 0.256
	n = 1710	ΣP(x) = 1	ΣxP(x) = 1.862			Σ(x − μ)²P(x) = 1.701

(b) $\mu = \Sigma x P(x) \approx 1.9$

(c) $\sigma^2 = \Sigma (x - \mu)^2 P(x) \approx 1.7$

(d) $\sigma = \sqrt{\sigma^2} \approx 1.3$

(e) The average number of accidents per student is 1.9 with a standard deviation of 1.3.

32. (a)

x	f	P(x)	xP(x)	(x − μ)	(x − μ)²	(x − μ)²P(x)
0	19	0.059	0.000	−3.350	11.223	(11.223)(0.059) = 0.666
1	39	0.122	0.122	−2.350	5.523	(5.523)(0.122) = 0.673
2	52	0.163	0.325	−1.350	1.823	(1.823)(0.163) = 0.296
3	57	0.178	0.534	−0.350	0.123	(0.123)(0.178) = 0.022
4	68	0.213	0.850	0.650	0.423	(0.423)(0.213) = 0.090
5	41	0.128	0.641	1.650	2.723	(2.723)(0.128) = 0.349
6	27	0.084	0.506	2.650	7.023	(7.023)(0.084) = 0.590
7	17	0.053	0.372	3.650	13.323	(13.323)(0.053) = 0.706
	n = 320	ΣP(x) = 1	ΣxP(x) = 3.350			Σ(x − μ)²P(x) = 3.392

(b) $\mu = \Sigma x P(x) = 3.4$

(c) $\sigma^2 = \Sigma (x - \mu)^2 P(x) 3.4$

(d) $\sigma = \sqrt{\sigma^2} \approx 1.8$

(e) The average number of school related extracurricular activities per student is 3.4 with a standard deviation of 1.8.

34.

x	P(x)	xP(x)	(x − μ)	(x − μ)²	(x − μ)²P(x)
0	0.22	0.00	−1.87	3.50	0.769
1	0.28	0.28	−0.87	0.76	0.212
2	0.20	0.40	0.13	0.02	0.003
3	0.12	0.36	1.13	1.28	0.153
4	0.09	0.36	2.13	4.54	0.408
5	0.07	0.35	3.13	9.80	0.686
6	0.02	0.12	4.13	17.60	0.341
		ΣP(x) = 1.87			Σ(x − μ)²P(x) = 2.573

(a) $\mu = \Sigma x P(x) = 1.9$

(b) $\sigma^2 = \Sigma (x - \mu)^2 P(x) = 2.6$

(c) $\sigma = \sqrt{\sigma^2} = 1.6$

(d) $E[x] = \mu = \Sigma x P(x) = 1.9$

(e) The average number of ATM transactions made in one day is 1.9 with a standard deviation of 1.6.

36. (a) $\mu = \Sigma x P(x) = 1.7$ (b) $\sigma^2 = \Sigma(x-\mu)^2 P(x) \approx 1.0$

(c) $\sigma = \sqrt{\sigma^2} \approx 1.0$ (d) $E[x] = \mu = \Sigma x P(x) = 1.7$

(e) The average car occupancy is 1.7 with a standard deviation of 1.0.

38. (a) $\mu = \Sigma x P(x) = 1.6$ (b) $\sigma^2 = \Sigma(x-\mu)^2 P(x) \approx 1.9$

(c) $\sigma = \sqrt{\sigma^2} \approx 1.4$ (d) $E[x] = \mu = \Sigma x P(x) = 1.6$

(e) The average car occupancy was 1.6 with a standard deviation of 1.4.

40. (a) $P(0) = 0.432$

(b) $P(x \geq 1) = 1 - P(0) = 1 - 0.432 = 0.568$

(c) $P(1 \leq x \leq 3) = 0.568$

42. A household with no computers is not unusual since the probability is 0.432.

44. $E(x) = \mu = \Sigma x P(x)$

$$= (3146) \cdot \left(\frac{1}{5000}\right) + (446) \cdot \left(\frac{1}{5000}\right) + (21) \cdot \left(\frac{15}{5000}\right) + (-4) \cdot \left(\frac{4983}{5000}\right) \approx -\$3.21$$

4.2 BINOMIAL DISTRIBUTIONS

4.2 EXERCISE SOLUTIONS

2. (a) $p = 0.75$ (graph is skewed left $\rightarrow p > 0.5$)

(b) $p = 0.50$ (graph is symmetric)

(c) $p = 0.25$ (graph is skewed right $\rightarrow p < 0.5$)

4. (a) $n = 10, (x = 0, 1, 2, \ldots, 10)$

(b) $n = 15, (x = 0, 1, 2, \ldots, 15)$

(c) $n = 5, (x = 0, 1, 2, 3, 4, 5)$

As n increases the probability distribution becomes more symmetric.

6. Is a binomial experiment.
Success: person does not make a purchase
$n = 18, p = 0.74, q = 0.26, x = 0, 1, 2, \ldots, 18$

8. Is not a binomial experiment because the probability of a success is not the same for each trial.

10. $\mu = np = (76)(0.8) = 60.8$

$\sigma^2 = npq = (76)(0.8)(0.2) = 12.16$

$\sigma = \sqrt{\sigma^2} = 3.487$

12. $\mu = np = (342)(0.63) = 215.46$

$\sigma^2 = npq = (342)(0.63)(0.37) = 79.720$

$\sigma = \sqrt{\sigma^2} = 8.929$

14. $n = 7, p = 0.70$

(a) $P(5) = 0.318$

(b) $P(x \geq 5) = P(5) + P(6) + P(7) = 0.318 + 0.247 + 0.082 = 0.647$

(c) $P(x < 5) = 1 - P(x \geq 5) = 1 - 0.647 = 0.353$

232 CHAPTER 4 | DISCRETE PROBABILITY DISTRIBUTIONS

16. $n = 12$, $p = 0.10$

(a) $P(4) \approx 0.021$

(b) $P(x \geq 4) = 1 - P(x < 4) = 1 - P(0) - P(1) - P(2) - P(3)$
$= 1 - 0.282 - 0.377 - 0.230 - 0.085 = 0.026$

(c) $P(x < 4) = 1 - P(x \geq 4) = 1 - 0.026 = 0.974$

18. $n = 20$, $p = 0.70$ (use binomial formula)

(a) $P(1) \approx 1.627 \times 10^{-9}$

(b) $P(x > 1) = P(0) - P(1) \approx 1 - 3.487 \times 10^{-11} - 1.627 \times 10^{-9} = 0.9999999983 \approx 1$

(c) $P(x \leq 1) = 1 - P(x > 1) \approx 1 - 0.9999999983 \approx 1.662 \times 10^{-9}$

20. (a) $n = 5$, $p = 0.25$ (b) 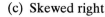 No Trouble Sleeping at Night   (c) Skewed right

x	P(x)
0	0.237305
1	0.395508
2	0.263672
3	0.087891
4	0.014648
5	0.000977

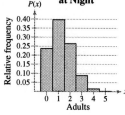

(d) $\mu = np = (5)(0.25) = 1.3$

(e) $\sigma^2 = npq = (5)(0.25)(0.75) = 0.9$

(f) $\sigma = \sqrt{\sigma^2} \approx 0.9$

(g) On average 1.3 adults, out of every 5, have difficulty sleeping at night. The standard deviation is 0.9 adults. Four or five would be uncommon due to their low probabilities.

22. (a) $n = 5$, $p = 0.38$ (b) Blood Type   (c) Skewed right

x	P(x)
0	0.092
1	0.281
2	0.344
3	0.211
4	0.065
5	0.008

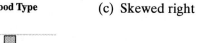

(d) $\mu = np = (5)(0.38) = 1.9$

(e) $\sigma^2 = npq = (5)(0.38)(0.62) = 1.2$

(f) $\sigma = \sqrt{\sigma^2} \approx 1.1$

(g) On average 1.9 adults, out of every 5, has O+ blood. The standard deviation is 1.1 adults. Four or five would be uncommon due to their low probabilities.

24. (a) $n = 5$, $p = 0.41$ (b) $P(2) = 0.345$ (c) $P(\text{less than } 4) = 1 - P(4) - P(5)$
$= 1 - 0.083 - 0.012$
$= 0.905$

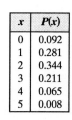

x	P(x)
0	0.071
1	0.248
2	0.345
3	0.240
4	0.083
5	0.012

26. $\mu = np = (5)(0.41) = 2.1$

$\sigma^2 = npq = (5)(0.41)(0.59) = 1.2$

$\sigma = \sqrt{\sigma^2} = 1.1$

If 5 employees are randomly selected, on average 2.1 employees would claim lack of time was a barrier. It would be unusual for all 5 employees to claim lack of time was a barrier since $P(5) = 0.012$.

28. $P(5, 2, 2, 1) = \dfrac{10!}{5!2!2!1!}\left(\dfrac{5}{16}\right)^5\left(\dfrac{4}{16}\right)^2\left(\dfrac{1}{16}\right)^2\left(\dfrac{6}{16}\right)^1 \approx 0.002$

4.3 MORE DISCRETE PROBABILITY DISTRIBUTIONS

4.3 EXERCISE SOLUTIONS

2. $P(1) = (0.25)(0.75)^0 = 0.25$

4. $P(4) = (0.62)(0.38)^3 = 0.034$

6. $P(5) = \dfrac{(6)^5(e^{-6})}{5!} = 0.161$

8. $P(4) = \dfrac{(8.1)^4(e^{-8.1})}{4!} = 0.054$

10. Poisson. We are interested in counting the number of occurrences that takes place within a given unit of time.

12. Binomial. We are interested in counting the number of successes out of n trials.

14. Geometric. We are interested in counting the number of trials until the first success.

16. $p = 0.541$

(a) $P(2) = (0.541)(0.459)^1 \approx 0.248$

(b) $P(\text{makes 1st or 2nd shot}) = P(1) + P(2) = (0.541)(0.459)^0 + (0.541)(0.459)^1 \approx 0.789$

(c) (Binomial: $n = 2, p = 0.541$)

$P(0) = \dfrac{2!}{0!2!}(0.541)^0(0.459)^2 \approx 0.211$

18. $p = 0.25$

(a) $P(4) = (0.25)(0.75)^3 \approx 0.105$

(b) $P(\text{prize with 1st, 2nd, or 3rd purchase}) = P(1) + P(2) + P(3)$
$= (0.25)(0.75)^0 + (0.25)(0.75)^1 + (0.25)(0.75)^2 \approx 0.578$

(c) $P(x > 4) = 1 - P(x \leq 4) = 1 - [P(1) + P(2) + P(3) + P(4)] = 1 - [0.684] \approx 0.317$

20. $\mu = 4$

(a) $P(3) = 0.195$

(b) $P(x \leq 3) = P(0) + P(1) + P(2) + P(3) = 0.0183 + 0.0733 + 0.1465 + 0.1954 = 0.433$

(c) $P(x > 3) = 1 - P(x \leq 3) = 1 - 0.4335 = 0.567$

22. $\mu = 8.7$

(a) $P(9) = \dfrac{8.7^9 e^{-8.7}}{9!} \approx 0.131$

(b) $P(x \leq 9) = 0.627$

(c) $P(x > 9) = 1 - P(x \leq 9) = 1 - 0.627 = 0.373$

24. (a) $P(0) = \dfrac{{}_2C_0\,{}_{13}C_3}{{}_{15}C_3} = \dfrac{(1)(286)}{(455)} \approx 0.629$ (b) $P(1) = \dfrac{{}_2C_1\,{}_{13}C_2}{{}_{15}C_3} = \dfrac{(2)(78)}{(455)} \approx 0.343$

(c) $P(2) = \dfrac{{}_2C_2\,{}_{13}C_1}{{}_{15}C_3} = \dfrac{(1)(13)}{(455)} \approx 0.029$

26. $p = 0.005$

(a) $\mu = \dfrac{1}{p} = \dfrac{1}{0.005} = 200$

$\sigma^2 = \dfrac{q}{p^2} = \dfrac{0.995}{(0.005)^2} = 39{,}800$

$\sigma = \sqrt{\sigma^2} \approx 199.5$

On average 200 records will be examined before finding one that has been miscalculated. The standard deviation is 199.5.

(b) 200

28. $\mu = 4$

(a) $\sigma^2 = 4$

$\sigma = \sqrt{\sigma^2} \approx 2$

The standard deviation is 2 inches.

(b) $P(X > 7) = 1 - P(X \le 7) \approx 1 - 0.9489 \approx 0.05$

CHAPTER 4 REVIEW EXERCISE SOLUTIONS

2. Continuous   4. Discrete   6. Yes

8. No, $P(5) > 1$ and $\Sigma P(x) \ne 1$.   10. No, $\Sigma P(x) \ne 1$.

12. (a)

x	Frequency	$P(x)$	$xP(x)$	$x - \mu$	$(x - \mu)^2$	$(x - \mu)^2 P(x)$
0	7	0.219	0.000	−1.688	2.848	0.623
1	8	0.250	0.250	−0.688	0.473	0.118
2	10	0.313	0.625	0.313	0.098	0.031
3	3	0.094	0.281	1.313	1.723	0.161
4	3	0.094	0.375	2.313	5.348	0.501
5	1	0.031	0.156	3.313	10.973	0.343
	$n = 32$	$\Sigma P(x) = 1$	$\Sigma xP(x) = 1.688$			$\Sigma(x - \mu)^2 P(x) = 1.777$

(b)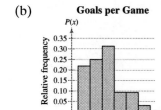

(c) $\mu = \Sigma xP(x) \approx 1.7$

$\sigma^2 = \Sigma(x - \mu)^2 P(x) \approx 1.8$

$\sigma = \sqrt{\sigma^2} \approx 1.3$

14. (a)

x	Frequency	$P(x)$	$xP(x)$	$x - \mu$	$(x - \mu)^2$	$(x - \mu)^2 P(x)$
15	76	0.134	2.014	−16.802	282.311	37.908
30	445	0.786	23.587	−1.802	3.248	2.553
60	30	0.053	3.180	28.198	795.120	42.144
90	3	0.005	0.477	58.198	3386.993	17.952
120	12	0.021	2.544	88.198	7778.866	164.923
	$n = 566$	$\Sigma P(x) = 1$	$\Sigma xP(x) = 31.802$			$\Sigma(x - \mu)^2 P(x) = 265.480$

(b)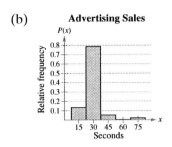

(c) $\mu = \Sigma x P(x) \approx 31.8$

$\sigma^2 = \Sigma(x - \mu)^2 P(x) \approx 265.5$

$\sigma = \sqrt{\sigma^2} \approx 16.3$

16. $E(x) = \mu = \Sigma x P(x) = 2.5$

18. No, the experiment is not repeated for a fixed number of trials.

20. $n = 12$, $p = 0.2$

(a) $P(4) = 0.133$

(b) $P(x \geq 4) = 1 - P(0) - P(1) - P(2) - P(3) = 1 - 0.069 - 0.206 - 0.283 - 0.236 = 0.205$

(c) $P(x > 4) = 1 - P(0) - P(1) - P(2) - P(3) - P(4)$
$= 1 - 0.069 - 0.206 - 0.283 - 0.236 - 0.133 = 0.073$

22. $n = 5$, $p = 0.26$

(a) $P(2) = 0.274$

(b) $P(x \geq 2) = P(2) + P(3) + P(4) + P(5) = 0.274 + 0.096 + 0.017 + 0.001 \approx 0.388$

(c) $P(x > 2) = P(3) + P(4) + P(5) \approx 0.096 + 0.017 + 0.001 \approx 0.114$

24. (a)

x	P(x)
0	0.006
1	0.050
2	0.167
3	0.294
4	0.293
5	0.155
6	0.034

(b)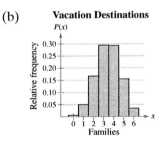

(c) $\mu = np = (6)(0.57) = 3.4$

$\sigma^2 = npq = (6)(0.57)(0.43) = 1.5$

$\sigma = \sqrt{\sigma^2} \approx 1.2$

26. $n = 5$, $p = 0.17$

(a)

x	P(x)
0	0.394
1	0.403
2	0.165
3	0.0338
4	0.00347
5	0.000142

(b)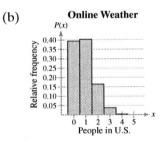

(c) $\mu = np = (5)(0.17) = 0.9$

$\sigma^2 = npq = (5)(0.17)(0.83) = 0.7$

$\sigma = \sqrt{\sigma^2} \approx 0.8$

28. $p = \dfrac{73}{153} = 0.477$

(a) $P(1) \approx 0.477$ (b) $P(2) \approx 0.249$

(c) $P(1 \text{ or } 2) = P(1) + P(2) \approx 0.477 + 0.249 = 0.726$

(d) $P(X \leq 3) = P(1) + P(2) + P(3) = 0.477 + 0.249 + 0.130 = 0.856$

30. (a) $\mu = 10$

$$P(x \geq 3) = 1 - P(x < 3)$$
$$= 1 - [P(0) + P(1) + P(2)] = 1 - [0.0000 + 0.0005 + 0.0023] = 0.9972$$

(b) $\mu = 5$

$$P(x \geq 3) = 1 - P(x < 3)$$
$$= 1 - [P(0) + P(1) + P(2)] = 1 - [0.0067 + 0.0337 + 0.0842] = 0.8754$$

(c) $\mu = 15$

$$P(x \geq 3) = 1 - P(x < 3) = 1 - [P(0) + P(1) + P(2)] \approx 1 - [0.0000 + 0.0000 + 0.0000] = 1$$

CHAPTER 4 QUIZ SOLUTIONS

1. (a) Discrete because the random variable is countable.

(b) Continuous because the random variable has an infinite number of possible outcomes and cannot be counted.

2. (a)

x	Frequency	$P(x)$	$xP(x)$	$x - \mu$	$(x - \mu)^2$	$(x - \mu)^2 P(x)$
1	61	0.370	0.370	−1.145	1.312	0.485
2	39	0.236	0.473	−0.145	0.021	0.005
3	48	0.291	0.873	0.855	0.730	0.212
4	14	0.085	0.339	1.855	3.439	0.292
5	3	0.018	0.091	2.855	8.148	0.149
	$n = 165$	$\Sigma P(x) = 1$	$\Sigma xP(x) = 2.145$			$\Sigma(x - \mu)^2 P(x) = 1.142$

(b)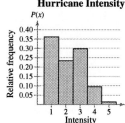

(c) $\mu = \Sigma xP(x) \approx 2.1$

$\sigma^2 = \Sigma(x - \mu)^2 P(x) \approx 1.1$

$\sigma = \sqrt{\sigma^2} \approx 1.1$

On average the intensity of a hurricane will be 2.1. The standard deviation is 1.1.

(d) $P(x \geq 4) = P(4) + P(5) = 0.085 + 0.018 = 0.103$

3. $n = 8, p = 0.80$

(a)

x	$P(x)$
0	0.000003
1	0.000082
2	0.001147
3	0.009175
4	0.045875
5	0.146801
6	0.293601
7	0.335544
8	0.167772

(b)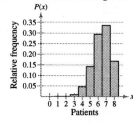

(c) $\mu = np = (8)(0.80) = 6.4$

$\sigma^2 = npq = (8)(0.80)(0.20) = 1.3$

$\sigma = \sqrt{\sigma^2} \approx 1.1$

(d) $P(2) = 0.001$

(e) $P(x < 2) = P(0) + P(1) = 0.000003 + 0.000082 = 0.000085$

4. $\mu = 5$

(a) $P(5) = 0.1755$

(b) $P(x < 5) = P(0) + P(1) + P(2) + P(3) + P(4)$
$= 0.0067 + 0.0337 + 0.0842 + 0.1404 + 0.1755 = 0.4405$

(c) $P(0) = 0.0067$

Normal Probability Distributions

CHAPTER 5

5.1 INTRODUCTION TO NORMAL DISTRIBUTIONS AND THE STANDARD NORMAL DISTRIBUTION

5.1 EXERCISE SOLUTIONS

2. 1000

4. Answers will vary.
Similarities: Both curves will have the same shape (i.e., equal standard deviations)
Differences: The two curves will have different lines of symmetry.

6. Transform each data value x into a z-score. This is done by subtracting the mean from x and dividing by the standard deviation. In symbols,

$$z = \frac{x - \mu}{\sigma}.$$

8. (c) is true since a z-score equal to zero indicates that the corresponding x-value is equal to the mean. (a) and (b) are not true since it is possible to have a z-score equal to zero and the mean is not zero or the corresponding x-value is not zero.

10. No, the graph is not symmetric.   **12.** No, the graph is skewed left.

14. No, the graph is not bell-shaped.

16. The histogram does not represent data from a normal distribution since it's not bell-shaped.

18. (Area left of $z = 0$) − (Area left of $z = -2.25$) = 0.5 − 0.0122 = 0.4878

20. (Area right of $z = 2$) = 1 − (Area left of $z = 2$) = 1 − 0.9772 = 0.0228

22. 0.5987   **24.** 0.8997   **26.** 1 − 0.0401 = 0.9599

28. 1 − 0.9901 = 0.0099   **30.** 0.0010   **32.** 1 − 0.9940 = 0.006

34. 0.999 − 0.5 = 0.499   **36.** 0.5 − 0.3050 = 0.195   **38.** 0.9901 − 0.0099 = 0.9802

40. 0.0250 + 0.0250 = 0.05

42. (a)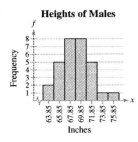

It is reasonable to assume that the heights are normally distributed since the histogram is nearly symmetric and bell-shaped.

(b) $\bar{x} \approx 68.75$, $s \approx 2.847$

(c) The mean of your sample is 0.45 inch less than that of the previous study so the average height from the sample is less than in the previous study. The standard deviation is about 0.053 less than that of the previous study, so the heights are slightly less spread out than in the previous study.

44. (a) A = 24,750   B = 30,000   C = 33,000   D = 35,150

(b) $x = 35{,}150 \Rightarrow z = \dfrac{x-\mu}{\sigma} = \dfrac{35{,}150 - 30{,}000}{2500} = 2.06$

$x = 24{,}750 \Rightarrow z = \dfrac{x-\mu}{\sigma} = \dfrac{24{,}750 - 30{,}000}{2500} = -2.1$

$x = 30{,}000 \Rightarrow z = \dfrac{x-\mu}{\sigma} = \dfrac{30{,}000 - 30{,}000}{2500} = 0$

$x = 33{,}000 \Rightarrow z = \dfrac{x-\mu}{\sigma} = \dfrac{33{,}000 - 30{,}000}{2500} = 1.2$

(c) 2.06 and −2.1 are unusual.

46. (a) A = 14   B = 18   C = 25   D = 32

(b) $x = 18 \Rightarrow z = \dfrac{x-\mu}{\sigma} = \dfrac{18 - 20.8}{4.8} = -0.58$

$x = 32 \Rightarrow z = \dfrac{x-\mu}{\sigma} = \dfrac{32 - 20.8}{4.8} = 2.33$

$x = 14 \Rightarrow z = \dfrac{x-\mu}{\sigma} = \dfrac{14 - 20.8}{4.8} = -1.42$

$x = 25 \Rightarrow z = \dfrac{x-\mu}{\sigma} = \dfrac{25 - 20.8}{4.8} = 0.86$

(c) 2.33 is unusual.

48. 0.1587

50. 1 − 0.1003 = 0.8997

52. 0.9772 − 0.6915 = 0.2857

54. $P(z < 0.45) = 0.6736$

56. $P(z > -0.25) = 1 - P(z < -0.25) = 1 - 0.4013 = 0.5987$

58. $P(-2.08 < z < 0) = 0.5 - 0.0188 = 0.4812$

60. $P(-1.96 < z < 1.96) = 0.9750 - 0.0250 = 0.95$

62. $P(z < -1.96 \text{ or } z > 1.96) = 2(0.025) = 0.05$

64.

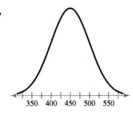

The normal distribution curve is centered at its mean (450) and has 2 points of inflection (400 and 500) representing $\mu \pm \sigma$.

66. (a)

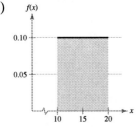

Area under curve = Area of rectangle
= (base)(height)
= (20 − 10) · (0.10)
= 1

(b) $P(12 < x < 15) = $ (base)(height) $= (3)(0.1) = 0.3$

(c) $P(13 < x < 18) = $ (base)(height) $= (5)(0.1) = 0.5$

5.2 NORMAL DISTRIBUTIONS: FINDING PROBABILITIES

5.2 EXERCISE SOLUTIONS

2. $P(x < 100) = P(z < 2.8) = 0.9974$

4. $P(x > 75) = P(z > -2.2) = 1 - 0.0139 = 0.9861$

6. $P(85 < x < 95) = P(-0.2 < z < 1.8) = 0.9641 - 0.4207 = 0.5434$

8. $P(670 < x < 800) = P(1.31 < z < 2.44) = 0.9927 - 0.9049 = 0.0878$

10. $P(200 < x < 239) = P(-0.53 < z < 0.37) = 0.6443 - 0.2981 = 0.3462$

12. $P(160 < x < 168) = P(-1.44 < z < 0) = 0.5 - 0.0749 = 0.4251$

14. (a) $P(x < 7) = P(z < -1.5) = 0.0668$

　　(b) $P(7 < x < 15) = P(-1.5 < z < 2.5) = 0.9938 - 0.0668 = 0.9270$

　　(c) $P(x > 15) = P(z > 2.5) = 1 - P(z < 2.5) = 1 - 0.9938 = 0.0062$

16. (a) $P(x < 13) = P(z < -0.67) = 0.2514$

　　(b) $P(13 < x < 17) = P(-0.67 < z < 0.67) = 0.7486 - 0.2514 = 0.4972$

　　(c) $P(x > 17) = P(z > 0.67) = 1 - P(z < 0.67) = 1 - 0.7486 = 0.2514$

18. (a) $P(x < 80) = P(z < -1.67) = 0.0475$

　　(b) $P(80 < x < 115) = P(-1.67 < z < 1.25) = 0.8944 - 0.0475 = 0.8469$

　　(c) $P(x > 115) = P(z > 1.25) = 1 - P(z < 1.25) = 1 - 0.8944 = 0.1056$

20. (a) $P(x < 17) = P(z < -0.6) = 0.2743$

　　(b) $P(17 < x < 22) = P(-0.6 < z < 0.4) = 0.6554 - 0.2743 = 0.3811$

　　(c) $P(x > 22) = P(z > 0.4) = 1 - 0.6554 = 0.3446$

22. (a) $P(x < 500) = P(z < -0.17) = 0.4325 \Rightarrow 43.25\%$

　　(b) $P(x > 600) = P(z > 0.70) = 1 - P(z < 0.70) = 1 - 0.7580 = 0.2420$
　　　　$(1500)(0.2420) = 363$

24. (a) $P(x < 239) = P(z < 0.37) = 0.6443$

　　(b) $P(x > 200) = P(z > -0.53) = 1 - P(z < -0.53) = 1 - 0.2981 = 0.7019$
　　　　$(200)(0.7019) = 140.38 \Rightarrow 140$

26. (a) $P(x > 20) = P(z > 1.67) = 1 - P(z < 1.67) = 1 - 0.9525 = 0.0475 \Rightarrow 4.75\%$

　　(b) $P(x < 12) = P(z < -1) = 0.1587$
　　　　$(50)(0.1587) = 7.935 \Rightarrow 7$

28. (a) $P(x > 125) = P(z > 2.08) = 1 - P(z < 2.08) = 1 - 0.9812 = 0.0188 \rightarrow 1.88\%$

　　(b) $P(x < 90) = P(z < -0.83) = 0.2033$
　　　　$(300)(0.2033) = 60.99 \Rightarrow 60$

240 CHAPTER 5 | NORMAL PROBABILITY DISTRIBUTIONS

30. $P(x < 3.1) = P(z < -1.56) = 0.0594 \Rightarrow 5.94\%$

It is not unusual for a person to consume less than 3.1 pounds of peanuts because the z-score is within 2 standard deviations of the mean.

32. Out of control, since two out of three consecutive points lie more than 2 standard deviations from the mean. (8th and 10th observations.)

34. In control, since none of the three warning signals detected a change.

5.3 NORMAL DISTRIBUTIONS: FINDING VALUES

5.3 SECTION EXERCISES

2. $z = -0.81$ **4.** $z = 0.37$ **6.** $z = -2.41$

8. $z = 0.73$ **10.** $z = 1.04$ **12.** $z = -2.325$

14. $z = -1.04$ **16.** $z = 0.125$ **18.** $z = 1.75$

20. $z = 0$ **22.** $z = -1.28$ **24.** $z = 0.385$

26. $z = 0.25$ **28.** $z = 1.99$ **30.** $z = \pm 1.96$

32. 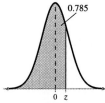 $\Rightarrow z = 0.79$ **34.** 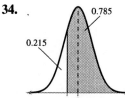 $\Rightarrow z = -0.79$

36. 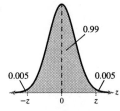 $\Rightarrow z = 2.575$ **38.** 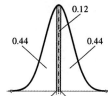 $\Rightarrow z = \pm 0.15$

40. (a) 90th percentile $\Rightarrow$ Area $= 0.90 \Rightarrow z = 1.28$
$x = \mu + z\sigma = 69.2 + (1.28)(2.9) \approx 72.91$

(b) 1st quartile $\Rightarrow$ Area $= 0.25 \Rightarrow z = -0.675$
$x = \mu + z\sigma = 69.2 + (-0.675)(2.9) \approx 67.24$

42. (a) 5th percentile $\Rightarrow$ Area $= 0.05 \Rightarrow z = -1.645$
$x = \mu + z\sigma = 11.7 + (-1.645)(3) = 6.765$

(b) 3rd quartile $\Rightarrow$ Area $= 0.75 \Rightarrow z = 0.67$
$x = \mu + z\sigma = 11.7 + (0.675)(3) = 13.725$

44. (a) Top 25% $\Rightarrow$ Area $= 0.75 \Rightarrow z = 0.675$
$x = \mu + z\sigma = 16.5 + (0.675)(2.5) = 18.188$

(b) Bottom 15% $\Rightarrow$ Area $= 0.15 \Rightarrow z = -1.035$
$x = \mu + z\sigma = 16.5 + (-1.035)(2.5) = 13.913$

46. Upper 7.5% $\Rightarrow$ Area = 0.925 $\Rightarrow$ z = 1.44

$x = \mu + z\sigma = 15 + (1.44)(0.085) = 15.122$

48. A: Top 10% $\Rightarrow$ Area = 0.90 $\Rightarrow$ z = 1.28

$x = \mu + z\sigma = 72 + (1.28)(9) = 83.52$

B: Top 30% $\Rightarrow$ Area = 0.70 $\Rightarrow$ z = 0.52

$x = \mu + z\sigma = 72 + (0.52)(9) = 76.68$

C: Top 70% $\Rightarrow$ Area = 0.30 $\Rightarrow$ z = -0.52

$x = \mu + z\sigma = 72 + (-0.52)(9) = 67.32$

D: Top 90% $\Rightarrow$ Area = 0.10 $\Rightarrow$ z = -1.28

$x = \mu + z\sigma = 72 + (-1.28)(9) = 60.48$

5.4 SAMPLING DISTRIBUTIONS AND THE CENTRAL LIMIT THEOREM

5.4 SECTION EXERCISES

2. $\mu_{\bar{x}} = \mu = 100$

$\sigma_{\bar{x}} = \dfrac{\sigma}{\sqrt{n}} = \dfrac{15}{\sqrt{100}} = 1.5$

4. $\mu_{\bar{x}} = \mu = 100$

$\sigma_{\bar{x}} = \dfrac{\sigma}{\sqrt{n}} = \dfrac{15}{\sqrt{1000}} = 0.47$

6. False. As the size of a sample increases the standard deviation of the distribution of sample means decreases.

8. True.

10. {120 120, 120 140, 120 180, 120 220, 140 120, 140 140, 140 180, 140 220, 180 120, 180 140, 180 180, 180 220, 220 120, 220 140, 220 180, 220 220}

$\mu_{\bar{x}} = 165$, $\sigma_{\bar{x}} = 27.157$

$\mu = 165$, $\sigma = 38.406$

12. (b) since $\mu_{\bar{x}} = 5.8$, $\sigma_{\bar{x}} = \dfrac{\sigma}{\sqrt{n}} = \dfrac{2.3}{\sqrt{100}} \approx 0.23$ and the graph approximates a normal curve.

14. $\mu_{\bar{x}} = 800$, $\sigma_{\bar{x}} = \dfrac{\sigma}{\sqrt{n}} = \dfrac{100}{\sqrt{15}} \approx 25.820$

16. $\mu_{\bar{x}} = 47.2$, $\sigma_{\bar{x}} = \dfrac{\sigma}{\sqrt{n}} = \dfrac{3.6}{\sqrt{36}} = 0.6$

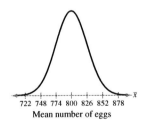
722 748 774 800 826 852 878
Mean number of eggs

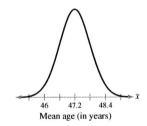
46 47.2 48.4
Mean age (in years)

18. $\mu_{\bar{x}} = 49.3$, $\sigma_{\bar{x}} = \frac{\sigma}{\sqrt{n}} = \frac{17.1}{\sqrt{25}} \approx 3.42$

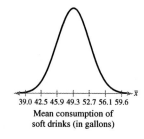

Mean consumption of soft drinks (in gallons)

20. $\mu_{\bar{x}} = 800$, $\sigma_{\bar{x}} = \frac{\sigma}{\sqrt{n}} = \frac{100}{\sqrt{30}} \approx 18.257$

$\mu_{\bar{x}} = 800$, $\sigma_{\bar{x}} = \frac{\sigma}{\sqrt{n}} = \frac{100}{\sqrt{45}} \approx 14.907$

As the sample size increases, the standard error decreases.

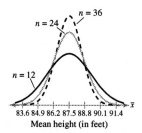

Mean height (in feet)

22. $z = \frac{\bar{x} - \mu}{\frac{\sigma}{\sqrt{n}}} = \frac{42{,}000 - 45{,}500}{\frac{1700}{\sqrt{35}}} = \frac{-3500}{207.35} \approx -16.88$

$P(\bar{x} < 42{,}000) = P(z < -16.88) \approx 0$

24. $z = \frac{\bar{x} - \mu}{\frac{\sigma}{\sqrt{n}}} = \frac{2.034 - 2.029}{\frac{0.049}{\sqrt{38}}} = \frac{0.005}{0.00794} \approx 0.63$

$z = \frac{\bar{x} - \mu}{\frac{\sigma}{\sqrt{n}}} = \frac{2.044 - 2.029}{\frac{0.049}{\sqrt{38}}} = \frac{0.015}{0.00794} \approx 1.89$

$P(2.034 < \bar{x} < 2.044) = P(0.63 < z < 1.89) = 0.9706 - 0.7357 = 0.2349$

26. $z = \frac{\bar{x} - \mu}{\frac{\sigma}{\sqrt{n}}} = \frac{70 - 69.2}{\frac{2.9}{\sqrt{60}}} = \frac{0.8}{0.374} \approx 2.14$

$P(\bar{x} > 70) = P(z > 2.14) = 1 - P(z < 2.14) = 1 - 0.9838 = 0.0162$

28. $z = \frac{x - \mu}{\sigma} = \frac{65 - 69.2}{2.9} \approx -1.45$

$P(x < 65) = P(z < -1.45) = 0.0735$

$z = \frac{\bar{x} - \mu}{\frac{\sigma}{\sqrt{n}}} = \frac{65 - 69.2}{\frac{2.9}{\sqrt{15}}} = \frac{-4.2}{0.749} \approx -5.61$

$P(\bar{x} < 65) = P(z < -5.61) \approx 0$

It is more likely to select one man with a height less than 65 inches because the probability is greater.

30. $z = \frac{\bar{x} - \mu}{\frac{\sigma}{\sqrt{n}}} = \frac{64.05 - 64}{\frac{0.11}{\sqrt{40}}} = \frac{0.05}{0.017} \approx 2.87$

$P(\bar{x} > 64.05) = P(z > 2.87) = 1 - P(z < 2.87) = 1 - 0.9979 = 0.0021$

Yes, it is very unlikely that we would have randomly sampled 40 containers with a mean equal to 64.05 ounces.

CHAPTER 5 | NORMAL PROBABILITY DISTRIBUTIONS **243**

32. (a) $\mu = 10$

$\sigma = 0.5$

$z = \dfrac{\bar{x} - \mu}{\dfrac{\sigma}{\sqrt{n}}} = \dfrac{10.21 - 10}{\dfrac{0.5}{\sqrt{25}}} = \dfrac{0.21}{0.1} = 2.1$

$P(\bar{x} \geq 10.21) = P(z \geq 2.1) = 1 - P(z \leq 2.1) = 1 - 0.9821 = 0.0179$

(b) Claim is inaccurate.

(c) Assuming the distribution is normally distributed:

$z = \dfrac{x - \mu}{\sigma} = \dfrac{10.21 - 10}{0.5} = 0.42$

$P(x \geq 10.21) = P(z \geq 0.42) = 1 - P(z \leq 0.42) = 1 - 0.6628 = 0.3372$

Assuming the manufacturer's claim is true, an individual carton with a weight of 10.21 would not be unusual because it is less than σ away from the mean for an individual ice cream carton.

34. (a) $\mu = 38{,}000$

$\sigma = 1000$

$z = \dfrac{\bar{x} - \mu}{\dfrac{\sigma}{\sqrt{n}}} = \dfrac{37{,}650 - 38{,}000}{\dfrac{1000}{\sqrt{50}}} = \dfrac{-350}{141.42} = -2.47$

$P(\bar{x} \leq 37{,}650) = P(z \leq -2.47) = 0.0068$

(b) Claim is inaccurate.

(c) Assuming the distribution is normally distributed:

$z = \dfrac{x - \mu}{\sigma} = \dfrac{37{,}650 - 38{,}000}{1000} = -0.35$

$P(x < 37{,}650) = P(z < -0.35) = 0.3632$

Assuming the manufacturer's claim is true, an individual brake pad lasting less than 37,650 miles would not be unusual, because it is within one standard deviation of the mean for an individual brake pad.

36. $\mu = 4$

$\sigma = 0.5$

$z = \dfrac{\bar{x} - \mu}{\sigma/\sqrt{n}} = \dfrac{4.2 - 4}{0.5/\sqrt{100}} = \dfrac{0.2}{0.05} = 4$

$P(\bar{x} \geq 4.2) = P(z \geq 4) = 1 - P(z \leq 4) \approx 1 - 1 \approx 0$

It is very unlikely the machine is calibrated to produce a bolt with a mean of 4 inches.

244 CHAPTER 5 | NORMAL PROBABILITY DISTRIBUTIONS

38. Use the finite correction factor since $n = 30 > 25 = 0.05N$.

$$z = \frac{\bar{x} - \mu}{\frac{\sigma}{\sqrt{n}}\sqrt{\frac{N-n}{N-1}}} = \frac{2.5 - 3.32}{\frac{1.09}{\sqrt{30}}\sqrt{\frac{500-30}{500-1}}} = \frac{-0.82}{(0.199)\sqrt{0.9419}} \approx -4.25$$

$$z = \frac{\bar{x} - \mu}{\frac{\sigma}{\sqrt{n}}\sqrt{\frac{N-n}{N-1}}} = \frac{4 - 3.32}{\frac{1.09}{\sqrt{30}}\sqrt{\frac{500-30}{500-1}}} = \frac{0.68}{(0.199)\sqrt{0.9419}} \approx 3.52$$

$$P(2.5 < \bar{x} < 4) = P(-4.25 < z < 3.52) \approx 1 - 0 = 1$$

5.5 NORMAL APPROXIMATIONS TO BINOMIAL DISTRIBUTIONS

5.5 EXERCISE SOLUTIONS

2. $np = (12)(0.60) = 7.2 \geq 5$
$nq = (12)(0.40) = 4.8 < 5$

Cannot use the normal distribution.

4. $np = (18)(0.85) = 15.3 \geq 5$
$nq = (18)(0.15) = 2.7 < 5$

Cannot use the normal distribution.

6. $n = 20, p = 0.63, q = 0.37$
$np = 12.6 \geq 5, nq = 7.4 \geq 5$

Use the normal distribution.

$\mu = np = (20)(0.66) = 13.2$
$\sigma = \sqrt{npq} = \sqrt{(20)(0.66)(0.34)} \approx 2.12$

8. $n = 30, p = 0.086, q = 0.914$
$np = 2.58 < 5, nq = 27.42 \geq 5$

Cannot use the normal distribution.

10. b **12.** c **14.** d **16.** b

18. Binomial: $P(2 \leq x \leq 4) = P(2) + P(3) + P(4) \approx 0.02 + 0.05 + 0.12 = 0.19$

Normal: $P(1.5 \leq x \leq 4.5) = P(-2.60 \leq z \leq -0.87) = 0.1922 - 0.0047 = 0.1875$

20. $n = 32, p = 0.34 \rightarrow np = 10.88$ and $nq = 21.12$

Use normal distribution.

(a) $z = \frac{x - \mu}{\sigma} = \frac{11.5 - 10.88}{2.68} \approx 0.23$

$z = \frac{x - \mu}{\sigma} = \frac{12.5 - 10.88}{2.68} \approx 0.60$

$P(x = 12) = P(11.5 < x < 12.5)$
$= P(0.23 < z < 0.60)$
$= 0.7257 - 0.5910 = 0.1347$

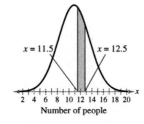

(b) $P(x \geq 12) = P(x > 11.5)$
$= P(z > 0.23)$
$= 1 - P(z < 0.23)$
$= 1 - 0.5910 = 0.4090$

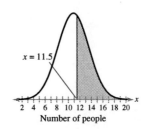

(c) $P(x < 12) = P(x < 11.5)$

$\qquad = P(z < 0.23) = 0.5910$

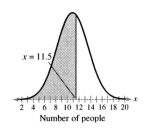

(d) $n = 150, p = 0.34 \rightarrow np = 51$ and $nq = 99$

$z = \dfrac{x - \mu}{\sigma} = \dfrac{59.5 - 51}{5.80} \approx 1.47$

$P(x < 60) = P(x < 59.5)$

$\qquad = P(z < 1.47)$

$\qquad = 0.9292$

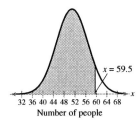

22. $n = 50, p = 0.31, q = 0.69$

$np = 15.5 \geq 5, nq = 34.5 \geq 5$

Use normal distribution.

(a) $z = \dfrac{x - \mu}{\sigma} = \dfrac{13.5 - 15.5}{3.27} = -0.61$

$z = \dfrac{x - \mu}{\sigma} = \dfrac{14.5 - 15.5}{3.27} = -0.31$

$P(x = 14) \approx P(13.5 \leq x \leq 14.5)$

$\qquad = P(-0.61 < z < -0.31)$

$\qquad = 0.3783 - 0.2709 = 0.1074$

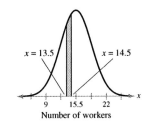

(b) $P(x \geq 14) \approx P(x \geq 13.5)$

$\qquad = P(z \geq -0.61)$

$\qquad = 1 - P(z \leq -0.61)$

$\qquad = 1 - 0.2709 = 0.7291$

$\qquad = 0.7291$

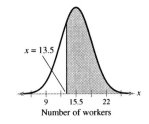

(c) $P(x < 14) \approx P(x \leq 13.5) = P(z \leq -0.61) = 0.2709$

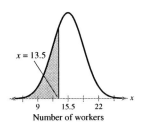

(d) $n = 150, p = 0.31, q = 0.69$

$np = 46.5, nq = 103.5$

Use normal distribution.

$z = \dfrac{x - \mu}{\sigma} = \dfrac{29.5 - 46.5}{5.66} = -3.00$

$P(x < 30) \approx P(x \leq 29.5) = P(z \leq -3.00) = 0.0013$

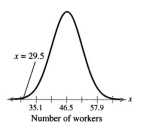

24. $n = 10, p = 0.029 \rightarrow np = 0.29$ and $nq = 9.71$

Cannot use normal distribution because $np < 5$.

(a) $P(x \leq 3) = P(x = 0) + P(x = 1) + P(x = 2) + P(x = 3)$

$= {}_{10}C_0(0.029)^0(0.971)^{10} + {}_{10}C_1(0.029)^1(0.971)^9 + {}_{10}C_2(0.029)^2(0.971)^8$

$+ {}_{10}C_3(0.029)^3(0.971)^7$

≈ 0.99987

(b) $P(x \geq 3) = 1 - P(x < 3) = 1 - P(x = 0) - P(x = 1) - P(x = 2) \approx 0.00251$

(c) $P(x > 3) = 1 - P(x \leq 3) \approx 1 - 0.99987 = 0.00013$

(d) $n = 50, p = 0.029 \rightarrow np = 1.45$ and $nq = 48.55$

Cannot use normal distribution.

$P(x = 0) = {}_{50}C_0(0.029)^0(0.971)^{50} \approx 0.230$

26. (a) $n = 40, p = 0.80, q = 0.20$

$np = 32 \geq 5, nq = 8 \geq 5$

Use normal distribution.

(b) $z = \dfrac{x - \mu}{\sigma} = \dfrac{26.5 - 32}{2.53} = -2.17$

$P(x \leq 26) \approx P(x \leq 26.5) = P(z \leq -2.17) = 0.0150$

(c) $z = \dfrac{x - \mu}{\sigma} = \dfrac{25.5 - 32}{2.53} = -2.57$

$z = \dfrac{x - \mu}{\sigma} = \dfrac{26.5 - 32}{2.53} = -2.17$

$P(x = 26) \approx P(25.5 \leq x \leq 26.5) = P(-2.57 \leq z \leq -2.17) = 0.0150 - 0.0051 = 0.0099$

Yes, because the z-score is more than 2 standard deviations from the mean.

28. $n = 200, p = 0.11$

9% of 200 = 18 people

$z = \dfrac{x - \mu}{\sigma} \approx \dfrac{17.5 - 22}{4.42} \approx -1.02$

$P(x < 18) = P(x < 17.5) = P(z < -1.02) = 0.1539$

It is probable that 18 of the 200 people responded that they participate in hiking. Answers will vary.

30. $n = 100, p = 0.65$

$$z = \frac{x - \mu}{\sigma} = \frac{69.5 - 65}{4.77} \approx 0.94$$

$P(\text{accept claim}) = P(x \geq 70)$
$= P(x > 69.5) = P(z > 0.94) = 1 - P(z < 0.94) = 1 - 0.8264 = 0.1736$

CHAPTER 5 REVIEW EXERCISE SOLUTIONS

2. $\mu = -3, \sigma = 5$

4. 1.32 and 1.78 are unusual.

6. 0.9946   **8.** 0.8962   **10.** $1 - 0.2033 = 0.7967$

12. $0.3336 - 0.1112 = 0.2224$ **14.** $0.9750 - 0.0250 = 0.95$

16. $0.5478 + 0.0427 = 0.5905$ **18.** $P(z > -0.74) = 0.7704$

20. $P(0.42 < z < 3.15) = 0.9992 - 0.6628 = 0.3364$

22. $P(z < 0 \text{ or } z > 1.68) = 0.5 + 0.0465 = 0.5465$

24. (a) $z = \dfrac{x - \mu}{\sigma} = \dfrac{1 - 1.5}{0.25} = -2$

$z = \dfrac{x - \mu}{\sigma} = \dfrac{2 - 1.5}{0.25} = 2$

$P(1 < x < 2) = P(-2 < z < 2) = 0.9772 - 0.0228 = 0.9544$

(b) $z = \dfrac{x - \mu}{\sigma} = \dfrac{1.6 - 1.5}{0.25} = 0.4$

$z = \dfrac{x - \mu}{\sigma} = \dfrac{2.2 - 1.5}{0.25} = 2.8$

$P(1.6 < x < 2.2) = P(0.4 < z < 2.8) = 0.9974 - 0.6554 = 0.3420$

(c) $z = \dfrac{x - \mu}{\sigma} = \dfrac{2.2 - 1.5}{0.25} = 2.8$

$P(x > 2.2) = P(z > 2.8) = 0.0026$

26. $z = -1.28$ **28.** $z = -2.05$ **30.** $z = -0.84$

32. $x = \mu + z\sigma = 45.1 + (1.2)(0.5) = 45.7$ meters

34. 3rd Quartile $\Rightarrow$ Area $= 0.75 \Rightarrow z = 0.67$

$x = \mu + z\sigma = 45.1 + (0.67)(0.5) = 45.435$

36. Bottom 5% $\Rightarrow$ Area $= 0.05 \Rightarrow z = -1.645$

$x = \mu + z\sigma = 45.1 + (-1.645)(0.5) \approx 44.28$

38. {00, 01, 02, 03, 10, 11, 12, 13, 20, 21, 22, 23, 30, 31, 32, 33}

$\mu = 1.5, \sigma \approx 1.118$

$\mu_{\bar{x}} = 1.5, \sigma_{\bar{x}} \approx 0.791$

40. $\mu_{\bar{x}} = 226.6$, $\sigma_{\bar{x}} = \frac{\sigma}{\sqrt{n}} = \frac{68.1}{\sqrt{40}} \approx 10.77$

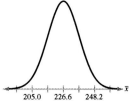

Mean consumption (in pounds)

42. (a) $z = \frac{\bar{x} - \mu}{\frac{\sigma}{\sqrt{n}}} = \frac{1.0 - 1.5}{\frac{0.25}{\sqrt{7}}} = \frac{-0.5}{0.0945} \approx -5.29$

$z = \frac{\bar{x} - \mu}{\frac{\sigma}{\sqrt{n}}} = \frac{2.0 - 1.5}{\frac{0.25}{\sqrt{7}}} = \frac{0.5}{0.0945} \approx 5.29$

$P(1.0 < \bar{x} < 2.0) = P(-5.29 < z < 5.29) \approx 1$

(b) $z = \frac{\bar{x} - \mu}{\frac{\sigma}{\sqrt{n}}} = \frac{1.6 - 1.5}{\frac{0.25}{\sqrt{7}}} = \frac{0.1}{0.0945} \approx 1.06$

$z = \frac{\bar{x} - \mu}{\frac{\sigma}{\sqrt{n}}} = \frac{2.2 - 1.5}{\frac{0.25}{\sqrt{7}}} = \frac{1.7}{0.0945} \approx 7.41$

$P(1.6 < \bar{x} < 2.2) = P(1.06 < z < 7.41) \approx 1 - 0.8554 = 0.1446$

(c) $z = \frac{\bar{x} - \mu}{\frac{\sigma}{\sqrt{n}}} = \frac{2.2 - 1.5}{\frac{0.25}{\sqrt{7}}} = \frac{1.7}{0.0945} \approx 7.41$

$P(\bar{x} > 2.2) = P(z > 7.41) \approx 0$

(a) is larger and (b) and (c) are smaller.

44. (a) $z = \frac{\bar{x} - \mu}{\frac{\sigma}{\sqrt{n}}} = \frac{1400 - 1300}{\frac{250}{\sqrt{36}}} = \frac{100}{41.67} \approx 2.4$

$P(x < 1400) = P(z < 2.4) = 0.9918$

(b) $z = \frac{\bar{x} - \mu}{\frac{\sigma}{\sqrt{n}}} = \frac{1150 - 1300}{\frac{250}{\sqrt{36}}} = \frac{-150}{41.67} = -3.6$

$P(x > 1150) = P(z > -3.6) = 1 - 0.0002 = 0.9998$

46. Assuming the distribution is normally distributed:

$z = \frac{\bar{x} - \mu}{\frac{\sigma}{\sqrt{n}}} = \frac{5250 - 5000}{\frac{300}{\sqrt{15}}} = \frac{250}{77.46} \approx 3.23$

$P(\bar{x} > 5250) = P(z > 3.23) = 1 - P(z < 3.23) = 1 - 0.9994 = .0006$

48. $n = 15, p = 0.59, q = 0.41$

$np = 8.85 > 5$ and $nq = 6.15 > 5$

Use the normal distribution.

$\mu = np = 8.85, \sigma = \sqrt{npq} = \sqrt{15(0.59)(0.41)} \approx 1.905$

50. $P(x \leq 36) = P(x < 36.5)$

52. $P(x = 50) = P(49.5 < x < 50.5)$

54. $n = 12, p = 0.33 \rightarrow np = 3.96 < 5$

Cannot use normal distribution.

$P(x > 5) = 1 - P(x \leq 5)$
$= 1 - [P(x = 0) + P(x = 1) + \cdots + P(x = 5)]$
$= 1 - [{}_{12}C_0(0.33)^0(0.67)^{12} + {}_{12}C_1(0.33)^1(0.67)^{11} + {}_{12}C_2(0.33)^2(0.67)^{10}$
$\quad + {}_{12}C_3(0.33)^3(0.67)^9 + {}_{12}C_4(0.33)^4(0.67)^8 + {}_{12}C_5(0.33)^5(0.67)^7]$
$= 1 - 0.829 \approx 0.171$

CHAPTER 5 QUIZ SOLUTIONS

1. (a) $P(z > -2.10) = 0.9821$

(b) $P(z < 3.22) = 0.9994$

(c) $P(-2.33 < z < 2.33) = 0.9901 - 0.0099 = 0.9802$

(d) $P(z < -1.75 \text{ or } z > -0.75) = 0.0401 + 0.7734 = 0.8135$

2. (a) $z = \dfrac{x - \mu}{\sigma} = \dfrac{5.36 - 5.5}{0.08} \approx -1.75$

$z = \dfrac{x - \mu}{\sigma} = \dfrac{5.64 - 5.5}{0.08} \approx 1.75$

$P(5.36 < x < 5.64) = P(-1.75 < z < 1.75) = 0.9599 - 0.0401 = 0.9198$

(b) $z = \dfrac{x - \mu}{\sigma} = \dfrac{-5.00 - (-8.2)}{7.84} \approx 0.41$

$z = \dfrac{x - \mu}{\sigma} = \dfrac{0 - (-8.2)}{7.84} \approx 1.05$

$P(-5.00 < x < 0) = P(0.41 < z < 1.05) = 0.8531 - 0.6591 = 0.1940$

(c) $z = \dfrac{x - \mu}{\sigma} = \dfrac{0 - 18.5}{9.25} = -2$

$z = \dfrac{x - \mu}{\sigma} = \dfrac{37 - 18.5}{9.25} = 2$

$P(x < 0 \text{ or } x > 37) = P(z < -2 \text{ or } z > 2) = 2(0.0228) = 0.0456$

3. $z = \dfrac{x - \mu}{\sigma} = \dfrac{315 - 281}{34.4} \approx 0.99$

$P(x > 315) = P(z > 0.99) = 0.1611$

4. $z = \dfrac{x - \mu}{\sigma} = \dfrac{250 - 281}{34.4} \approx -0.90$

$z = \dfrac{x - \mu}{\sigma} = \dfrac{305 - 281}{34.4} \approx 0.70$

$P(250 < x < 305) = P(-0.90 < z < 0.70) = 0.7580 - 0.1841 = 0.5739$

5. $P(x > 250) = P(z > -0.90) = 0.8159 \rightarrow 81.59\%$

6. $z = \dfrac{x - \mu}{\sigma} = \dfrac{300 - 281}{34.4} \approx 0.55$

$P(x < 300) = P(z < 0.55) = 0.7088$

$(2000)(0.7088) = 1417.6$

7. top 5% $\rightarrow z \approx 1.645$

$\mu + z\sigma = 281 + (1.645)(34.4) = 337.588$

8. bottom 25% $\rightarrow z \approx -0.67$

$\mu + z\sigma = 281 + (-0.67)(34.4) = 257.952$

9. $z = \dfrac{\bar{x} - \mu}{\dfrac{\sigma}{\sqrt{n}}} = \dfrac{300 - 281}{\dfrac{34.4}{\sqrt{60}}} = \dfrac{19}{4.44} \approx 4.28$

$P(\bar{x} > 300) = P(z > 4.28) \approx 0$

10. $z = \dfrac{x - \mu}{\sigma} = \dfrac{300 - 281}{34.4} \approx 0.55$

$P(\bar{x} > 300) = P(z > 0.55) = 0.2912$

$z = \dfrac{\bar{x} - \mu}{\dfrac{\sigma}{\sqrt{n}}} = \dfrac{305 - 281}{\dfrac{34.4}{\sqrt{15}}} = \dfrac{24}{8.88} \approx 2.70$

$P(\bar{x} > 300) = P(z > 2.70) = 0.0035$

You are more likely to select one student with a test score greater than 300 because the standard error of the mean is less than the standard deviation.

11. $n = 24, p = 0.68 \rightarrow np = 16.32, nq = 7.68$

Use normal distribution.

$\mu = np = 16.32 \qquad \sigma = \sqrt{npq} \approx 2.285$

12. $z = \dfrac{x - \mu}{\sigma} = \dfrac{15.5 - 16.32}{2.285} \approx -0.36$

$P(x \leq 15) = P(x < 15.5) = P(z < -0.36) = 0.3594$

Confidence Intervals

CHAPTER 6

6.1 CONFIDENCE INTERVALS FOR THE MEAN (LARGE SAMPLES)

6.1 EXERCISE SOLUTIONS

2. b

4. b; As n increases, E decreases because $\sqrt{n}$ is in the denominator of the formula for E. Therefore, the intervals become narrower.

6. 1.44

8. 2.17

10. $|\bar{x} - \mu| = |9.5 - 8.76| = 0.74$

12. $|\bar{x} - \mu| = |46.56 - 48.12| = 1.56$

14. $|\bar{x} - \mu| = |86.4 - 80.9| = 5.5$

16. $E = z_c \dfrac{s}{\sqrt{n}} = 1.96 \dfrac{3.0}{\sqrt{60}} \approx 0.759$

18. $E = z_c \dfrac{s}{\sqrt{n}} = 2.24 \dfrac{3.4}{\sqrt{100}} = 0.762$

20. $c = 0.90 \Rightarrow z_c = 1.645$

$\bar{x} = 57.2, s = 7.1, n = 50$

$\bar{x} \pm z_c \dfrac{s}{\sqrt{n}} = 57.2 \pm 1.645 \dfrac{7.1}{\sqrt{50}}$

$= 57.2 \pm 1.652 = (55.5, 58.9)$

Answer: (d)

22. $c = 0.98 \Rightarrow z_c = 2.33$

$\bar{x} = 57.2, s = 7.1, n = 50$

$\bar{x} \pm z_c \dfrac{s}{\sqrt{n}} = 57.2 \pm 2.33 \dfrac{7.1}{\sqrt{50}}$

$= 57.2 \pm 2.340 = (54.9, 59.5)$

Answer: (a)

24. $\bar{x} \pm z_c \dfrac{s}{\sqrt{n}} = 31.39 \pm 1.96 \dfrac{0.8}{\sqrt{82}} = 31.3 \pm 0.173 \approx (31.22, 31.56)$

26. $\bar{x} \pm z_c \dfrac{s}{\sqrt{n}} = 13.5 \pm 2.575 \dfrac{1.5}{\sqrt{100}} = 13.5 \pm 0.386 \approx (13.1, 13.9)$

28. $c = 0.95 \Rightarrow z_c = 1.96$

$n = \left(\dfrac{z_c \sigma}{E}\right)^2 = \left(\dfrac{(1.96)(2.5)}{1}\right)^2 = 24.01 \Rightarrow 25$

30. $c = 0.98 \Rightarrow z_c = 2.33$

$n = \left(\dfrac{z_c \sigma}{E}\right)^2 = \left(\dfrac{(2.33)(10.1)}{2}\right)^2 = 138.45 \Rightarrow 139$

32. $(44.07)(80.97) \Rightarrow 2E = 80.97 - 44.07 = 36.9 \Rightarrow E = 18.45$ and $\bar{x} = 44.07 + E$

$\phantom{(44.07)(80.97) \Rightarrow 2E = 80.97 - 44.07 = 36.9 \Rightarrow E = 18.45 \text{ and } \bar{x}} = 44.07 + 18.45 = 62.52$

34. 90% CI: $\bar{x} \pm z_c \dfrac{s}{\sqrt{n}} = 19.31 \pm 1.645 \dfrac{2.37}{\sqrt{36}} = 19.31 \pm 0.650 \approx (18.66, 19.96)$

95% CI: $\bar{x} \pm z_c \dfrac{s}{\sqrt{n}} = 19.31 \pm 1.96 \dfrac{2.37}{\sqrt{36}} = 19.31 \pm 0.774 \approx (18.54, 20.08)$

The 95% CI is wider.

36. 90% CI: $\bar{x} \pm z_c \dfrac{s}{\sqrt{n}} = 23 \pm 1.645 \dfrac{6.7}{\sqrt{36}} = 23 \pm 1.837 \approx (21, 25)$

95% CI: $\bar{x} \pm z_c \dfrac{s}{\sqrt{n}} = 23 \pm 1.96 \dfrac{6.7}{\sqrt{36}} = 23 \pm 2.189 \approx (21, 25)$

The 95% CI is wider.

38. $\bar{x} \pm z_c \dfrac{s}{\sqrt{n}} = 75 \pm 2.575 \dfrac{12.5}{\sqrt{50}} = 75 \pm 4.552 \approx (70, 80)$

40. $\bar{x} \pm z_c \dfrac{s}{\sqrt{n}} = 75 \pm 2.575 \dfrac{12.5}{\sqrt{90}} = 75 \pm 3.393 \approx (72, 78)$

$n = 50$ CI is wider because we have taken a smaller sample giving us less information about the population.

42. $\bar{x} \pm z_c \dfrac{s}{\sqrt{n}} = 0.20 \pm 1.96 \dfrac{0.06}{\sqrt{56}} = 0.20 \pm 0.016 \approx (0.18, 0.22)$

44. $\bar{x} \pm z_c \dfrac{s}{\sqrt{n}} = 0.20 \pm 1.96 \dfrac{0.1}{\sqrt{56}} = 0.20 \pm 0.026 \approx (0.17, 0.23)$

$s = 0.1$ CI is wider because of the increased variability within the population.

46. Answers will vary.

48. 90% CI: $\bar{x} \pm z_c = \dfrac{\sigma}{\sqrt{n}} = 1.7 \pm 1.645 \dfrac{0.6}{\sqrt{21}} = 1.7 \pm 0.215 \approx (1.5, 1.9)$

99% CI: $\bar{x} \pm z_c = \dfrac{\sigma}{\sqrt{n}} = 1.7 \pm 2.575 \dfrac{0.6}{\sqrt{21}} = 1.7 \pm 0.337 \approx (1.4, 2.0)$

99% CI is wider.

50. $n = \left(\dfrac{z_c \sigma}{E}\right)^2 = \left(\dfrac{2.575 \cdot 1.4}{2}\right)^2 \approx 3.249 \rightarrow 4$

52. (a) $n = \left(\dfrac{z_c \sigma}{E}\right)^2 = \left(\dfrac{1.645 \cdot 1.2}{1}\right)^2 \approx 3.897 \rightarrow 4$

(b) $n = \left(\dfrac{z_c \sigma}{E}\right)^2 = \left(\dfrac{2.575 \cdot 1.2}{1}\right)^2 \approx 9.548 \rightarrow 10$

99% CI requires larger sample because more information is needed from the population to be 99% confident.

54. (a) $n = \left(\dfrac{z_c \sigma}{E}\right)^2 = \left(\dfrac{1.96 \cdot 3}{1}\right)^2 \approx 34.574 \rightarrow 35$

(b) $n = \left(\dfrac{z_c \sigma}{E}\right)^2 = \left(\dfrac{1.96 \cdot 3}{2}\right)^2 \approx 8.644 \rightarrow 9$

$E = 1$ requires a larger sample size. As the error size decreases, a larger sample must be taken to obtain enough information from the population to ensure desired accuracy.

56. (a) $n = \left(\dfrac{z_c \sigma}{E}\right)^2 = \left(\dfrac{1.645 \cdot 0.25}{0.0625}\right)^2 \approx 43.296 \rightarrow 44$

(b) $n = \left(\dfrac{z_c \sigma}{E}\right)^2 = \left(\dfrac{1.645 \cdot 0.25}{0.03125}\right)^2 \approx 173.186 \rightarrow 174$

$E = 0.03125$ requires a larger sample size. As the error size decreases, a larger sample must be taken to obtain enough information from the population to ensure desired accuracy.

58. (a) $n = \left(\dfrac{z_c \sigma}{E}\right)^2 = \left(\dfrac{2.575 \cdot 0.20}{0.15}\right)^2 \approx 11.788 \to 12$

(b) $n = \left(\dfrac{z_c \sigma}{E}\right)^2 = \left(\dfrac{2.575 \cdot 0.10}{0.15}\right)^2 \approx 2.947 \to 3$

$\sigma = 0.20$ requires a larger sample size. Due to the decreased variability in the population, a smaller sample is needed to ensure desired accuracy.

60. A 99% CI may not be practical to use in all situations. It may produce a CI so wide that it has no practical application.

62. $\bar{x} = 101.8, s \approx 6.7, n = 35$

$\bar{x} \pm z_c \dfrac{s}{\sqrt{n}} = 101.8 \pm 1.96 \dfrac{6.7}{\sqrt{35}} = 6.7 \pm 2.217 \approx (99.6, 104.0)$

64. $\bar{x} = 18.137, s \approx 4.319, n = 33$

$\bar{x} \pm z_c \dfrac{s}{\sqrt{n}} = 18.137 \pm 1.96 \dfrac{4.319}{\sqrt{33}} = 18.137 \pm 1.474 \approx (16.663, 19.611)$

66. (a) $\sqrt{\dfrac{N-n}{N-1}} = \sqrt{\dfrac{100-50}{100-1}} \approx 0.711$ (b) $\sqrt{\dfrac{N-n}{N-1}} = \sqrt{\dfrac{400-50}{400-1}} \approx 0.937$

(c) $\sqrt{\dfrac{N-n}{N-1}} = \sqrt{\dfrac{700-50}{700-1}} \approx 0.964$ (d) $\sqrt{\dfrac{N-n}{N-1}} = \sqrt{\dfrac{2000-50}{2000-1}} \approx 0.988$

(e) The finite population correction factor approaches 1 as the population size increases while the sample size remains the same.

6.2 CONFIDENCE INTERVALS FOR THE MEAN (SMALL SAMPLES)

6.2 EXERCISE SOLUTIONS

2. 2.201   **4.** 2.539

6. (a) $E = z_c \dfrac{s}{\sqrt{n}} = 2.575 \dfrac{3}{\sqrt{6}} \approx 3.154$ (b) $E = t_c \dfrac{s}{\sqrt{n}} = 4.032 \dfrac{3}{\sqrt{6}} \approx 4.938$

8. (a) $\bar{x} \pm t_c \dfrac{s}{\sqrt{n}} = 13.4 \pm 2.365 \dfrac{0.85}{\sqrt{8}} = 13.4 \pm 0.711 \approx (12.7, 14.1)$

(b) $\bar{x} \pm z_c \dfrac{s}{\sqrt{n}} = 13.4 \pm 1.96 \dfrac{0.85}{\sqrt{8}} = 13.4 \pm 0.589 \approx (12.8, 14.0)$

t-CI is wider.

10. (a) $\bar{x} \pm t_c \dfrac{s}{\sqrt{n}} = 14 \pm 3.250 \dfrac{2.0}{\sqrt{10}} = 14 \pm 2.055 \approx (12, 16)$

(b) $\bar{x} \pm z_c \dfrac{s}{\sqrt{n}} = 14 \pm 2.575 \dfrac{2.0}{\sqrt{10}} = 14 \pm 1.625 \approx (12, 16)$

t-CI is wider.

12. $\bar{x} \pm t_c \dfrac{s}{\sqrt{n}} = 100 \pm 2.447 \dfrac{42.50}{\sqrt{7}} = 100 \pm 39.307 \approx (60.69, 139.31)$

$E = t_c \dfrac{s}{\sqrt{n}} = 2.447 \dfrac{42.50}{\sqrt{7}} \approx 39.307$

14. $\bar{x} \pm t_c \dfrac{\sigma}{\sqrt{n}} = 100 \pm 1.96 \dfrac{50}{\sqrt{7}} = 100 \pm 37.041 \approx (62.96, 137.04)$

$E = z_c \dfrac{\sigma}{\sqrt{n}} = 1.96 \dfrac{50}{\sqrt{7}} \approx 37.041$

t-CI is wider.

16. (a) $\bar{x} \pm t_c \dfrac{s}{\sqrt{n}} = 1.4 \pm 1.796 \dfrac{0.3}{\sqrt{12}} = 1.4 \pm 0.156 \approx (1.2, 1.6)$

(b) $\bar{x} \pm z_c \dfrac{s}{\sqrt{n}} = 1.4 \pm 1.645 \dfrac{0.3}{\sqrt{600}} = 1.4 \pm 0.020 \approx (1.38, 1.42)$

t-CI is wider.

18. (a) $\bar{x} \approx 4418.817$ (b) $s \approx 389.294$

(c) $\bar{x} \pm t_c \dfrac{s}{\sqrt{n}} = 4418.817 \pm 3.012 \dfrac{389.294}{\sqrt{14}} = 4418.817 \pm 313.378 \approx (4105.439, 4732.195)$

20. (a) $\bar{x} \approx 2.35$ (b) $s \approx 1.03$

(c) $\bar{x} \pm t_c \dfrac{s}{\sqrt{n}} = 2.35 \pm 2.977 \dfrac{1.03}{\sqrt{15}} = 2.35 \pm 0.793 \approx (1.56, 3.14)$

22. $\bar{x} = 61.12, s = 24.62, n < 30, \sigma$ unknown, and pop normally distributed $\rightarrow$ use t-distribution

$\bar{x} \pm t_c \dfrac{s}{\sqrt{n}} = 61.12 \pm 2.201 \dfrac{24.62}{\sqrt{12}} = 61.12 \pm 15.643 \approx (45.48, 76.76)$

24. $\bar{x} = 20.8, s \approx 4.5, n < 30, \sigma$ known, and pop normally distributed $\rightarrow$ use normal distribution

$\bar{x} \pm z_c \dfrac{\sigma}{\sqrt{n}} = 20.8 \pm 1.96 \dfrac{4.7}{\sqrt{20}} = 20.8 \pm 2.06 \approx (18.7, 22.9)$

26. $n < 30, \sigma$ unknown, and pop normally distributed $\rightarrow$ use t-distribution

$\bar{x} \pm t_c \dfrac{s}{\sqrt{n}} = 28.13 \pm 2.210 \dfrac{12.05}{\sqrt{17}} = 28.13 \pm 6.459 \approx (21.93, 34.33)$

28. $n = 16, \bar{x} = 1015, s = 25$

$\pm t_{0.99} \rightarrow 99\%$ t-CI

$\bar{x} \pm t_c \dfrac{s}{\sqrt{n}} = 1015 \pm 2.947 \dfrac{25}{\sqrt{16}} = 1015 \pm 18.419 \approx (997, 1033)$

They are making good light bulbs since the desired bulb life of 1000 hours is contained between 997 and 1033 hours.

6.3 CONFIDENCE INTERVALS FOR POPULATION PROPORTIONS

6.3 EXERCISE SOLUTIONS

2. True

4. $\hat{p} = \dfrac{x}{n} = \dfrac{501}{1001} \approx 0.500$

$\hat{q} = 1 - \hat{p} \approx 0.500$

6. $\hat{p} = \dfrac{x}{n} = \dfrac{800}{2008} \approx 0.398$

$\hat{q} = 1 - \hat{p} \approx 0.602$

8. $\hat{p} = \dfrac{x}{n} = \dfrac{151}{1004} \approx 0.150$

$\hat{q} = 1 - \hat{p} \approx 0.850$

10. $\hat{p} = \dfrac{x}{n} = \dfrac{662}{1003} \approx 0.660$

$\hat{q} = 1 - \hat{p} = 0.340$

12. $\hat{p} = \dfrac{x}{n} = \dfrac{1515}{3224} = 0.470$

$\hat{q} = 1 - \hat{p} = 1 - 0.470 = 0.530$

14. $\hat{p} = 0.15, E = 0.052$

$\hat{p} \pm E = 0.51 \pm 0.052 = (0.458, 0.562)$

16. 95% CI: $\hat{p} \pm z_c \sqrt{\dfrac{\hat{p}\hat{q}}{n}} = 0.500 \pm 1.96 \sqrt{\dfrac{0.500 \cdot 0.500}{1001}} = 0.500 \pm 0.031 \approx (0.469, 0.531)$

99% CI: $\hat{p} \pm z_c \sqrt{\dfrac{\hat{p}\hat{q}}{n}} = 0.500 \pm 2.575 \sqrt{\dfrac{0.500 \cdot 0.500}{1001}} = 0.500 \pm 0.041 \approx (0.459, 0.541)$

99% CI is wider.

18. 95% CI: $\hat{p} \pm z_c \sqrt{\dfrac{\hat{p}\hat{q}}{n}} = 0.398 \pm 1.96 \sqrt{\dfrac{0.398 \cdot 0.602}{2008}} = 0.398 \pm 0.021 \approx (0.377, 0.419)$

99% CI: $\hat{p} \pm z_c \sqrt{\dfrac{\hat{p}\hat{q}}{n}} = 0.398 \pm 2.575 \sqrt{\dfrac{0.398 \cdot 0.602}{2008}} = 0.398 \pm 0.028 \approx (0.370, 0.426)$

99% CI is wider.

20. 95% CI: $\hat{p} \pm z_c \sqrt{\dfrac{\hat{p}\hat{q}}{n}} = 0.150 \pm 1.96 \sqrt{\dfrac{0.150 \cdot 0.850}{1004}} = 0.150 \pm 0.022 \approx (0.128, 0.172)$

99% CI: $\hat{p} \pm z_c \sqrt{\dfrac{\hat{p}\hat{q}}{n}} = 0.150 \pm 2.575 \sqrt{\dfrac{0.150 \cdot 0.850}{1004}} = 0.150 \pm 0.029 \approx (0.121, 0.179)$

99% CI is wider.

22. (a) $n = \hat{p}\hat{q}\left(\dfrac{z_c}{E}\right)^2 = 0.5 \cdot 0.5 \left(\dfrac{2.326}{0.04}\right)^2 \approx 845.356 \rightarrow 846$

(b) $n = \hat{p}\hat{q}\left(\dfrac{z_c}{E}\right)^2 = 0.2 \cdot 0.8 \left(\dfrac{2.326}{.04}\right)^2 \approx 541.028 \rightarrow 542$

(c) Having an estimate of the proportion reduces the minimum sample size needed.

24. (a) $n = \hat{p}\hat{q}\left(\dfrac{z_c}{E}\right)^2 = 0.5 \cdot 0.5 \left(\dfrac{2.17}{0.035}\right)^2 \approx 961.000$

(b) $n = \hat{p}\hat{q}\left(\dfrac{z_c}{E}\right)^2 = 0.19 \cdot 0.81 \left(\dfrac{2.17}{0.035}\right)^2 \approx 591.592 \rightarrow 592$

(c) Having an estimate of the proportion reduces the minimum sample size needed.

26. (a) $\hat{p} = 0.55, n = 350$

$\hat{p} \pm z_c \sqrt{\dfrac{\hat{p}\hat{q}}{n}} = 0.55 \pm 2.575 \sqrt{\dfrac{0.55 \cdot 0.45}{350}} = 0.55 \pm 0.069 \approx (0.482, 0.618)$

256 CHAPTER 6 | CONFIDENCE INTERVALS

(b) $\hat{p} = 0.40, n = 350$

$$\hat{p} \pm z_c \sqrt{\frac{\hat{p}\hat{q}}{n}} = 0.40 \pm 2.575 \sqrt{\frac{0.40 \cdot 0.60}{350}} = 0.40 \pm 0.067 \approx (0.333, 0.467)$$

It is unlikely that the two proportions are equal because the confidence intervals estimating the proportions do not overlap.

28. (a) We do not have enough evidence to say the two proportions are unequal since the 2 CI's (0.195, 0.285) and (0.252, 0.348) overlap.

(b) We do not have enough evidence to say the two proportions are unequal since the 2 CI's (0.252, 0.348) and (0.310, 0.430) overlap.

(c) We do have enough evidence to conclude the two proportions are unequal since the 2 CI's (0.195, 0.285) and (0.310, 0.430) do not overlap.

30. $27\% \pm 3\% \rightarrow (24\%, 30\%)$

$$E = z_c \sqrt{\frac{\hat{p}\hat{q}}{n}} \rightarrow z_c = E\sqrt{\frac{n}{\hat{p}\hat{q}}} = 0.03 \sqrt{\frac{1001}{0.27 \cdot 0.73}} \approx 2.138 \rightarrow z_c = 2.14 \rightarrow c = 0.968$$

(24%, 30%) is approximately a 96.8% CI.

32. $E = z_c \sqrt{\frac{\hat{p}\hat{q}}{n}} \rightarrow \frac{E}{z_c} = \sqrt{\frac{\hat{p}\hat{q}}{n}} \rightarrow \left(\frac{E}{z_c}\right)^2 = \frac{\hat{p}\hat{q}}{n} \rightarrow n = \hat{p}\hat{q}\left(\frac{z_c}{E}\right)^2$

6.4 CONFIDENCE INTERVALS FOR VARIANCE AND STANDARD DEVIATION

6.4 EXERCISE SOLUTIONS

2. $\chi_R^2 = 28.299, \chi_L^2 = 3.074$ **4.** $\chi_R^2 = 44.314, \chi_L^2 = 11.524$ **6.** $\chi_R^2 = 37.916, \chi_L^2 = 18.939$

8. (a) $s = 0.0321$

$$\left(\frac{(n-1)s^2}{\chi_R^2}, \frac{(n-1)s^2}{\chi_L^2}\right) \approx \left(\frac{14 \cdot (0.0321)^2}{23.685}, \frac{14 \cdot (0.0321)^2}{6.571}\right) \approx (0.000609, 0.00220)$$

(b) $\left(\sqrt{0.000609}, \sqrt{0.00220}\right) \approx (0.0247, 0.0469)$

10. (a) $s = 0.0918$

$$\left(\frac{(n-1)s^2}{\chi_R^2}, \frac{(n-1)s^2}{\chi_L^2}\right) = \left(\frac{16 \cdot (0.0918)^2}{28.845}, \frac{16 \cdot (0.0918)^2}{6.908}\right) \approx (0.00467, 0.0195)$$

(b) $\left(\sqrt{0.00467}, \sqrt{0.0195}\right) \approx (0.0683, 0.140)$

12. (a) $\left(\frac{(n-1)s^2}{\chi_R^2}, \frac{(n-1)s^2}{\chi_L^2}\right) = \left(\frac{16 \cdot (37)^2}{28.845}, \frac{16 \cdot (37)^2}{6.908}\right) \approx (759, 3171)$

(b) $\left(\sqrt{759}, \sqrt{3171}\right) \approx (28, 56)$

14. (a) $\left(\frac{(n-1)s^2}{\chi_R^2}, \frac{(n-1)s^2}{\chi_L^2}\right) = \left(\frac{15(6.4)^2}{27.488}, \frac{15(6.4)^2}{6.262}\right) = (22.4, 98.1)$

(b) $\left(\sqrt{22.4}, \sqrt{98.1}\right) = (4.7, 9.9)$

16. (a) $\left(\frac{(n-1)s^2}{\chi_R^2}, \frac{(n-1)s^2}{\chi_L^2}\right) = \left(\frac{29(3600)^2}{42.557}, \frac{29(3600)^2}{17.708}\right) = (8{,}831{,}450, 21{,}224{,}305)$

(b) $\left(\sqrt{8{,}831{,}450}, \sqrt{21{,}224{,}305}\right) = (2971, 4607)$

18. (a) $\left(\dfrac{(n-1)s^2}{\chi_R^2}, \dfrac{(n-1)s^2}{\chi_L^2}\right) = \left(\dfrac{12 \cdot (6.7)^2}{26.217}, \dfrac{12 \cdot (6.7)^2}{3.571}\right) \approx (20.5, 150.8)$

(b) $\left(\sqrt{20.5}, \sqrt{150.8}\right) \approx (4.5, 12.3)$

20. (a) $\left(\dfrac{(n-1)s^2}{\chi_R^2}, \dfrac{(n-1)s^2}{\chi_L^2}\right) = \left(\dfrac{17(6400)^2}{27.587}, \dfrac{17(6400)^2}{8.672}\right) = (25{,}240{,}874, 80{,}295{,}203)$

(b) $\left(\sqrt{25{,}240{,}874}, \sqrt{80{,}295{,}203}\right) = (5024, 8961)$

22. 90% CI for σ: $(0.0247, 0.0469)$ No, because the majority of the confidence interval is above 0.025.

CHAPTER 6 REVIEW EXERCISE SOLUTIONS

2. (a) $\bar{x} \approx 9.5$

(b) $s \approx 7.1$

$E = z_c \dfrac{s}{\sqrt{n}} = 1.645 \dfrac{7.1}{\sqrt{32}} \approx 2.064$

4. $\bar{x} \pm z_c \dfrac{s}{\sqrt{n}} = 0.0925 \pm 1.645 \dfrac{0.0013}{\sqrt{45}} = 0.0925 \pm 0.0003 \approx (0.0922, 0.0928)$

6. $n = \left(\dfrac{z_c \sigma}{E}\right)^2 = \left(\dfrac{2.575 \cdot 34.663}{2}\right)^2 \approx 1991.713 \to 1992$

8. $n = \left(\dfrac{z_c \sigma}{E}\right)^2 = \left(\dfrac{(2.326)(7.098)}{0.5}\right)^2 = 1090.31 \Rightarrow 1091$

10. $t_c = 1.323$

12. $t_c = 2.756$

14. $E = t_c \dfrac{s}{\sqrt{n}} = 2.064 \dfrac{0.05}{\sqrt{25}} \approx 0.021$

16. $E = t_c \dfrac{s}{\sqrt{n}} = 2.861 \left(\dfrac{16.5}{\sqrt{20}}\right) = 10.6$

18. $\bar{x} \pm t_c \dfrac{s}{\sqrt{n}} = 3.5 \pm 2.064 \dfrac{0.05}{\sqrt{25}} = 3.5 \pm 0.021 \approx (3.479, 3.521)$

20. $\bar{x} \pm t_c \dfrac{s}{\sqrt{n}} = 25.2 \pm 2.861 \left(\dfrac{6.5}{\sqrt{20}}\right) = 25.2 \pm 10.556 = (14.6, 35.8)$

22. $\bar{x} \pm t_c \dfrac{s}{\sqrt{n}} = 80 \pm 2.977 \dfrac{14}{\sqrt{15}} = 80 \pm 10.761 \approx (69, 91)$

24. $\hat{p} = \dfrac{x}{n} = \dfrac{1004}{3462} = 0.290, \hat{q} = 0.710$

26. $\hat{p} = \dfrac{x}{n} = \dfrac{125}{1000} = 0.125, \hat{q} = 0.875$

28. $\hat{p} = \dfrac{x}{n} = \dfrac{668}{957} = 0.698, \hat{q} = 0.302$

30. $\hat{p} = \dfrac{x}{n} = \dfrac{1230}{2365} = 0.520, \hat{q} = 0.480$

32. $\hat{p} \pm z_c \sqrt{\dfrac{\hat{p}\hat{q}}{n}} = 0.290 \pm 2.575 \sqrt{\dfrac{0.290 \cdot 0.710}{3462}} = 0.290 \pm 0.020 \approx (0.270, 0.310)$

34. $\hat{p} \pm z_c \sqrt{\dfrac{\hat{p}\hat{q}}{n}} = 0.125 \pm 2.326 \sqrt{\dfrac{0.125 \cdot 0.875}{1000}} = 0.125 \pm 0.024 \approx (0.101, 0.149)$

36. $\hat{p} \pm z_c \sqrt{\dfrac{\hat{p}\hat{q}}{n}} = 0.698 \pm 1.645 \sqrt{\dfrac{0.698 \cdot 0.302}{957}} = (0.674, 0.722)$

38. $\hat{p} \pm z_c \sqrt{\dfrac{\hat{p}\hat{q}}{n}} = 0.52 \pm 2.326 \sqrt{\dfrac{(0.52)(0.48)}{2365}} = 0.52 \pm 0.024 = (0.496, 0.544)$

40. $n = \hat{p}\hat{q}\left(\dfrac{z_c}{E}\right)^2 = 0.23 \cdot 0.77\left(\dfrac{2.575}{0.025}\right)^2 \approx 1878.854 \to 1879$

The sample size is larger.

42. $\chi_R^2 = 42.980$, $\chi_L^2 = 10.856$ **44.** $\chi_R^2 = 23.589$, $\chi_L^2 = 1.735$

46. 99% CI for σ^2: $\left(\dfrac{(n-1)s^2}{\chi_R^2}, \dfrac{(n-1)s^2}{\chi_L^2}\right) = \left(\dfrac{15 \cdot (0.073)^2}{32.801}, \dfrac{15 \cdot (0.073)^2}{4.601}\right) \approx (0.002, 0.017)$

99% CI for σ: $\left(\sqrt{0.002}, \sqrt{0.017}\right) \approx (0.045, 0.130)$

48. 95% CI for σ^2: $\left(\dfrac{(n-1)s^2}{\chi_R^2}, \dfrac{(n-1)s^2}{\chi_L^2}\right) = \left(\dfrac{(23)(1.12)^2}{38.076}, \dfrac{(23)(1.12)^2}{11.689}\right) = (0.76, 2.47)$

95% CI for σ: $\left(\sqrt{0.76}, \sqrt{2.47}\right) \approx (0.87, 0.157)$

CHAPTER 6 QUIZ SOLUTIONS

1. (a) $\bar{x} \approx 98.110$

(b) $s \approx 24.722$

$E = t_c \dfrac{s}{\sqrt{n}} = 1.960 \dfrac{24.722}{\sqrt{30}} \approx 8.847$

(c) $\bar{x} \pm t_c \dfrac{s}{\sqrt{n}} = 98.110 \pm 1.960 \dfrac{24.722}{\sqrt{30}} = 98.110 \pm 8.847 \approx (89.263, 106.957)$

You are 95% confident that the population mean repair costs is contained between \$89.26 and \$106.96.

2. $n = \left(\dfrac{z_c \sigma}{E}\right)^2 = \left(\dfrac{2.575 \cdot 22.50}{10}\right)^2 \approx 33.568 \to$ sample size must be 34

3. (a) $\bar{x} = 6.61$ (b) $s \approx 3.38$

(c) $\bar{x} \pm t_c \dfrac{s}{\sqrt{n}} = 6.61 \pm 1.833 \dfrac{3.38}{\sqrt{10}} = 6.61 \pm 1.957 \approx (4.65, 8.57)$

(d) $\bar{x} \pm z_c \dfrac{\sigma}{\sqrt{n}} = 6.61 \pm 1.645 \dfrac{3.5}{\sqrt{10}} = 6.61 \pm 1.821 \approx (4.79, 8.43)$

The t-CI is wider since less information is available.

4. $\bar{x} \pm t_c \dfrac{s}{\sqrt{n}} = 4744 \pm 2.365 \dfrac{580}{\sqrt{8}} = 4744 \pm 484.97 \approx (4259, 5229)$

5. (a) $\hat{p} = \dfrac{x}{n} = \dfrac{643}{1037} = 0.620$

(b) $\hat{p} \pm z_c \sqrt{\dfrac{\hat{p}\hat{q}}{n}} = 0.620 \pm 1.645 \sqrt{\dfrac{0.620 \cdot 0.38}{1037}} = 0.620 \pm 0.025 \approx (0.595, 0.645)$

(c) $n = \hat{p}\hat{q}\left(\dfrac{z_c}{E}\right)^2 = 0.620 \cdot 0.38 \left(\dfrac{2.575}{0.04}\right)^2 \approx 976.36 \to 977$

6. (a) $\left(\dfrac{(n-1)s^2}{\chi_R^2}, \dfrac{(n-1)s^2}{\chi_L^2}\right) = \left(\dfrac{29 \cdot (24.722)^2}{45.722}, \dfrac{29 \cdot (24.722)^2}{16.047}\right) \approx (387.650, 1104.514)$

(b) $\left(\sqrt{387.650}, \sqrt{1104.514}\right) \approx (19.689, 33.234)$

CHAPTER 7

Hypothesis Testing with One Sample

7.1 INTRODUCTION TO HYPOTHESIS TESTING

7.1 EXERCISE SOLUTIONS

2. Type I Error: The null hypothesis is rejected when it is true.

   Type II Error: The null hypothesis is not rejected when it is false.

4. False. A statistical hypothesis is a statement about a population.

6. True

8. False. If you want to support a claim, write it as your alternative hypothesis.

10. H_0: $\mu \geq 128$; H_a: $\mu < 128$

12. H_0: $\sigma^2 \geq 1.2$; H_a: $\sigma^2 < 1.2$

14. H_0: $p = 0.21$; H_a: $p \neq 0.21$

16. d, H_a: $\mu \geq 3$

18. a, H_a: $\mu \leq 2$

20. Left-tailed

22. Two-tailed

24. $\sigma < 3$

    H_0: $\sigma \geq 3$; H_a: $\sigma < 3$ (claim)

26. $p = 0.33$

    H_0: $p = 0.33$ (claim); H_a: $p \neq 0.33$

28. $\mu = 24$

    H_0: $\mu = 24$ (claim); H_a: $\mu \neq 24$

30. Type I: Rejecting H_0: $p = 0.21$ when actually $p = 0.21$.
    Type II: Not rejecting H_0: $p = 0.21$ when actually $p \neq 0.21$.

32. Type I: Rejecting H_0: $p = 0.02$ when actually $p = 0.02$.
    Type II: Not rejecting H_0: $p = 0.02$ when actually $p \neq 0.02$.

34. Type I: Rejecting H_0: $\sigma \geq 5$ when actually $\sigma \geq 5$.
    Type II: Not rejecting H_0: $\sigma \geq 5$ when actually $\sigma < 5$.

36. The null hypothesis is H_0: $\mu \leq 0.02$, the alternative hypothesis is H_a: $\mu > 0.02$. Therefore, because the alternative hypothesis contains >, the test is a right-tailed test.

38. The null hypothesis is H_0: $\mu \geq 50{,}000$, the alternative hypothesis is H_a: $\mu < 50{,}000$. Therefore, because the alternative hypothesis contains <, the test is a left-tailed test.

40. The null hypothesis is H_0: $\mu \leq 10$, the alternative hypothesis is H_a: $\mu > 10$. Therefore, because the alternative hypothesis contains >, the test is a right-tailed test.

42. (a) There is enough evidence to reject the Postal Service's claim.

    (b) There is not enough evidence to reject the Postal Service's claim.

44. (a) There is enough evidence to reject the manufacturer's claim.

    (b) There is not enough evidence to reject the manufacturer's claim.

46. (a) There is enough evidence to reject the soft-drink maker's claim.

(b) There is not enough evidence to reject the soft-drink maker's claim.

48. $H_0: \mu = 21; H_a: \mu \neq 21$

50. (a) $H_0: \mu \leq 28; H_a: \mu > 28$

(b) $H_0: \mu \geq 28; H_a: \mu < 28$

52. There are no z-values that correspond to $\alpha = 0$. If $\alpha = 0$, the null hypothesis cannot be rejected and the hypothesis test is useless.

54. (a) Reject H_0 since the CI is located above 54.

(b) Do not reject H_0 since the CI includes values below 54.

(c) Do not reject H_0 since the CI includes values below 54.

56. (a) Reject H_0 since the CI is located below 0.73.

(b) Do not reject H_0 since the CI includes values above 0.73.

(c) Do not reject H_0 since the CI includes values above 0.73.

7.2 HYPOTHESIS TESTING FOR THE MEAN (LARGE SAMPLES)

7.2 EXERCISE SOLUTIONS

2.
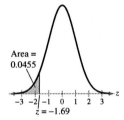

$P = 0.0455$

Reject H_0 since $P = 0.0455 < 0.05 = \alpha$.

4.
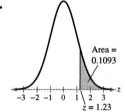

$P = 0.1093$

Fail to reject H_0 since $P = 0.1093 > 0.10 = \alpha$.

6.
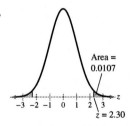

$P = 2 \text{ (Area)} = 2(0.0107) = 0.0214$; Fail to reject H_0 since $P = 0.0214 > 0.01 = \alpha$.

8. d **10.** f **12.** a

14. (a) Fail to reject H_0 ($P = 0.0691 > 0.01 = \alpha$).

(b) Fail to reject H_0 ($P = 0.0691 > 0.05 = \alpha$).

16. 1.41 **18.** -1.34 **20.** ± 1.645

22. Two-tailed ($\alpha = 0.05$) **24.** Left-tailed ($\alpha = 0.05$)

26. (a) Reject H_0 since $z > 1.96$.

(b) Fail to reject H_0 since $-1.96 < z < 1.96$.

(c) Fail to reject H_0 since $-1.96 < z < 1.96$.

(d) Reject H_0 since $z < -1.96$.

28. (a) Fail to reject H_0 since $-2.575 < z < 2.575$.

(b) Reject H_0 since $z < -2.575$.

(c) Reject H_0 since $z > 2.575$.

(d) Fail to reject H_0 since $-2.575 < z < 2.575$.

30. $H_0: \mu \leq 1030$ and $H_a: \mu > 1030$

$\alpha = 0.05 \rightarrow z_0 = 1.645$

$z = \dfrac{\bar{x} - \mu}{\frac{s}{\sqrt{n}}} = \dfrac{1035 - 1030}{\frac{23}{\sqrt{50}}} = \dfrac{5}{3.253} \approx 1.537$

Fail to reject H_0. There is not enough evidence to support the claim.

32. $H_0: \mu \leq 22{,}500$ and $H_a: \mu > 22{,}500$

$\alpha = 0.01 \rightarrow z_0 = 2.33$

$z = \dfrac{\bar{x} - \mu}{\frac{s}{\sqrt{n}}} = \dfrac{23{,}250 - 22{,}500}{\frac{1200}{\sqrt{45}}} = \dfrac{750}{178.885} \approx 4.193$

Reject H_0. There is enough evidence to reject the claim.

34. (a) $H_0: \mu \leq 1.5$; $H_a: \mu > 1.5$ (claim)

(b) $z = \dfrac{\bar{x} - \mu}{\frac{s}{\sqrt{n}}} = \dfrac{1.55 - 1.5}{\frac{0.32}{\sqrt{50}}} = \dfrac{0.05}{0.0453} \approx 1.105$

Area $= 0.8654$

(c) P-value $= \{\text{Area to right of } z = 1.11\} = 0.1335$

(d) Fail to reject H_0.

(e) There is insufficient evidence at the 10% level of significance to support the claim that the mean reserve capacity of the battery is greater than 1.5 hours.

36. (a) $H_0: \mu = 3.4$ (claim); $H_a: \mu \neq 3.4$

(b) $z = \dfrac{\bar{x} - \mu}{\frac{s}{\sqrt{n}}} = \dfrac{3.2 - 3.4}{\frac{1.13}{\sqrt{60}}} = \dfrac{-0.2}{0.146} \approx -1.371$

Area $= 0.0853$

(c) P-value $= 2\{\text{Area to left of } z = -1.37\} = 2\{0.0853\} = 0.1706$

(d) Fail to reject H_0.

(e) There is insufficient evidence at the 8% level to reject the claim that the mean tuna consumed by a person in the U.S. is 3.4 pounds per year.

38. (a) $H_0: \mu \geq \$90{,}900$ (claim); $H_a: \mu < \$90{,}900$

(b) $\bar{x} \approx 91{,}316.794, \quad s \approx 10{,}202.570$

$$z = \frac{\bar{x} - \mu}{\frac{s}{\sqrt{n}}} = \frac{91{,}316.794 - 90{,}900}{\frac{10{,}202.570}{\sqrt{34}}} = \frac{416.794}{1749.726} \approx 0.238$$

Area = 0.5948

(c) P-value = {Area to left of $z = 0.24$} = 0.5948

(d) Fail to reject H_0.

(e) There is insufficient evidence at the 3% level to reject the claim that the mean annual salary for engineering managers in Alabama is at least $69,000.

40. (a) $H_0: \mu = 80$ (claim); $H_a: \mu \neq 80$

(b) $z_0 = \pm 1.96$; Rejection regions: $z < -1.96$ and $z > 1.96$

(c) $z = \dfrac{\bar{x} - \mu}{\frac{s}{\sqrt{n}}} = \dfrac{83 - 80}{\frac{35}{\sqrt{42}}} = \dfrac{3}{5.401} \approx 0.555$

(d) Fail to reject H_0.

(e) There is insufficient evidence at the 5% level to reject the claim that the mean caffeine content is 80 milligrams per five ounces.

42. (a) $H_0: \mu \leq 230$ (claim); $H_a: \mu > 230$

(b) $z_0 = 1.75$; Rejection region: $z > 1.75$

(c) $z = \dfrac{\bar{x} - \mu}{\frac{s}{\sqrt{n}}} = \dfrac{232 - 230}{\frac{10}{\sqrt{52}}} = \dfrac{2}{1.387} \approx 1.442$

(d) Fail to reject H_0.

(e) There is insufficient evidence at the 4% level to reject the claim that the mean sodium content per serving of cereal is no more than 230 milligrams.

44. (a) $H_0: \mu \geq 10{,}000$ (claim); $H_a: \mu < 10{,}000$

(b) $z_0 = -1.34$; Rejection region: $z < -1.34$

(c) $\bar{x} = 9580.9, \quad s = 1722.4$

$$z = \dfrac{\bar{x} - \mu}{\frac{s}{\sqrt{n}}} = \dfrac{9580.9 - 10{,}000}{\frac{1722.4}{\sqrt{32}}} = \dfrac{-419.1}{304.5} = -1.377$$

(d) Reject H_0.

(e) There is sufficient evidence at the 9% level to reject the claim that the mean life of fluorescent lamps is at least 10,000 hours.

46. (a) $H_0: \mu \leq 10$; $H_a: \mu > 10$ (claim)   (b) $z_0 = 1.48$; Rejection region: $z > 1.48$

(c) $\bar{x} = 10.49, \quad s = 2.36$

$$z = \dfrac{\bar{x} - \mu}{\frac{s}{\sqrt{n}}} = \dfrac{10.49 - 10}{\frac{2.36}{\sqrt{45}}} = \dfrac{4.9}{3.5} = 1.393$$

(d) Fail to reject H_0.

(e) There is insufficient evidence at the 7% level to support the claim that the mean life span of a microwave oven for that company is more than 10 years.

48. $z = \dfrac{\bar{x} - \mu}{\dfrac{s}{\sqrt{n}}} = \dfrac{22{,}200 - 22{,}000}{\dfrac{775}{\sqrt{36}}} = \dfrac{200}{129.2} = 1.548$

P-value = {Area right of $z = 1.59$} = 0.0559

Fail to reject H_0 because P-value = $0.0559 > 0.05 = \alpha$.

50. (a) $\alpha = 0.06$; Reject H_0.

(b) $\alpha = 0.07$; Reject H_0.

(c) $z = \dfrac{\bar{x} - \mu}{\dfrac{s}{\sqrt{n}}} = \dfrac{22{,}200 - 20{,}000}{\dfrac{775}{\sqrt{40}}} = \dfrac{200}{122.5} = 1.63$

P-value = {Area right of $z = 1.63$} = 0.0516; Fail to reject H_0.

(d) $z = \dfrac{\bar{x} - \mu}{\dfrac{s}{\sqrt{n}}} = \dfrac{22{,}200 - 20{,}000}{\dfrac{775}{\sqrt{80}}} = \dfrac{200}{86.6} \approx 2.31$

P-value = {Area right of $z = 2.31$} = 0.0104; Reject H_0.

7.3 HYPOTHESIS TESTING FOR THE MEAN (SMALL SAMPLES)

7.3 EXERCISE SOLUTIONS

2. Identify the claim. State H_0 and H_a. Specify the level of significance. Identify the degrees of freedom and sketch the sampling distribution. Determine the critical value(s) and rejection region(s). Find the standardized test statistic. Make a decision and interpret it in the context of the original claim. The population must be normal or nearly normal.

4. $t_0 = 2.764$

6. $t_0 = -1.771$

8. $t_0 = \pm 2.262$

10. 1.761

12. -3.106

14. ± 1.721

16. (a) Fail to reject H_0 since $-1.372 < t < 1.372$.

(b) Reject H_0 since $t < -1.372$.

(c) Reject H_0 since $t > 1.372$.

(d) Fail to reject H_0 since $-1.372 < t < 1.372$.

18. (a) Fail to reject H_0 since $-1.725 < t < 1.725$.

(b) Reject H_0 since $t < -1.725$.

(c) Fail to reject H_0 since $-1.725 < t < 1.725$.

(d) Reject H_0 since $t > 1.725$.

20. H_0: $\mu \le 25$; H_a: $\mu > 25$ (claim)

$\alpha = 0.05$ and d.f. $= n - 1 = 16$

$t_0 = 1.746$

$t = \dfrac{\bar{x} - \mu}{\dfrac{s}{\sqrt{n}}} = \dfrac{26.2 - 25}{\dfrac{2.32}{\sqrt{17}}} = \dfrac{1.2}{0.563} \approx 2.133$

Reject H_0. There is enough evidence to support the claim.

22. H_0: $\mu = 52{,}200$; H_a: $\mu \ne 52{,}200$ (claim)

$\alpha = 0.05$ and d.f. $= n - 1 = 3$

$t_0 = \pm 3.182$

$t = \dfrac{\bar{x} - \mu}{\dfrac{s}{\sqrt{n}}} = \dfrac{53{,}220 - 52{,}200}{\dfrac{1200}{\sqrt{4}}} = \dfrac{1020}{600} = 1.7$

Fail to reject H_0. There is not enough evidence to support the claim.

24. (a) H_0: $\mu \le 95$; H_a: $\mu > 95$ (claim)

(b) $t_0 = 3.143$; Reject H_0 if $t > 3.143$.

(c) $t = \dfrac{\bar{x} - \mu}{\dfrac{s}{\sqrt{n}}} = \dfrac{100 - 95}{\dfrac{42.50}{\sqrt{7}}} = \dfrac{5}{16.0635} \approx 0.311$

(d) Fail to reject H_0.

(e) There is not enough evidence at the 1% level to support the claim that the mean repair cost for damaged computers is more than $95.

26. (a) H_0: $\mu \le 4$; H_a: $\mu > 4$ (claim)

(b) $t_0 = 1.833$; Reject H_0 if $t > 1.833$.

(c) $t = \dfrac{\bar{x} - \mu}{\dfrac{s}{\sqrt{n}}} = \dfrac{4.3 - 4}{\dfrac{1.2}{\sqrt{10}}} = \dfrac{0.3}{0.379} = 0.791$

(d) Fail to reject H_0.

(e) There is not enough evidence at the 5% level to support the claim that the mean waste generated by adults in the U.S. is more than 4 pounds per day.

28. (a) H_0: $\mu = \$17{,}100$ (claim); H_a: $\mu \ne \$17{,}100$

(b) $t_0 = \pm 2.201$; Reject H_0 if $t < -2.201$ or $t > 2.201$.

(c) $\bar{x} \approx \$17{,}551.750$, $s \approx \$1650.830$

$t = \dfrac{\bar{x} - \mu}{\dfrac{s}{\sqrt{n}}} = \dfrac{17{,}551.750 - 17{,}100}{\dfrac{1650.830}{\sqrt{12}}} = \dfrac{451.75}{47.871} \approx 0.948$

(d) Fail to reject H_0.

(e) There is not enough evidence at the 5% level to reject the claim that the mean annual pay for full-time female workers over age 25 without high school diplomas is $17,100.

30. (a) H_0: $\mu \le \$580$; H_a: $\mu > \$580$ (claim)

(b) $\bar{x} = 635.8$, $s = 146.8$

$$t = \frac{\bar{x} - \mu}{\frac{s}{\sqrt{n}}} = \frac{635.8 - 580}{\frac{146.8}{\sqrt{24}}} = \frac{55.8}{29.9} \approx 1.86$$

P-value = {Area right of $t = 1.86$} = 0.0378

(c) Reject H_0.

(d) There is sufficient evidence at the 5% level to support the claim that teachers spend a mean of more than $580 of their own money on school supplies in a year.

32. (a) H_0: $\mu = 11.0$ (claim); H_a: $\mu \neq 11.0$

(b) $\bar{x} = 10.050$, $s = 2.485$

$$t = \frac{\bar{x} - \mu}{\frac{s}{\sqrt{n}}} = \frac{10.050 - 11.0}{\frac{2.485}{\sqrt{8}}} = \frac{-.95}{0.879} \approx -1.081$$

P-value = 2{area left of $t = -1.081$} = 2(0.158) = 0.316

(c) Fail to reject H_0.

(d) There is not enough evidence at the 1% level to reject the claim that the mean number of classroom hours per week for full-time faculty is 11:0.

34. (a) H_0: $\mu = \$132$ (claim); H_a: $\mu \neq \$132$

(b) $\bar{x} = \$142.8$, $s = \$37.52$

$$t = \frac{\bar{x} - \mu}{\frac{s}{\sqrt{n}}} = \frac{142.8 - 132}{\frac{37.52}{\sqrt{10}}} = \frac{10.8}{11.865} = 0.91$$

P-value = 2{Area right of $t = 0.91$}
= 2{0.193}
= 0.386

(c) Fail to reject H_0.

(d) There is insufficient evidence at the 2% level to reject the claim that the daily lodging costs for a family in the U.S. is $132.

36. (a) Since $0.096 > 0.05 = \alpha$, fail to reject H_0.

(b) Since $0.096 < 0.10 = \alpha$, reject H_0.

(c) $t = \dfrac{\bar{x} - \mu}{\frac{s}{\sqrt{n}}} = \dfrac{2494 - 2294}{\frac{325}{\sqrt{12}}} = \dfrac{200}{93.819} = 2.132$

P-value = {Area right of $t = 2.132$} ≈ 0.028

Since $0.028 > 0.01 = \alpha$, fail to reject H_0.

(d) $t = \dfrac{\bar{x} - \mu}{\frac{s}{\sqrt{n}}} = \dfrac{2494 - 2294}{\frac{325}{\sqrt{24}}} = \dfrac{200}{66.340} = 3.015$

P-value = {Area right of $t = 3.015$} ≈ 0.003

Since $0.003 < 0.01 = \alpha$, reject H_0.

266 CHAPTER 7 | HYPOTHESIS TESTING WITH ONE SAMPLE

38. Since σ is unknown and $n \geq 30$, use the z-distribution.

$H_0: \mu = 337$ (claim); $H_a: \mu \neq 337$

$$z = \frac{\bar{x} - \mu}{\frac{s}{\sqrt{n}}} = \frac{332 - 337}{\frac{10}{\sqrt{50}}} = \frac{-5}{1.414} \approx -3.536$$

P-value $= 2\{$Area left of $z = -3.536\} = 2(0.0002) = 0.0004$

Reject H_0. There is enough evidence at the 1% level to reject the claim that the mean time full-time faculty spend with students outside of class is 337 hours per week.

7.4 HYPOTHESIS TESTING FOR PROPORTIONS

7.4 EXERCISE SOLUTIONS

2. If $np \geq 5$ and $nq \geq 5$, the normal distribution can be used.

4. $np = (500)(0.30) = 150 \geq 5$

$nq = (500)(0.70) = 350 \geq 5 \rightarrow$ use normal distribution

$H_0: p \leq 0.30$ (claim); $H_a: p > 0.30$

$z_0 = 1.645$

$$z = \frac{\hat{p} - p}{\sqrt{\frac{pq}{n}}} = \frac{0.35 - 0.30}{\sqrt{\frac{(0.30)(0.70)}{500}}} = \frac{0.05}{0.0205} \approx 2.440$$

Reject H_0. There is enough evidence to reject the claim.

6. $np = (45)(0.125) = 5.625 \geq 5$

$nq = (45)(0.875) = 39.375 \geq 5 \rightarrow$ use normal distribution

$H_0: p \leq 0.125$; $H_a: p > 0.125$ (claim)

$z_0 = 2.33$

$$z = \frac{\hat{p} - p}{\sqrt{\frac{pq}{n}}} = \frac{0.2325 - 0.125}{\sqrt{\frac{(0.125)(0.875)}{45}}} = \frac{0.1075}{0.0493} \approx 2.180$$

Fail to reject H_0. There is not enough evidence to reject the claim.

8. $np = (16)(0.80) = 12.8 \geq 5$

$nq = (16)(0.20) = 3.2 < 5 \rightarrow$ cannot use normal distribution

10. (a) $H_0: p \leq 0.55$ (claim); $H_a: p > 0.55$

(b) $z_0 = 2.33$; Reject H_0 if $z > 2.33$.

(c) $z = \dfrac{\hat{p} - p}{\sqrt{\frac{pq}{n}}} = \dfrac{0.564 - 0.55}{\sqrt{\frac{(0.55)(0.45)}{250}}} = \dfrac{0.014}{0.03146} \approx 0.445$

(d) Fail to reject H_0.

(e) There is not enough evidence at the 1% level to reject the claim that no more than 55% of U.S. adults eat breakfast every day.

12. (a) $H_0: p \leq 0.58$; $H_a: p > 0.58$ (claim)

    (b) $z_0 = 1.28$; Reject H_0 if $z > 1.28$.

    (c) $z = \dfrac{\hat{p} - p}{\sqrt{\dfrac{pq}{n}}} = \dfrac{0.61 - 0.58}{\sqrt{\dfrac{(0.58)(0.42)}{100}}} = \dfrac{0.03}{0.05} = 0.6$

    (d) Fail to reject H_0.

    (e) There is not enough evidence at the 10% level to support the claim that more than 58% of British consumers want supermarkets to stop selling genetically modified foods.

14. (a) $H_0: p = 0.44$ (claim); $H_a: p \neq 0.44$

    (b) $z_0 = \pm 1.96$; Reject H_0 if $z < -1.96$ or $z > 1.96$.

    (c) $\hat{p} = \dfrac{556}{1165} \approx 0.477$

    $z = \dfrac{\hat{p} - p}{\sqrt{\dfrac{pq}{n}}} = \dfrac{0.477 - 0.44}{\sqrt{\dfrac{(0.44)(0.56)}{1165}}} = \dfrac{0.037}{0.015} \approx 2.54$

    (d) Reject H_0.

    (e) There is sufficient evidence at the 5% level to reject the claim that 44% of adults in the United States do volunteer work.

16. The company should continue the use of giveaways since there is not enough evidence to say that less than 52% of the attendees would be more likely to stop at the exhibit.

18. $z = \dfrac{\hat{p} - p}{\sqrt{\dfrac{pq}{n}}} \Rightarrow \dfrac{\left(\dfrac{x}{n}\right) - p}{\sqrt{\dfrac{pq}{n}}} \Rightarrow \dfrac{\left(\dfrac{x}{n}\right) - p}{\dfrac{\sqrt{pq}}{\sqrt{n}}} \Rightarrow \dfrac{\left[\left(\dfrac{x}{n}\right) - p\right]\sqrt{n}}{\sqrt{pq}} \cdot \dfrac{\sqrt{n}}{\sqrt{n}} \Rightarrow \dfrac{\left[\left(\dfrac{x}{n}\right) - p\right]n}{\sqrt{pqn}} \Rightarrow \dfrac{x - np}{\sqrt{pqn}}$

7.5 HYPOTHESIS TESTING FOR VARIANCE AND STANDARD DEVIATION

7.5 EXERCISE SOLUTIONS

2. State H_0 and H_a. Specify the level of significance. Determine the degrees of freedom. Determine the critical value(s) and rejection region(s). Find the standardized test statistic. Make a decision and interpret in the context of the original claim.

4. $\chi_0^2 = 14.684$

6. $\chi_0^2 = 13.091$

8. $\chi_L^2 = 12.461$, $\chi_R^2 = 50.993$

10. (a) Fail to reject H_0.
    (b) Fail to reject H_0.
    (c) Reject H_0.
    (d) Reject H_0.

12. (a) Fail to reject H_0.
    (b) Fail to reject H_0.
    (c) Fail to reject H_0.
    (d) Reject H_0.

14. $H_0: \sigma \geq 40$; $H_a: \sigma < 40$ (claim)

    $\chi_0^2 = 3.053$

    $\chi^2 = \dfrac{(n-1)s^2}{\sigma^2} = \dfrac{(11)(40.8)^2}{(40)^2} \approx 11.444$

    Fail to reject H_0. There is insufficient evidence to support the claim.

16. (a) $H_0: \sigma^2 = 5$ (claim); $H_a: \sigma^2 \neq 5$

    (b) $\chi_L^2 = 10.982$, $\chi_R^2 = 36.781$; Reject H_0 if $\chi^2 > 36.781$ or $\chi^2 < 10.982$.

    (c) $\chi^2 = \dfrac{(n-1)s^2}{\sigma^2} = \dfrac{(22)(4.5)}{5} = 19.800$

    (d) Fail to reject H_0.

    (e) There is not enough evidence at the 5% level to reject the claim that the variance of the gas mileage is 5.

18. (a) $H_0: \sigma \geq 30$; $H_a: \sigma < 30$ (claim)

    (b) $\chi_0^2 = 2.088$; Reject H_0 if $\chi^2 < 2.088$.

    (c) $\chi^2 = \dfrac{(n-1)s^2}{\sigma^2} = \dfrac{(9)(28.8)^2}{(30)^2} \approx 8.294$

    (d) Fail to reject H_0.

    (e) There is not enough evidence at the 1% level to support the claim that the standard deviation of test scores for eighth grade students who took a life-science assessment test is less than 30.

20. (a) $H_0: \sigma = 6.14$ (claim); $H_a: \sigma \neq 6.14$

    (b) $\chi_L^2 = 8.907$, $\chi_R^2 = 32.852$; Reject H_0 if $\chi^2 < 8.907$ or $\chi^2 > 32.852$.

    (c) $\chi^2 = \dfrac{(n-1)s^2}{\sigma^2} = \dfrac{(19)(6.5)^2}{(6.14)^2} = 21.293$

    (d) Fail to reject H_0.

    (e) There is not enough evidence to reject the claim.

22. (a) $H_0: \sigma \leq \$25$ (claim); $H_a: \sigma > \$25$

    (b) $\chi_0^2 = 26.217$; Reject H_0 if $\chi^2 < 26.217$.

    (c) $\chi^2 = \dfrac{(n-1)s^2}{\sigma^2} = \dfrac{(12)(27.50)^2}{(25)^2} = 14.520$

    (d) Fail to reject H_0.

    (e) There is not enough evidence at the 1% level to reject the claim that the standard deviation of the room rates of hotels in the city is no more than $25.

24. (a) H_0: $\sigma \geq 14{,}500$ (claim); H_a: $\sigma < 14{,}500$

(b) $\chi_0^2 = 10.085$; Reject H_0 if $\chi^2 < 10.085$.

(c) $s = 14{,}060.664$

$$\chi^2 = \frac{(n-1)s^2}{\sigma^2} = \frac{(17)(14{,}060.664)^2}{(14{,}500)^2} \approx 15.985$$

(d) Fail to reject H_0.

(e) There is not enough evidence at the 10% level to reject the claim that the standard deviation of the annual salaries for public relations managers is at least $14,500.

26. $\chi^2 = 14.520$

P-value = {Area right of $\chi^2 = 14.520$} = 0.269

Fail to reject H_0 because P-value = $0.269 > 0.01 = \alpha$.

28. $\chi^2 = 15.985$

P-value = {Area left of $\chi^2 = 15.985$} = 0.475

Fail to reject H_0 because P-value = $0.475 > 0.10 = \alpha$.

CHAPTER 7 REVIEW EXERCISE SOLUTIONS

2. H_0: $\mu = 35$ (claim); H_a: $\mu \neq 35$

4. H_0: $\mu = 150{,}020$; H_a: $\mu \neq 150{,}020$ (claim)

6. H_0: $p \geq 0.70$ (claim); H_a: $p < 0.70$

8. (a) H_0: $\mu \geq 30{,}000$ (claim); H_a: $\mu < 30{,}000$

(b) Type I error will occur if H_0 is rejected when the actual mean tire life is at least 30,000 miles.
Type II error if H_0 is not rejected when the actual mean tire life is less than 30,000 miles.

(c) Left-tailed, since hypothesis compares "$\geq$ vs $<$".

(d) There is enough evidence to reject the claim.

(e) There is not enough evidence to reject the claim.

10. (a) H_0: $\mu \geq 20$; H_a: $\mu < 20$ (claim)

(b) Type I error will occur if H_0 is rejected when the actual mean number of fat calories is at least 20.
Type II error if H_0 is not rejected when the actual mean number of fat calories is less than 20.

(c) Left-tailed, since hypothesis compares "$\geq$ vs $<$".

(d) There is enough evidence to support the claim.

(e) There is not enough evidence to support the claim.

12. $z_0 = \pm 2.81$

14. $z_0 = \pm 1.75$

270 CHAPTER 7 | HYPOTHESIS TESTING WITH ONE SAMPLE

16. H_0: $\mu = 0$; H_a: $\mu \neq 0$ (claim)

$z_0 = \pm 1.96$

$z = \dfrac{\bar{x} - \mu}{\frac{s}{\sqrt{n}}} = \dfrac{-0.69 - 0}{\frac{2.62}{\sqrt{60}}} = \dfrac{-0.69}{0.338} \approx -2.040$

Reject H_0. There is enough evidence to support the claim.

18. H_0: $\mu = 7450$ (claim); H_a: $\mu \neq 7450$

$z_0 = \pm 1.96$

$z = \dfrac{\bar{x} - \mu}{\frac{s}{\sqrt{n}}} = \dfrac{7512 - 7450}{\frac{243}{\sqrt{57}}} = \dfrac{62}{32.186} \approx 1.926$

Fail to reject H_0. There is not enough evidence to reject the claim.

20. H_0: $\mu = 230$; H_a: $\mu \neq 230$ (claim)

$z = \dfrac{\bar{x} - \mu}{\frac{s}{\sqrt{n}}} = \dfrac{216.5 - 230}{\frac{17.3}{\sqrt{48}}} = \dfrac{-13.5}{2.497} \approx -5.406$

P-value $= 2\{\text{Area left of } z = -5.406\} \approx 0$

$\alpha = 0.10$; Reject H_0.

$\alpha = 0.05$; Reject H_0.

$\alpha = 0.01$; Reject H_0.

22. H_0: $\mu \leq \$533$ (claim); H_a: $\mu > \$533$

$z = \dfrac{\bar{x} - \mu}{\frac{s}{\sqrt{n}}} = \dfrac{542 - 533}{\frac{40}{\sqrt{45}}} = \dfrac{9}{5.963} \approx -1.509$

P-value $= \{\text{Area right of } z = 1.509\} = 0.066$

Fail to reject H_0. There is not enough evidence to reject the claim.

24. $t_0 = 2.998$ **26.** $t_0 = \pm 2.201$

28. H_0: $\mu \leq 12{,}700$; H_a: $\mu > 12{,}700$ (claim)

$t_0 = 1.725$

$t = \dfrac{\bar{x} - \mu}{\frac{s}{\sqrt{n}}} = \dfrac{12{,}804 - 12{,}700}{\frac{248}{\sqrt{21}}} = \dfrac{104}{54.118} \approx 1.922$

Reject H_0. There is enough evidence to support the claim.

30. H_0: $\mu = 4.20$ (claim); H_a: $\mu \neq 4.20$

$t_0 = \pm 2.896$

$t = \dfrac{\bar{x} - \mu}{\frac{s}{\sqrt{n}}} = \dfrac{4.41 - 4.20}{\frac{0.26}{\sqrt{9}}} = \dfrac{0.21}{0.0867} \approx 2.423$

Fail to reject H_0. There is not enough evidence to reject the claim.

32. H_0: $\mu \geq 850$; H_a: $\mu < 850$ (claim)

$t_0 = -2.160$

$t = \dfrac{\bar{x} - \mu}{\dfrac{s}{\sqrt{n}}} = \dfrac{875 - 850}{\dfrac{25}{\sqrt{14}}} = \dfrac{25}{6.682} \approx 3.742$

Fail to reject H_0. There is not enough evidence to support the claim.

34. H_0: $\mu \leq 10$ (claim); H_a: $\mu > 10$

$t_0 = 1.397$

$t = \dfrac{\bar{x} - \mu}{\dfrac{s}{\sqrt{n}}} = \dfrac{13.5 - 10}{\dfrac{5.8}{\sqrt{9}}} = \dfrac{3.5}{1.933} \approx 1.810$

Reject H_0. There is enough evidence to reject the claim.

36. H_0: $\mu \leq 9$; H_a: $\mu > 9$ (claim)

$\bar{x} = 9.982$, $s = 2.125$

$t = \dfrac{\bar{x} - \mu}{\dfrac{s}{\sqrt{n}}} = \dfrac{9.982 - 9}{\dfrac{2.125}{\sqrt{11}}} = \dfrac{0.982}{0.641} \approx 1.532$

P-value $= 0.078$

Fail to reject H_0. There is not enough evidence to support the claim.

38. H_0: $p \geq 0.70$; H_a: $p < 0.70$ (claim)

$z_0 = -2.33$

$z = \dfrac{\hat{p} - p}{\sqrt{\dfrac{pq}{n}}} = \dfrac{0.50 - 0.70}{\sqrt{\dfrac{(0.70)(0.30)}{68}}} = \dfrac{-0.2}{0.0556} \approx -3.599$

Reject H_0. There is enough evidence to support the claim.

40. H_0: $p = 0.50$ (claim); H_a: $p \neq 0.50$

$z_0 = \pm 1.645$

$z = \dfrac{\hat{p} - p}{\sqrt{\dfrac{pq}{n}}} = \dfrac{0.71 - 0.50}{\sqrt{\dfrac{(0.50)(0.50)}{129}}} = \dfrac{0.21}{0.0440} \approx 4.770$

Reject H_0. There is enough evidence to reject the claim.

42. H_0: $p = 0.34$; H_a: $p \neq 0.34$ (claim)

$z_0 = \pm 2.575$

$z = \dfrac{\hat{p} - p}{\sqrt{\dfrac{pq}{n}}} = \dfrac{0.29 - 0.34}{\sqrt{\dfrac{(0.34)(0.66)}{60}}} = \dfrac{-0.05}{0.061} \approx 0.820$

Fail to reject H_0. There is not enough evidence to support the claim.

44. H_0: $p \leq 0.80$ (claim); H_a: $p > 0.80$

$z_0 = 1.28$

$z = \dfrac{\hat{p} - p}{\sqrt{\dfrac{pq}{n}}} = \dfrac{0.85 - 0.80}{\sqrt{\dfrac{(0.80)(0.20)}{43}}} = \dfrac{0.05}{0.061} \approx 0.820$

Fail to reject H_0. There is not enough evidence to reject the claim.

46. H_0: $p = 0.02$ (claim); H_a: $p \neq 0.02$

$z_0 = \pm 1.96$

$\hat{p} = \dfrac{x}{n} = \dfrac{4}{121} \approx 0.033$

$z = \dfrac{\hat{p} - p}{\sqrt{\dfrac{pq}{n}}} = \dfrac{0.033 - 0.02}{\sqrt{\dfrac{(0.02)(0.98)}{121}}} = \dfrac{0.013}{0.013} \approx 1.00$

Fail to reject H_0. There is not enough evidence to reject the claim.

48. $\chi_L^2 = 3.565$, $\chi_R^2 = 29.819$

50. $\chi_0^2 = 1.145$

52. H_0: $\sigma^2 \leq 60$ (claim); H_a: $\sigma^2 > 60$

$\chi_0^2 = 26.119$

$\chi^2 = \dfrac{(n-1)s^2}{\sigma^2} = \dfrac{(14)(72.7)}{(60)} \approx 16.963$

Fail to reject H_0. There is not enough evidence to support the claim.

54. H_0: $\sigma = 0.035$; H_a: $\sigma \neq 0.035$ (claim)

$\chi_L^2 = 4.601$, $\chi_R^2 = 32.801$

$\chi^2 = \dfrac{(n-1)s^2}{\sigma^2} = \dfrac{(15)(0.026)^2}{(0.035)^2} \approx 8.278$

Fail to reject H_0. There is not enough evidence to support the claim.

56. H_0: $\sigma \leq 0.0025$ (claim); H_a: $\sigma > 0.0025$

$\chi_0^2 = 27.688$

$\chi^2 = \dfrac{(n-1)s^2}{\sigma^2} = \dfrac{(13)(0.0031)^2}{(0.0025)^2} \approx 19.989$

Fail to reject H_0. There is not enough evidence to reject the claim.

CHAPTER 7 QUIZ SOLUTIONS

1. (a) $H_0: \mu \geq 94$ (claim); $H_a: \mu < 94$

   (b) "$\geq$ vs $<$" $\rightarrow$ Left-tailed

   σ is unknown and $n \geq 30 \rightarrow z$-test.

   (c) $z_0 = -2.05$; Reject H_0 if $z < -2.05$.

   (d) $z = \dfrac{\bar{x} - \mu}{\frac{s}{\sqrt{n}}} = \dfrac{93.5 - 94}{\frac{30}{\sqrt{103}}} = \dfrac{-0.5}{2.956} \approx -0.169$

   (e) Fail to reject H_0. There is insufficient evidence at the 2% level to reject the claim that the mean consumption of fresh citrus fruits by people in the U.S. is at least 94 pounds per year.

2. (a) $H_0: \mu \geq 25$ (claim); $H_a: \mu < 25$

   (b) "$\geq$ vs $<$" $\rightarrow$ Left-tailed

   σ is unknown, the population is normal, and $n < 30 \rightarrow t$-test.

   (c) $t_0 = -1.895$; Reject H_0 if $t < -1.895$.

   (d) $z = \dfrac{\bar{x} - \mu}{\frac{s}{\sqrt{n}}} = \dfrac{23 - 25}{\frac{5}{\sqrt{8}}} = \dfrac{-2}{1.768} \approx -1.131$

   (e) Fail to reject H_0. There is insufficient evidence at the 5% level to reject the claim that the mean gas mileage is at least 25 miles per gallon.

3. (a) $H_0: p \leq 0.10$ (claim); $H_a: p > 0.10$

   (b) "$\leq$ vs $>$" $\rightarrow$ Right-tailed

   $np \geq 5$ and $nq \geq 5 \rightarrow z$-test

   (c) $z_0 = 1.75$; Reject H_0 if $z > 1.75$.

   (d) $z = \dfrac{\hat{p} - p}{\sqrt{\frac{pq}{n}}} = \dfrac{0.13 - 0.10}{\sqrt{\frac{(0.10)(0.90)}{57}}} = \dfrac{0.03}{0.0397} \approx 0.755$

   (e) Fail to reject H_0. There is insufficient evidence at the 4% level to reject the claim that no more than 10% of microwaves need repair during the first five years of use.

4. (a) $H_0: \sigma = 105$ (claim); $H_a: \sigma \neq 105$

   (b) "$=$ vs $\neq$" $\rightarrow$ Two-tailed

   Assuming the scores are normally distributed and you are testing the hypothesized standard deviation $\rightarrow \chi^2$ test.

   (c) $\chi_L^2 = 3.565$, $\chi_R^2 = 29.819$; Reject H_0 if $\chi^2 < 3.565$ or if $\chi^2 > 29.819$.

   (d) $\chi^2 = \dfrac{(n-1)s^2}{\sigma^2} = \dfrac{(13)(113)^2}{(105)^2} \approx 15.056$

   (e) Fail to reject H_0. There is insufficient evidence at the 1% level to reject the claim that the standard deviation of the SAT verbal scores for the state is 105.

5. (a) $H_0: \mu = \$59{,}482$ (claim); $H_a: \mu \neq \$59{,}482$

   (b) "= vs $\neq$" $\rightarrow$ Two-tailed

   σ is unknown, $n < 30$, and assuming the salaries are normally distributed $\rightarrow$ t-test.

   (c) not applicable

   (d) $t = \dfrac{\bar{x} - \mu}{\frac{s}{\sqrt{n}}} = \dfrac{58{,}581 - 59{,}482}{\frac{6500}{\sqrt{12}}} = \dfrac{-901}{1876.388} \approx -0.480$

   P-value = 2{Area left of $t = -0.480$} = 2(0.320) = 0.640

   (e) Fail to reject H_0. There is insufficient evidence at the 5% level to reject the claim that the mean annual salary for full-time male workers age 25 to 34 with a bachelor's degree is $59,482.

6. (a) $H_0: \mu = \$276$ (claim); $H_a: \mu \neq \$276$

   (b) "= vs $\neq$" $\rightarrow$ Two-tailed

   σ is unknown, $n \geq 30$ $\rightarrow$ z-test.

   (c) not applicable

   (d) $z = \dfrac{\bar{x} - \mu}{\frac{s}{\sqrt{n}}} = \dfrac{285 - 276}{\frac{30}{\sqrt{35}}} = \dfrac{9}{5.071} \approx 1.775$

   P-value = 2{Area right of $z = 1.775$} = 2{0.038} = 0.0759

   (e) Fail to reject H_0. There is insufficient evidence at the 5% level to reject the claim that the mean daily cost of meals and lodging for a family of four traveling in Massachusetts is $276.

Hypothesis Testing with Two Samples

CHAPTER 8

8.1 TESTING THE DIFFERENCE BETWEEN MEANS (LARGE INDEPENDENT SAMPLES)

8.1 EXERCISE SOLUTIONS

2. (1) The samples must be independent.

(2) $\{n_1 \geq 30$ and $n_2 \geq 30\}$ or $\{$each population must be normally distributed with known standard deviations$\}$

4. H_0: $\mu_1 \leq \mu_2$; H_1: $\mu_1 > \mu_2$ (claim)

Rejection region: $z_0 > 1.28$ (Right-tailed test)

(a) $\bar{x}_1 - \bar{x}_2 = 500 - 510 = -10$

(b) $z = \dfrac{(\bar{x}_1 - \bar{x}_2) - (\mu_1 - \mu_2)}{\sqrt{\dfrac{s_1^2}{n_1} + \dfrac{s_2^2}{n_2}}} = \dfrac{(500 - 510) - (0)}{\sqrt{\dfrac{(30)^2}{100} + \dfrac{(15)^2}{75}}} = \dfrac{-10}{\sqrt{12}} \approx -2.887$

(c) z is not in the rejection region because $-2.887 < 1.28$.

(d) Fail to reject H_0. There is not enough evidence to support the claim.

6. H_0: $\mu_1 \leq \mu_2$ (claim); H_1: $\mu_1 > \mu_2$

Rejection region: $z_0 > 1.88$ (Right-tailed test)

(a) $\bar{x}_1 - \bar{x}_2 = 5004 - 4895 = 109$

(b) $z = \dfrac{(\bar{x}_1 - \bar{x}_2) - (\mu_1 - \mu_2)}{\sqrt{\dfrac{s_1^2}{n_1} + \dfrac{s_2^2}{n_2}}} = \dfrac{(5004 - 4895) - (0)}{\sqrt{\dfrac{(136)^2}{144} + \dfrac{(215)^2}{156}}} = \dfrac{109}{\sqrt{424.759}} \approx 5.289$

(c) z is in the rejection region because $5.289 > 1.88$.

(d) Reject H_0. There is enough evidence to reject the claim.

8. H_0: $\mu_1 = \mu_2$ (claim); H_a: $\mu_1 \neq \mu_2$

$z_0 = \pm 1.96$

$z = \dfrac{(\bar{x}_1 - \bar{x}_2) - (\mu_1 - \mu_2)}{\sqrt{\dfrac{s_1^2}{n_1} + \dfrac{s_2^2}{n_2}}} = \dfrac{(52 - 45) - (0)}{\sqrt{\dfrac{(2.5)^2}{70} + \dfrac{(5.5)^2}{60}}} = \dfrac{7}{\sqrt{0.59345}} \approx 9.087$

Reject H_0. There is enough evidence to reject the claim.

10. (a) H_0: $\mu_1 \leq \mu_2$; H_a: $\mu_1 > \mu_2$ (claim)

(b) $z_0 = 1.28$; Reject H_0 if $z > 1.28$.

(c) $z = \dfrac{(\bar{x}_1 - \bar{x}_2) - (\mu_1 - \mu_2)}{\sqrt{\dfrac{s_1^2}{n_1} + \dfrac{s_2^2}{n_2}}} = \dfrac{(55 - 51) - (0)}{\sqrt{\dfrac{(5.3)^2}{50} + \dfrac{(4.9)^2}{50}}} = \dfrac{4}{\sqrt{1.042}} \approx 3.919$

(d) Reject H_0.

(e) There is sufficient evidence at the 10% level to support the claim that the mean braking distance for Type C is greater than for Type D.

12. (a) H_0: $\mu_1 = \mu_2$ (claim); H_a: $\mu_1 \neq \mu_2$

    (b) $z_0 = \pm 2.575$; Reject H_0 if $z < -2.575$ or $z > 2.575$.

    (c) $z = \dfrac{(\bar{x}_1 - \bar{x}_2) - (\mu_1 - \mu_2)}{\sqrt{\dfrac{s_1^2}{n_1} + \dfrac{s_2^2}{n_2}}} = \dfrac{(50 - 60) - (0)}{\sqrt{\dfrac{(10)^2}{34} + \dfrac{(18)^2}{46}}} = \dfrac{-10}{\sqrt{9.985}} \approx -3.165$

    (d) Reject H_0.

    (e) There is sufficient evidence at the 1% level to reject the claim that the mean repair costs for Model A and Model B are the same.

14. (a) H_0: $\mu_1 \leq \mu_2$; H_a: $\mu_1 > \mu_2$ (claim)

    (b) $z_0 = 1.28$; Reject H_0 if $z > 1.28$.

    (c) $z = \dfrac{(\bar{x}_1 - \bar{x}_2) - (\mu_1 - \mu_2)}{\sqrt{\dfrac{s_1^2}{n_1} + \dfrac{s_2^2}{n_2}}} = \dfrac{(22.1 - 19.8) - (0)}{\sqrt{\dfrac{(4.8)^2}{49} + \dfrac{(5.4)^2}{44}}} = \dfrac{2.3}{\sqrt{1.133}} \approx 2.161$

    (d) Reject H_0.

    (e) There is sufficient evidence at the 10% level to support the claim that the ACT scores are higher for high school students in a college prep program.

16. (a) H_0: $\mu_1 \geq \mu_2$; H_a: $\mu_1 < \mu_2$ (claim)

    (b) $z_0 = -1.645$; Reject H_0 if $z < -1.645$.

    (c) $z = \dfrac{(\bar{x}_1 - \bar{x}_2) - (\mu_1 - \mu_2)}{\sqrt{\dfrac{s_1^2}{n_1} + \dfrac{s_2^2}{n_2}}} = \dfrac{(1526 - 2136) - (0)}{\sqrt{\dfrac{(225)^2}{30} + \dfrac{(350)^2}{30}}} = \dfrac{-610}{\sqrt{5770.833}} \approx -8.03$

    (d) Reject H_0.

    (e) There is sufficient evidence at the 5% level to support the claim that households in the U.S. headed by people under the age of 25 spend less on food away from home than households headed by people ages 55–64.

18. (a) H_0: $\mu_1 \geq \mu_2$; H_a: $\mu_1 < \mu_2$ (claim)

    (b) $z_0 = -1.645$; Reject H_0 if $z < -1.645$.

    (c) $z = \dfrac{(\bar{x}_1 - \bar{x}_2) - (\mu_1 - \mu_2)}{\sqrt{\dfrac{s_1^2}{n_1} + \dfrac{s_2^2}{n_2}}} = \dfrac{(1600 - 2040) - (0)}{\sqrt{\dfrac{(230)^2}{40} + \dfrac{(380)^2}{40}}} = \dfrac{-440}{\sqrt{4932.5}} \approx -6.26$

    (d) Reject H_0.

    (e) There is sufficient evidence at the 5% level to support the claim that households in the U.S. headed by people under the age of 25 spend less on food away from home than households headed by people ages 55–64. The new samples do not lead to a different conclusion.

20. (a) H_0: $\mu_1 \geq \mu_2$; H_a: $\mu_1 < \mu_2$ (claim)

    (b) $z_0 = -1.88$; Reject H_0 if $z < -1.88$.

(c) $\bar{x}_1 \approx 14.134$, $s_1 \approx 4.589$, $n_1 = 35$

$\bar{x}_2 = 21.140$, $s_2 \approx 4.586$, $n_2 = 35$

$$z = \frac{(\bar{x}_1 - \bar{x}_2) - (\mu_1 - \mu_2)}{\sqrt{\frac{s_1^2}{n_1} + \frac{s_2^2}{n_2}}} = \frac{(14.134 - 21.140) - (0)}{\sqrt{\frac{(4.589)^2}{35} + \frac{(4.586)^2}{35}}} = \frac{-7.006}{\sqrt{1.203}} \approx -6.388$$

(d) Reject H_0.

(e) There is sufficient evidence at the 3% significance level to support the sociologist's claim.

22. (a) H_0: $\mu_1 = \mu_2$ (claim); H_a: $\mu_1 \neq \mu_2$

(b) $z_0 \approx \pm 2.05$; Reject H_0 if $z < -2.05$ or $z > 2.05$.

(c) $\bar{x}_1 = 3.337$, $s_1 = 0.011$, $n_1 = 40$

$\bar{x}_2 = 3.500$, $s_2 = 0.010$, $n_2 = 40$

$$z = \frac{(\bar{x}_1 - \bar{x}_2) - (\mu_1 - \mu_2)}{\sqrt{\frac{s_1^2}{n_1} + \frac{s_2^2}{n_2}}} = \frac{(3.337 - 3.500) - (0)}{\sqrt{\frac{(0.011)^2}{40} + \frac{(0.010)^2}{40}}} = \frac{-0.163}{\sqrt{0.000005525}} \approx -69.346$$

(d) Reject H_0.

(e) There is sufficient evidence at the 4% level of significance to reject the claim.

24. They are equivalent through algebraic manipulation of the equation.

$\mu_1 \geq \mu_2 \rightarrow \mu_1 - \mu_2 \geq 0$

26. H_0: $\mu_1 - \mu_2 = -1$ (claim); H_a: $\mu_1 - \mu_2 \neq -1$

$z_0 = \pm 1.96$; Reject H_0 if $z < -1.96$ or $z > 1.96$.

$$z = \frac{(\bar{x}_1 - \bar{x}_2) - (\mu_1 - \mu_2)}{\sqrt{\frac{s_1^2}{n_1} + \frac{s_2^2}{n_2}}} = \frac{(12.63 - 13.60) - (-1)}{\sqrt{\frac{(4.21)^2}{48} + \frac{(4.53)^2}{56}}} = \frac{0.03}{\sqrt{0.7357}} \approx 0.035$$

Fail to reject H_0. There is not enough evidence to reject the claim.

28. H_0: $\mu_1 - \mu_2 \geq 30{,}000$; H_a: $\mu_1 - \mu_2 < 30{,}000$ (claim)

$z_0 = -1.645$; Reject H_0 if $z < -1.645$.

$$z = \frac{(\bar{x}_1 - \bar{x}_2) - (\mu_1 - \mu_2)}{\sqrt{\frac{s_1^2}{n_1} + \frac{s_2^2}{n_2}}} = \frac{(52{,}300 - 25{,}300) - (30{,}000)}{\sqrt{\frac{(8250)^2}{31} + \frac{(3200)^2}{33}}} = \frac{-3000}{\sqrt{2{,}505{,}867.546}} = -1.895$$

Reject H_0. There is enough evidence to support the claim.

30.
$$(\bar{x}_1 - \bar{x}_2) - z_c \sqrt{\frac{s_1^2}{n_1} + \frac{s_2^2}{n_2}} < \mu_1 - \mu_2 < (\bar{x}_1 - \bar{x}_2) + z_c \sqrt{\frac{s_1^2}{n_1} + \frac{s_2^2}{n_2}}$$

$$(10.3 - 8.5) - 1.96 \sqrt{\frac{(1.2)^2}{140} + \frac{(1.5)^2}{127}} < \mu_1 - \mu_2 < (10.3 - 8.5) + 1.96 \sqrt{\frac{(1.2)^2}{140} + \frac{(1.5)^2}{127}}$$

$$1.8 - 1.96\sqrt{0.028} < \mu_1 - \mu_2 < 1.8 + 1.96\sqrt{0.028}$$

$$1.47 < \mu_1 - \mu_2 < 2.13$$

278 CHAPTER 8 | HYPOTHESIS TESTING WITH TWO SAMPLES

32. $H_0: \mu_1 - \mu_2 \leq 0; H_a: \mu_1 - \mu_2 > 0$ (claim)

$z_0 = -1.645$; Reject H_0 if $z > -1.645$.

$$z = \frac{(\bar{x}_1 - \bar{x}_2) - (\mu_1 - \mu_2)}{\sqrt{\frac{s_1^2}{n_1} + \frac{s_2^2}{n_2}}} = \frac{(10.3 - 8.5) - (0)}{\sqrt{\frac{(1.2)^2}{140} + \frac{(1.5)^2}{127}}} = \frac{1.8}{\sqrt{0.028}} \approx 10.76$$

Reject H_0. There is enough evidence to support the claim. I would recommend using Irinotecan over Fluorouracil because the average number of months with no reported cancer related pain was significantly higher.

34. $H_0: \mu_1 - \mu_2 \leq 0; H_1: \mu_1 - \mu_2 > 0$ (claim)

The 95% CI for $\mu_1 - \mu_2$ in Exercise 30 contained only values greater than zero and, as found in Exercise 32, there was sufficient evidence at the 5% level of significance to support the claim. If the CI for $\mu_1 - \mu_2$ contains only positive numbers, you reject H_0, because the null hypothesis states that $\mu_1 - \mu_2$ is less than or equal to zero, i.e., $\mu_1 - \mu_2$ is not positive.

8.2 TESTING THE DIFFERENCE BETWEEN MEANS (SMALL INDEPENDENT SAMPLES)

8.2 EXERCISE SOLUTIONS

2. (1) The samples must be independent.

(2) Each population must have a normal distribution.

(3) The size of at least one of the samples is less than 30.

4. (a) d.f. $= n_1 + n_2 - 2 = 25$ (b) d.f. $= \min\{n_1 - 1, n_2 - 1\} = 11$

$t_0 = 2.485$ $t_0 = 2.718$

6. (a) d.f. $= n_1 + n_2 - 2 = 39$ (b) d.f. $= \min\{n_1 - 1, n_2 - 1\} = 18$

$t_0 = \pm 1.96$ $t_0 = \pm 2.101$

8. (a) d.f. $= n_1 + n_2 - 2 = 7$ (b) d.f. $= \min\{n_1 - 1, n_2 - 1\} = 1$

$t_0 = -1.415$ $t_0 = -3.078$

10. (a) d.f. $= n_1 + n_2 - 2 = 14$ (b) d.f. $= \min\{n_1 - 1, n_2 - 1\} = 4$

$t_0 = 2.977$ $t_0 = 4.604$

12. $H_0: \mu_1 \geq \mu_2$ (claim); $H_a: \mu_1 < \mu_2$

d.f. $= n_1 + n_2 - 2 = 18$

$t_0 = -1.330$ (Left-tailed test)

(a) $\bar{x}_1 - \bar{x}_2 = 0.515 - 0.475 = 0.04$

(b) $t = \dfrac{(\bar{x}_1 - \bar{x}_2) - (\mu_1 - \mu_2)}{\sqrt{\dfrac{(n_1 - 1)s_1^2 + (n_2 - 1)s_2^2}{n_1 + n_2 - 2}}\sqrt{\dfrac{1}{n_1} + \dfrac{1}{n_2}}}$

$= \dfrac{(0.515 - 0.475) - (0)}{\sqrt{\dfrac{(11 - 1)(0.305)^2 + (9 - 1)(0.215)^2}{11 + 9 - 2}}\sqrt{\dfrac{1}{11} + \dfrac{1}{9}}} = \dfrac{0.4}{\sqrt{0.0722}\sqrt{0.202}} \approx 0.331$

(c) t is not in the rejection region.

(d) Fail to reject H_0. There is not enough evidence to reject the claim.

14. H_0: $\mu_1 \leq \mu_2$ (claim); H_a: $\mu_1 > \mu_2$

d.f. $= \min\{n_1 - 1, n_2 - 1\} = 13$

(a) $\bar{x}_1 - \bar{x}_2 = 45 - 50 = -5$

(b) $t_0 = 2.650$ (Right-tailed test)

$$t = \frac{(\bar{x}_1 - \bar{x}_2) - (\mu_1 - \mu_2)}{\sqrt{\frac{s_1^2}{n_1} + \frac{s_2^2}{n_2}}} = \frac{(45 - 50) - (0)}{\sqrt{\frac{(4.8)^2}{16} + \frac{(1.2)^2}{14}}} = \frac{-5}{\sqrt{1.543}} \approx -4.025$$

(c) t is not in the rejection region.

(d) Fail to reject H_0. There is not enough evidence to reject the claim.

16. (a) H_0: $\mu_1 = \mu_2$ (claim); H_a: $\mu_1 \neq \mu_2$

(b) d.f. $= n_1 + n_2 - 2 = 11$

$t_0 = \pm 2.201$; Reject H_0 if $t < -2.201$ or $t > 2.201$.

(c) $t = \dfrac{(\bar{x}_1 - \bar{x}_2) - (\mu_1 - \mu_2)}{\sqrt{\dfrac{(n_1 - 1)s_1^2 + (n_2 - 1)s_2^2}{n_1 + n_2 - 2}} \sqrt{\dfrac{1}{n_1} + \dfrac{1}{n_2}}}$

$= \dfrac{(20.1 - 12.0) - (0)}{\sqrt{\dfrac{(5 - 1)(5.03)^2 + (8 - 1)(7.38)^2}{5 + 8 - 2}} \sqrt{\dfrac{1}{5} + \dfrac{1}{8}}} = \dfrac{8.1}{\sqrt{43.859}\sqrt{0.325}} \approx 2.145$

(d) Fail to reject H_0.

(e) There is not enough evidence to reject the claim.

18. (a) H_0: $\mu_1 \leq \mu_2$; H_a: $\mu_1 > \mu_2$ (claim)

(b) d.f. $= n_1 + n_2 - 2 = 11$

$t_0 = 1.363$; Reject H_0 if $t > 1.363$.

(c) $t = \dfrac{(\bar{x}_1 - \bar{x}_2) - (\mu_1 - \mu_2)}{\sqrt{\dfrac{(n_1 - 1)s_1^2 + (n_2 - 1)s_2^2}{n_1 + n_2 - 2}} \sqrt{\dfrac{1}{n_1} + \dfrac{1}{n_2}}}$

$= \dfrac{(1090 - 485) - (0)}{\sqrt{\dfrac{(5 - 1)(403)^2 + (8 - 1)(382)^2}{5 + 8 - 2}} \sqrt{\dfrac{1}{5} + \dfrac{1}{8}}}$

$= \dfrac{605}{\sqrt{151{,}918.546}\sqrt{0.325}} \approx 2.723$

(d) Reject H_0.

(e) There is enough evidence to support the claim.

20. (a) H_0: $\mu_1 = \mu_2$; H_a: $\mu_1 \neq \mu_2$ (claim)

(b) d.f. $= n_1 + n_2 - 2 = 33$

$t_0 = \pm 2.576$; Reject H_0 if $t < -2.576$ or $t > 2.576$.

(c) $t = \dfrac{(\bar{x}_1 - \bar{x}_2) - (\mu_1 - \mu_2)}{\sqrt{\dfrac{(n_1-1)s_1^2 + (n_2-1)s_2^2}{n_1+n_2-2}}\sqrt{\dfrac{1}{n_1}+\dfrac{1}{n_2}}}$

$= \dfrac{(35{,}800 - 35{,}100) - (0)}{\sqrt{\dfrac{(17-1)(7800)^2 + (18-1)(7375)^2}{17+18-2}}\sqrt{\dfrac{1}{17}+\dfrac{1}{18}}}$

$= \dfrac{700}{\sqrt{57{,}517{,}594.70}\sqrt{0.114}} \approx 0.273$

(d) Fail to reject H_0.

(e) There is not enough evidence to support the claim.

22. (a) $H_0: \mu_1 \le \mu_2$; $H_a: \mu_1 > \mu_2$ (claim)

(b) d.f. $= \min\{n_1 - 1, n_2 - 1\} = 13$

$t_0 = 1.350$; Reject H_0 if $t > 1.350$.

(c) $\bar{x}_1 = 402.765$, $s_1 = 11.344$, $n_1 = 17$

$\bar{x}_2 = 384.000$, $s_2 = 17.698$, $n_2 = 14$

$t = \dfrac{(\bar{x}_1 - \bar{x}_2) - (\mu_1 - \mu_2)}{\sqrt{\dfrac{s_1^2}{n_1}+\dfrac{s_2^2}{n_2}}} = \dfrac{(402.765 - 384.000) - (0)}{\sqrt{\dfrac{(11.344)^2}{17}+\dfrac{(17.698)^2}{14}}} = \dfrac{18.765}{\sqrt{29.943}} \approx 3.429$

(d) Reject H_0.

(e) There is enough evidence to support the claim and to recommend using the experimental method.

24. (a) $H_0: \mu_1 \ge \mu_2$; $H_a: \mu_1 < \mu_2$ (claim)

(b) d.f. $= n_1 + n_2 - 2 = 39$

$t_0 = -1.645$; Reject H_0 if $t < -1.645$.

(c) $\bar{x}_1 = 79.091$, $s_1 = 6.900$, $n_1 = 22$

$\bar{x}_2 = 83.000$, $s_2 = 7.645$, $n_2 = 19$

$t = \dfrac{(\bar{x}_1 - \bar{x}_2) - (\mu_1 - \mu_2)}{\sqrt{\dfrac{(n_1-1)s_1^2 + (n_2-1)s_2^2}{n_1+n_2-2}}\sqrt{\dfrac{1}{n_1}+\dfrac{1}{n_2}}}$

$= \dfrac{(79.091 - 83.000) - (0)}{\sqrt{\dfrac{(22-1)(6.900)^2 + (19-1)(7.645)^2}{22+19-2}}\sqrt{\dfrac{1}{22}+\dfrac{1}{19}}} = \dfrac{-3.909}{\sqrt{52.611}\sqrt{0.0981}} \approx -1.721$

(d) Reject H_0.

(e) There is enough evidence to support the claim.

26. $\hat{\sigma} = \sqrt{\dfrac{(n_1-1)s_1^2 + (n_2-1)s_2^2}{n_1+n_2-2}} = \sqrt{\dfrac{(15-1)(2.1)^2 + (12-1)(1.8)^2}{15+12-2}} \approx 1.974$

$(\bar{x}_1 - \bar{x}_2) \pm t_c \hat{\sigma}\sqrt{\dfrac{1}{n_1}+\dfrac{1}{n_2}} \to (29 - 27) \pm 2.060 \cdot 1.974 \sqrt{\dfrac{1}{15}+\dfrac{1}{12}}$

$\to 2 \pm 1.575 \to 0.425 < \mu_1 - \mu_2 < 3.575$

28. $(\bar{x}_1 - \bar{x}_2) \pm t_c \sqrt{\dfrac{s_1^2}{n_1} + \dfrac{s_2^2}{n_2}} \to (39 - 37) \pm 1.345 \sqrt{\dfrac{(2.42)^2}{20} + \dfrac{(1.65)^2}{12}}$

$\to 2 \pm 0.926 \to 1.074 < \mu_1 - \mu_2 < 2.926$

8.3 TESTING THE DIFFERENCE BETWEEN MEANS (DEPENDENT SAMPLES)

8.3 EXERCISE SOLUTIONS

2. (1) Each sample must be randomly selected from a normal population.

    (2) Each member of the first sample must be paired with a member of the second sample.

4. Dependent because the samples can be paired with respect to each individual.

6. Independent; the data represent weights from different individuals.

8. Dependent because the samples can be paired with respect to each car.

10. Dependent because the samples can be paired with respect to each individual.

12. $H_0: \mu_d = 0$ (claim); $H_a: \mu_d \neq 0$

    $\alpha = 0.01$ and d.f. $= n - 1 = 7$

    $t_0 = \pm 3.499$ (Two-tailed)

    $t = \dfrac{\bar{d} - \mu_d}{\dfrac{s_d}{\sqrt{n}}} = \dfrac{3.2 - 0}{\dfrac{0.25}{\sqrt{8}}} = \dfrac{3.2}{0.0884} \approx 36.204$

    Reject H_0. There is enough evidence to reject the claim.

14. $H_0: \mu_d \leq 0$; $H_a: \mu_d > 0$ (claim)

    $\alpha = 0.05$ and d.f. $= n - 1 = 27$

    $t_0 = 1.703$ (Right-tailed)

    $t = \dfrac{\bar{d} - \mu_d}{\dfrac{s_d}{\sqrt{n}}} = \dfrac{5.3 - 0}{\dfrac{0.41}{\sqrt{28}}} = \dfrac{5.3}{0.077} \approx 68.40$

    Reject H_0. There is enough evidence to support the claim.

16. $H_0: \mu_d = 0$; $H_a: \mu_d \neq 0$ (claim)

    $\alpha = 0.10$ and d.f. $= n - 1 = 19$

    $t_0 = \pm 1.729$ (Two-tailed)

    $t = \dfrac{\bar{d} - \mu_d}{\dfrac{s_d}{\sqrt{n}}} = \dfrac{4.2 - 0}{\dfrac{0.8}{\sqrt{20}}} = \dfrac{4.2}{0.1789} \approx 23.479$

    Reject H_0. There is enough evidence to support the claim.

18. (a) $H_0: \mu_d \geq 0$; $H_a: \mu_d < 0$ (claim)

    (b) $t_0 = -2.821$; Reject H_0 if $t < -2.821$.

(c) $\bar{d} \approx -59.9$ and $s_d \approx 26.831$

(d) $t = \dfrac{\bar{d} - \mu_d}{\frac{s_d}{\sqrt{n}}} = \dfrac{-59.9 - 0}{\frac{26.831}{\sqrt{10}}} = \dfrac{-59.9}{8.485} \approx -7.060$

(e) Reject H_0.

(f) There is enough evidence to support the claim that the SAT prep course improves verbal SAT scores.

20. (a) H_0: $\mu_d \geq 0$; H_a: $\mu_d < 0$ (claim)

(b) $t_0 = -1.397$; Reject H_0 if $t < -1.397$.

(c) $\bar{d} \approx -1.611$ and $s_d \approx 0.770$

(d) $t = \dfrac{\bar{d} - \mu_d}{\frac{s_d}{\sqrt{n}}} = \dfrac{-1.611 - 0}{\frac{0.770}{\sqrt{9}}} = \dfrac{-1.611}{0.257} \approx -6.277$

(e) Reject H_0.

(f) There is enough evidence to support the claim that the fuel additive improved gas mileage.

22. (a) H_0: $\mu_d \leq 0$; H_a: $\mu_d > 0$ (claim)

(b) $t_0 = 1.350$; Reject H_0 if $t > 1.350$.

(c) $\bar{d} = 1.357$ and $s_d = 3.97$

(d) $t = \dfrac{\bar{d} - \mu_d}{\frac{s_d}{\sqrt{n}}} = \dfrac{1.357 - 0}{\frac{3.97}{\sqrt{14}}} = \dfrac{1.357}{1.06} = 1.278$

(e) Fail to reject H_0.

(f) There is not enough evidence to support the claim that the program helped adults lose weight.

24. (a) H_0: $\mu_d \leq 0$; H_a: $\mu_d > 0$ (claim)

(b) $t_0 = 2.681$; Reject H_0 if $t > 2.681$.

(c) $\bar{d} \approx 1.469$ and $s_d \approx 0.494$

(d) $t = \dfrac{\bar{d} - \mu_d}{\frac{s_d}{\sqrt{n}}} = \dfrac{1.469 - 0}{\frac{0.494}{\sqrt{13}}} = \dfrac{1.469}{0.137} \approx 10.722$

(e) Reject H_0.

(f) There is enough evidence to support the claim that the new drug helps to reduce the number of daily headache hours.

26. (a) H_0: $\mu_d \leq 0$; H_a: $\mu_d > 0$ (claim)

(b) $t_0 = 1.860$; Reject H_0 if $t > 1.860$.

(c) $\bar{d} \approx 6.333$ and $s_d \approx 5.362$

(d) $t = \dfrac{\bar{d} - \mu_d}{\dfrac{s_d}{\sqrt{n}}} = \dfrac{6.333 - 0}{\dfrac{5.362}{\sqrt{9}}} = \dfrac{6.333}{1.787} \approx 3.544$

(e) Reject H_0.

(f) There is enough evidence to support the claim that the new drug reduces diastolic blood pressure.

28. (a) H_0: $\mu_d = 0$; H_a: $\mu_d \neq 0$ (claim)

(b) $t_0 = \pm 2.571$; Reject H_0 if $t < -2.571$ or $t > 2.571$.

(c) $\bar{d} \approx -7.167$ and $s_d \approx 10.25$

(d) $t = \dfrac{\bar{x} - \mu_d}{\dfrac{s_d}{\sqrt{n}}} = \dfrac{-7.167 - 0}{\dfrac{10.25}{\sqrt{6}}} = \dfrac{-7.167}{4.18} \approx -1.715$

(e) Fail to reject H_0.

(f) There is not enough evidence to support the claim that the plant performance ratings have changed.

30. $\bar{d} = -0.436$ and $s_d \approx 0.677$

$$\bar{d} - t_{\alpha/2}\dfrac{s_d}{\sqrt{n}} < \mu_d < \bar{d} + t_{\alpha/2}\dfrac{s_d}{\sqrt{n}}$$

$$-0.436 - 2.160\left(\dfrac{0.677}{\sqrt{14}}\right) < \mu_d < -0.436 + 2.160\left(\dfrac{0.677}{\sqrt{14}}\right)$$

$$-0.436 - 0.391 < \mu_d < -0.436 + 0.391$$

$$-0.827 < \mu_d < -0.045$$

8.4 TESTING THE DIFFERENCE BETWEEN PROPORTIONS

8.4 EXERCISE SOLUTIONS

2. (1) The sample must be independent.

(2) $n_1 p_1 \geq 5$, $n_1 q_1 \geq 5$, $n_2 p_2 \geq 5$, and $n_2 q_2 \geq 5$

4. H_0: $p_1 \geq p_2$; H_a: $p_1 < p_2$ (claim)

$z_0 = -1.645$ (Left-tailed test)

$\bar{p} = \dfrac{x_1 + x_2}{n_1 + n_2} = \dfrac{471 + 372}{785 + 465} = 0.674$

$\bar{q} = 0.326$

$z = \dfrac{(\hat{p}_1 - \hat{p}_2) - (p_1 - p_2)}{\sqrt{\bar{p}\bar{q}\left(\dfrac{1}{n_1} + \dfrac{1}{n_2}\right)}} = \dfrac{(0.600 - 0.800) - (0)}{\sqrt{0.674 \cdot 0.326\left(\dfrac{1}{785} + \dfrac{1}{465}\right)}} = \dfrac{-0.2}{\sqrt{0.000752428}} \approx -7.291$

Reject H_0. There is enough evidence to support the claim.

284 CHAPTER 8 | HYPOTHESIS TESTING WITH TWO SAMPLES

6. (a) H_0: $p_1 = p_2$ (claim); H_a: $p_1 \neq p_2$

   (b) $z_0 = \pm 1.96$ (Two-tailed test)

   $\bar{p} = \dfrac{x_1 + x_2}{n_1 + n_2} = \dfrac{29 + 25}{45 + 30} = 0.72$

   $\bar{q} = 0.28$

   (c) $z = \dfrac{(\hat{p}_1 - \hat{p}_2) - (p_1 - p_2)}{\sqrt{\bar{p}\,\bar{q}\left(\dfrac{1}{n_1} + \dfrac{1}{n_2}\right)}} = \dfrac{(0.644 - 0.833) - (0)}{\sqrt{(0.72)(0.28)\left(\dfrac{1}{45} + \dfrac{1}{30}\right)}} = \dfrac{-0.189}{\sqrt{0.0112}} \approx -1.786$

   (d) Fail to reject H_0.

   (e) There is not enough evidence to reject the claim.

8. (a) H_0: $p_1 = p_2$ (claim); H_a: $p_1 \neq p_2$

   (b) $z_0 = \pm 1.645$; Reject H_0 if $z < -1.645$ or $z > 1.645$.

   $\bar{p} = \dfrac{x_1 + x_2}{n_1 + n_2} = \dfrac{5 + 19}{77 + 84} = 0.149$

   $\bar{q} = 0.851$

   (c) $z = \dfrac{(\hat{p}_1 - \hat{p}_2) - (p_1 - p_2)}{\sqrt{\bar{p}\,\bar{q}\left(\dfrac{1}{n_1} + \dfrac{1}{n_2}\right)}} = \dfrac{(0.065 - 0.226) - (0)}{\sqrt{0.149 \cdot 0.851\left(\dfrac{1}{77} + \dfrac{1}{84}\right)}} = \dfrac{-0.161}{\sqrt{0.00316}} \approx -2.866$

   (d) Reject H_0.

   (e) There is sufficient evidence at the 10% level to reject the claim that the proportions of patients suffering new bouts of depression are the same for both groups.

10. (a) H_0: $p_1 \geq p_2$; H_a: $p_1 < p_2$ (claim)

    (b) $z_0 = -2.33$; Reject H_0 if $z < -2.33$.

    $\bar{p} = \dfrac{x_1 + x_2}{n_1 + n_2} = \dfrac{361 + 341}{1245 + 1065} = 0.304$

    $\bar{q} = 0.696$

    (c) $z = \dfrac{(\hat{p}_1 - \hat{p}_2) - (p_1 - p_2)}{\sqrt{\bar{p}\,\bar{q}\left(\dfrac{1}{n_1} + \dfrac{1}{n_2}\right)}} = \dfrac{(0.290 - 0.320)}{\sqrt{(0.304)(0.696)\left(\dfrac{1}{1245} + \dfrac{1}{1065}\right)}} = \dfrac{-0.03}{\sqrt{0.0003686}} = -1.563$

    (d) Fail to reject H_0.

    (e) There is not enough evidence to support the claim that the proportion of male senior citizens eating the daily recommended number of servings of fruit is less than female senior citizens.

12. (a) H_0: $p_1 \geq p_2$; H_a: $p_1 < p_2$ (claim)

    (b) $z_0 = -1.645$; Reject H_0 if $z < -1.645$.

    $\bar{p} = \dfrac{x_1 + x_2}{n_1 + n_2} = \dfrac{246 + 224}{1500 + 1000} = 0.188$

    $\bar{q} = 0.812$

(c) $z = \dfrac{(\hat{p}_1 - \hat{p}_2) - (p_1 - p_2)}{\sqrt{\bar{p}\bar{q}\left(\dfrac{1}{n_1} + \dfrac{1}{n_2}\right)}} = \dfrac{(0.164 - 0.224) - (0)}{\sqrt{0.188 \cdot 0.812\left(\dfrac{1}{1500} + \dfrac{1}{1000}\right)}} = \dfrac{-0.06}{\sqrt{0.0002544}} \approx -3.762$

(d) Reject H_0.

(e) There is sufficient evidence at the 5% level to conclude that the proportion of adults who are smokers is lower in California than in Oregon.

14. (a) H_0: $p_1 = p_2$ (claim); H_a: $p_1 \neq p_2$

(b) $z_0 = \pm 1.645$; Reject H_0 if $z < -1.645$ or $z > 1.645$.

$\bar{p} = \dfrac{x_1 + x_2}{n_1 + n_2} = \dfrac{5335 + 3444}{15{,}200 + 12{,}900} = 0.312$

$\bar{q} = 0.688$

(c) $z = \dfrac{(\hat{p}_1 - \hat{p}_2) - (p_1 - p_2)}{\sqrt{\bar{p}\bar{q}\left(\dfrac{1}{n_1} + \dfrac{1}{n_2}\right)}} = \dfrac{(0.351 - 0.267) - (0)}{\sqrt{0.312 \cdot 0.688\left(\dfrac{1}{15{,}200} + \dfrac{1}{12.900}\right)}}$

$= \dfrac{0.084}{\sqrt{0.000030762}} \approx 15.145$

(d) Reject H_0.

(e) There is sufficient evidence at the 10% level to reject the claim that the proportion of college students who said they had smoked in the last 30 days has not changed.

16. (a) H_0: $p_1 = p_2$ (claim); H_a: $p_1 \neq p_2$

(b) $z_0 = \pm 1.96$; Reject H_0 if $z < -1.96$ or $z > 1.96$.

(c) $\bar{p} = \dfrac{x_1 + x_2}{n_1 + n_2} = \dfrac{2100 + 1482}{8750 + 12{,}350} = 0.170$

$\bar{q} = 0.830$

$z = \dfrac{(\hat{p}_1 - \hat{p}_2) - (p_1 - p_2)}{\sqrt{\bar{p}\bar{q}\left(\dfrac{1}{n_1} + \dfrac{1}{n_2}\right)}} = \dfrac{(0.24 - 0.12) - 0}{\sqrt{(0.17)(0.83)\left(\dfrac{1}{8750} + \dfrac{1}{12{,}350}\right)}} = \dfrac{0.12}{\sqrt{0.00002755}}$

$= 22.862$

(d) Reject H_0.

(e) There is sufficient evidence at the 5% level to reject the claim that the proportion of adults who have high cholesterol are the same for both groups.

18. H_0: $p_1 \leq p_2$; H_a: $p_1 > p_2$ (claim)

$z_0 = 2.33$

$\bar{p} = \dfrac{x_1 + x_2}{n_1 + n_2} = \dfrac{161 + 85}{700 + 500} = 0.205$

$\bar{q} = 0.795$

$z = \dfrac{(\hat{p}_1 - \hat{p}_2) - (p_1 - p_2)}{\sqrt{\bar{p}\bar{q}\left(\dfrac{1}{n_1} + \dfrac{1}{n_2}\right)}} = \dfrac{(0.23 - 0.17) - (0)}{\sqrt{0.205 \cdot 0.795\left(\dfrac{1}{700} + \dfrac{1}{500}\right)}} = \dfrac{0.05}{\sqrt{0.0005588}} \approx 2.115$

Fail to reject H_0. There is not sufficient evidence at the 1% level to support the representatives belief.

286 CHAPTER 8 | HYPOTHESIS TESTING WITH TWO SAMPLES

20. H_0: $p_1 = p_2$ (claim); H_a: $p_1 \neq p_2$

    $z_0 = \pm 1.96$

    $\bar{p} = \dfrac{x_1 + x_2}{n_1 + n_2} = \dfrac{224 + 145}{700 + 500} = 0.308$

    $\bar{q} = 0.692$

    $z = \dfrac{(\hat{p}_1 - \hat{p}_2) - (p_1 - p_2)}{\sqrt{\bar{p}\bar{q}\left(\dfrac{1}{n_1} + \dfrac{1}{n_2}\right)}} = \dfrac{(0.32 - 0.29) - (0)}{\sqrt{0.308 \cdot 0.692\left(\dfrac{1}{700} + \dfrac{1}{500}\right)}} = \dfrac{0.03}{\sqrt{0.00073075}} \approx 1.110$

    Fail to reject H_0. There is insufficient evidence at the 5% level to reject the claim.

22. H_0: $p_1 \leq p_2$; H_a: $p_1 > p_2$ (claim)

    $z_0 = 1.645$

    $\bar{p} = \dfrac{x_1 + x_2}{n_1 + n_2} = \dfrac{240 + 195.5}{500 + 425} = 0.471$

    $\bar{q} = 0.529$

    $z = \dfrac{(\hat{p}_1 - \hat{p}_2) - (p_1 - p_2)}{\sqrt{\bar{p}\bar{q}\left(\dfrac{1}{n_1} + \dfrac{1}{n_2}\right)}} = \dfrac{(0.48 - 0.46) - (0)}{\sqrt{(0.471)(0.529)\left(\dfrac{1}{500} + \dfrac{1}{425}\right)}} = \dfrac{0.02}{\sqrt{0.00108}} = 0.607$

    Fail to reject H_0. There is not sufficient evidence at the 5% level to reject the claim.

24. H_0: $p_1 = p_2$ (claim); H_a: $p_1 \neq p_2$

    $z_0 = \pm 2.576$

    $\bar{p} = \dfrac{x_1 + x_2}{n_1 + n_2} = \dfrac{275 + 195.5}{500 + 425} = 0.509$

    $\bar{q} = 0.491$

    $z = \dfrac{(\hat{p}_1 - \hat{p}_2) - (p_1 - p_2)}{\sqrt{\bar{p}\bar{q}\left(\dfrac{1}{n_1} + \dfrac{1}{n_2}\right)}} = \dfrac{(0.55 - 0.46) - (0)}{\sqrt{(0.509)(0.491)\left(\dfrac{1}{500} + \dfrac{1}{425}\right)}} = \dfrac{0.09}{\sqrt{0.001087}} = 2.729$

    Reject H_0. There is sufficient evidence at the 1% level to reject the claim.

26. $(\hat{p}_1 - \hat{p}_2) \pm z_c \sqrt{\dfrac{\hat{p}_1\hat{q}_1}{n_1} + \dfrac{\hat{p}_2\hat{q}_2}{n_2}} \rightarrow (0.075 - 0.113) \pm 1.96 \sqrt{\dfrac{0.075 \cdot 0.925}{997{,}000} + \dfrac{0.113 \cdot 0.887}{1{,}085{,}000}}$

    $\rightarrow -0.038 \pm 1.96\sqrt{0.000000161963}$

    $\rightarrow -0.039 < p_1 - p_2 < -0.037$

CHAPTER 8 REVIEW EXERCISE SOLUTIONS

2. H_0: $\mu_1 = \mu_2$ (claim); H_a: $\mu_1 \neq \mu_2$

    $z_0 = \pm 2.575$

    $z = \dfrac{(\bar{x}_1 - \bar{x}_2) - (\mu_1 - \mu_2)}{\sqrt{\dfrac{s_1^2}{n_1} + \dfrac{s_2^2}{n_2}}} = \dfrac{(5595 - 5575) - (0)}{\sqrt{\dfrac{(52)^2}{156} + \dfrac{(68)^2}{216}}} = \dfrac{20}{\sqrt{38.741}} \approx 3.213$

    Reject H_0. There is enough evidence to reject the claim.

4. H_0: $\mu_1 = \mu_2$; H_a: $\mu_1 \neq \mu_2$ (claim)

$z_0 = \pm 1.96$

$z = \dfrac{(\bar{x}_1 - \bar{x}_2) - (\mu_1 - \mu_2)}{\sqrt{\dfrac{s_1^2}{n_1} + \dfrac{s_2^2}{n_2}}} = \dfrac{(82 - 90) - (0)}{\sqrt{\dfrac{(11)^2}{410} + \dfrac{(10)^2}{340}}} = \dfrac{-8}{\sqrt{0.58924}} \approx -10.422$

Reject H_0. There is enough evidence to support the claim.

6. (a) H_0: $\mu_1 = \mu_2$; H_a: $\mu_1 \neq \mu_2$ (claim)

(b) $z_0 = \pm 1.645$; Reject H_0 if $z < -1.645$ or $z > 1.645$.

(c) $z = \dfrac{(\bar{x}_1 - \bar{x}_2) - (\mu_1 - \mu_2)}{\sqrt{\dfrac{s_1^2}{n_1} + \dfrac{s_2^2}{n_2}}} = \dfrac{(360 - 390) - (0)}{\sqrt{\dfrac{(50)^2}{38} + \dfrac{(45)^2}{35}}} = \dfrac{-30}{\sqrt{123.647}} \approx -2.698$

(d) Reject H_0.

(e) There is sufficient evidence at the 10% level to support the claim that the caloric content of the two types of french fries is different.

8. (a) H_0: $\mu_1 = \mu_2$ (claim); H_a: $\mu_1 \neq \mu_2$

(b) d.f. $= \min\{n_1 - 1, n_2 - 1\} = 5$

$t_0 = \pm 2.015$

(c) $t = \dfrac{(\bar{x}_1 - \bar{x}_2) - (\mu_1 - \mu_2)}{\sqrt{\dfrac{s_1^2}{n_1} + \dfrac{s_2^2}{n_2}}} = \dfrac{(0.015 - 0.019) - (0)}{\sqrt{\dfrac{(0.011)^2}{8} + \dfrac{(0.004)^2}{6}}} = \dfrac{-.004}{\sqrt{0.0000178}} \approx -0.948$

(d) Fail to reject H_0.

(e) There is not enough evidence to reject the claim.

10. H_0: $\mu_1 \geq \mu_2$ (claim); H_a: $\mu_1 < \mu_2$

d.f. $= n_1 + n_2 - 2 = 47$

$t_0 = -2.326$

$t = \dfrac{(\bar{x}_1 - \bar{x}_2) - (\mu_1 - \mu_2)}{\sqrt{\dfrac{(n_1 - 1)s_1^2 + (n_2 - 1)s_2^2}{n_1 + n_2 - 2}} \sqrt{\dfrac{1}{n_1} + \dfrac{1}{n_2}}}$

$= \dfrac{(25.6 - 22.4) - (0)}{\sqrt{\dfrac{(15-1)(8.25)^2 + (34-1)(7.85)^2}{15 + 34 - 2}} \sqrt{\dfrac{1}{15} + \dfrac{1}{34}}} = \dfrac{3.2}{\sqrt{63.541}\,\sqrt{0.0961}} \approx 1.295$

Fail to reject H_0. There is not enough evidence to reject the claim.

12. H_0: $\mu_1 \geq \mu_2$ (claim); H_a: $\mu_1 < \mu_2$

d.f. $= \min\{n_1 - 1, n_2 - 1\} = 5$

$t_0 = -1.476$

$t = \dfrac{(\bar{x}_1 - \bar{x}_2) - (\mu_1 - \mu_2)}{\sqrt{\dfrac{s_1^2}{n_1} + \dfrac{s_2^2}{n_2}}} = \dfrac{(520 - 500) - (0)}{\sqrt{\dfrac{(25)^2}{7} + \dfrac{(55)^2}{6}}} = \dfrac{20}{\sqrt{593.452}} \approx 0.821$

Fail to reject H_0. There is not enough evidence to reject the claim.

288 CHAPTER 8 | HYPOTHESIS TESTING WITH TWO SAMPLES

14. (a) H_0: $\mu_1 = \mu_2$; H_a: $\mu_1 \neq \mu_2$ (claim)

(b) d.f. $= n_1 + n_2 - 2 = 20$

$t_0 = \pm 2.086$; Reject H_0 if $t < -2.086$ or $t > 2.086$.

(c) $t = \dfrac{(\bar{x}_1 - \bar{x}_2) - (\mu_1 - \mu_2)}{\sqrt{\dfrac{(n_1 - 1)s_1^2 + (n_2 - 1)s_2^2}{n_1 + n_2 - 2}} \sqrt{\dfrac{1}{n_1} + \dfrac{1}{n_2}}}$

$= \dfrac{(18{,}250 - 17{,}500) - (0)}{\sqrt{\dfrac{(12-1)(1200)^2 + (10-1)(950)^2}{12 + 10 - 2}} \sqrt{\dfrac{1}{12} + \dfrac{1}{10}}}$

$= \dfrac{750}{\sqrt{1{,}198{,}125}\, \sqrt{0.183}} \approx 1.600$

(d) Fail to reject H_0. There is not enough evidence to support the claim.

16. Dependent, because the same mice were used for both experiments.

18. H_0: $\mu_d \geq 0$; H_a: $\mu_d < 0$ (claim)

$\alpha = 0.01$ and d.f. $= n - 1 = 24$

$t_0 = -2.492$ (Left-tailed test)

$t = \dfrac{\bar{d} - \mu_d}{\dfrac{s_d}{\sqrt{n}}} = \dfrac{3.2 - 0}{\dfrac{1.38}{\sqrt{25}}} = \dfrac{3.2}{0.276} \approx 11.594$

Fail to reject H_0. There is not enough evidence to reject the claim.

20. H_0: $\mu_d = 0$; H_a: $\mu_d \neq 15$ (claim)

$\alpha = 0.05$ and d.f. $= n - 1 = 36$

$t_0 = \pm 1.96$ (Two-tailed test)

$t = \dfrac{\bar{d} - \mu_d}{\dfrac{s_d}{\sqrt{n}}} = \dfrac{17.5 - 15}{\dfrac{4.05}{\sqrt{37}}} = \dfrac{2.5}{0.6691} \approx 3.755$

Reject H_0. There is enough evidence to support the claim.

22. (a) H_0: $\mu_d \leq 0$; H_a: $\mu_d > 0$ (claim)

(b) $t_0 = 1.372$; Reject H_0 if $t > 1.372$.

(c) $\bar{d} \approx -0.636$ and $s_d \approx 5.870$

(d) $t = \dfrac{\bar{d} - \mu_d}{\dfrac{s_d}{\sqrt{n}}} = \dfrac{-0.636 - 0}{\dfrac{5.870}{\sqrt{11}}} = \dfrac{-0.636}{1.7699} \approx -0.359$

(e) Fail to reject H_0.

(f) There is not enough evidence to support the claim.

24. H_0: $p_1 \leq p_2$ (claim); H_a: $p_1 > p_2$

$z_0 = 2.33$ (Right-tailed test)

$\bar{p} = \dfrac{x_1 + x_2}{n_1 + n_2} = \dfrac{15 + 42}{100 + 200} = 0.190$

$\bar{q} = 0.810$

$z = \dfrac{(\hat{p}_1 - \hat{p}_2) - (p_1 - p_2)}{\sqrt{\bar{p}\bar{q}\left(\dfrac{1}{n_1} + \dfrac{1}{n_2}\right)}} = \dfrac{(0.150 - 0.210) - (0)}{\sqrt{0.190 \cdot 0.810\left(\dfrac{1}{100} + \dfrac{1}{200}\right)}} = \dfrac{-.06}{\sqrt{0.00231}} \approx -1.249$

Fail to reject H_0. There is not enough evidence to reject the claim.

26. H_0: $p_1 \geq p_2$; H_a: $p_1 < p_2$ (claim)

$z_0 = -1.645$ (Left-tailed test)

$\bar{p} = \dfrac{x_1 + x_2}{n_1 + n_2} = \dfrac{86 + 107}{900 + 1200} \approx 0.092$

$\bar{q} = 0.908$

$z = \dfrac{(\hat{p}_1 - \hat{p}_2) - (p_1 - p_2)}{\sqrt{\bar{p}\bar{q}\left(\dfrac{1}{n_1} + \dfrac{1}{n_2}\right)}} = \dfrac{(0.096 - 0.089) - (0)}{\sqrt{0.092 \cdot 0.908\left(\dfrac{1}{900} + \dfrac{1}{1200}\right)}} = \dfrac{0.007}{\sqrt{0.00016243}} \approx 0.549$

Fail to reject H_0. There is not enough evidence to support the claim.

28. (a) H_0: $p_1 \leq p_2$; H_a: $p_1 > p_2$ (claim)

(b) $z_0 = 1.645$; Reject H_0 if $z > 1.645$.

$\bar{p} = \dfrac{x_1 + x_2}{n_1 + n_2} = \dfrac{61{,}745 + 33{,}039}{79{,}160 + 50{,}829} \approx 0.729$

$\bar{q} = 0.271$

(c) $z = \dfrac{(\hat{p}_1 - \hat{p}_2) - (p_1 - p_2)}{\sqrt{\bar{p}\bar{q}\left(\dfrac{1}{n_1} + \dfrac{1}{n_2}\right)}} = \dfrac{(0.780 - 0.650) - (0)}{\sqrt{0.729 \cdot 0.271\left(\dfrac{1}{79{,}160} + \dfrac{1}{50{,}829}\right)}}$

$= \dfrac{0.13}{\sqrt{0.0000063824}} \approx 51.458$

(d) Reject H_0.

(e) There is enough evidence to support the claim.

CHAPTER 8 QUIZ SOLUTIONS

1. (a) H_0: $\mu_1 \leq \mu_2$; H_a: $\mu_1 > \mu_2$ (claim)

(b) n_1 and $n_2 > 30$ and the samples are independent $\rightarrow$ Right tailed z-test

(c) $z_0 = 1.645$; Reject H_0 if $z > 1.645$.

(d) $z = \dfrac{(\bar{x}_1 - \bar{x}_2) - (\mu_1 - \mu_2)}{\sqrt{\dfrac{s_1^2}{n_1} + \dfrac{s_2^2}{n_2}}} = \dfrac{(300.4 - 290.6) - (0)}{\sqrt{\dfrac{(1.6)^2}{49} + \dfrac{(1.5)^2}{50}}} = \dfrac{9.8}{\sqrt{0.0972}} \approx 31.426$

(e) Reject H_0.

(f) There is sufficient evidence at the 5% level to support the claim that the mean score on the science assessment for male high school students was higher than for the female high school students.

2. (a) $H_0: \mu_1 = \mu_2$ (claim); $H_a: \mu_1 \neq \mu_2$

(b) n_1 and $n_2 < 30$, the samples are independent, and the populations are normally distributed. → Two-tailed t-test (assume vars are equal)

(c) d.f. $= n_1 + n_2 - 2 = 26$
$t_0 = \pm 2.779$; Reject H_0 if $t < -2.779$ or $t > 2.779$.

(d) $t = \dfrac{(\bar{x}_1 - \bar{x}_2) - (\mu_1 - \mu_2)}{\sqrt{\dfrac{(n_1-1)s_1^2 + (n_2-1)s_2^2}{n_1 + n_2 - 2}} \sqrt{\dfrac{1}{n_1} + \dfrac{1}{n_2}}} = \dfrac{(230.9 - 227.9) - (0)}{\sqrt{\dfrac{(13-1)(1.3)^2 + (15-1)(1.1)^2}{13 + 15 - 2}} \sqrt{\dfrac{1}{13} + \dfrac{1}{15}}}$

$= \dfrac{3.0}{\sqrt{1.43154}\sqrt{0.14359}} \approx 6.617$

(e) Reject H_0.

(f) There is sufficient evidence at the 1% level to reject the teacher's suggestion that the mean scores on the science assessment test are the same for nine-year old boys and girls.

3. (a) $H_0: p_1 \leq p_2$; $H_a: p_1 > p_2$ (claim)

(b) Testing 2 proportions, $n_1\bar{p}$, $n_1\bar{q}$, $n_2\bar{p}$, and $n_2\bar{q} \geq 5$, and the samples are independent → Right-tailed z-test

(c) $z_0 = 1.282$; Reject H_0 if $z > 1.282$.

(d) $\bar{p} = \dfrac{x_1 + x_2}{n_1 + n_2} = \dfrac{64{,}800 + 8560}{1{,}296{,}000 + 856{,}000} = 0.034$

$\bar{q} = 0.966$

$z = \dfrac{(\hat{p}_1 - \hat{p}_2) - (p_1 - p_2)}{\sqrt{\bar{p}\bar{q}\left(\dfrac{1}{n_1} + \dfrac{1}{n_2}\right)}} = \dfrac{(0.05 - 0.01) - (0)}{\sqrt{0.034 \cdot 0.966 \left(\dfrac{1}{1{,}296{,}000} + \dfrac{1}{856{,}000}\right)}}$

$= \dfrac{0.04}{\sqrt{0.0000000637118}} \approx 158.471$

(e) Reject H_0.

(f) There is sufficient evidence at the 10% level to support the claim that the proportion of accidents involving alcohol is higher for drivers in the 21 to 24 age group than for drivers aged 65 and older.

4. (a) $H_0: \mu_d \geq 0$; $H_a: \mu_d < 0$ (claim)

(b) Dependent samples and both populations are normally distributed. → Dependent t-test

(c) $t_0 = -1.796$; Reject H_0 if $t < -1.796$.

(d) $t = \dfrac{\bar{d} - \mu_d}{\dfrac{s_d}{\sqrt{n}}} = \dfrac{0 - 68.5}{\dfrac{26.318}{\sqrt{12}}} = \dfrac{-68.5}{7.597} \approx 9.016$

(e) Reject H_0.

(f) There is sufficient evidence at the 5% level to conclude that the students' SAT scores improved on the second test.

Correlation and Regression

CHAPTER 9

9.1 CORRELATION

9.1 EXERCISE SOLUTIONS

2. (b) Correlation coefficients must be contained in the interval $[-1, 1]$.

4. r = sample correlation coefficient
ρ = population correlation coefficient

6. No linear correlation

8. Negative linear correlation

10. (d) You would not expect age and height to be correlated.

12. (a) You would expect the relationship between age and body temperature to be fairly constant.

14. Explanatory variable: Hours of safety classes.
Response variable: Number of accidents.

16. (a)

[Scatter plot: Vocabulary size (y-axis, 0 to 2500) vs Age in years (x-axis, 1 to 6)]

(b)

x	y	xy	x^2	y^2
1	3	3	1	9
2	400	880	4	193,600
3	1200	3600	9	1,440,000
4	1500	6000	16	2,250,000
5	2100	10,500	25	4,410,000
6	2600	15,600	36	6,760,000
3	1100	3300	9	1,210,000
5	2000	10,000	25	4,000,000
2	500	1000	4	250,000
4	1525	6100	16	2,325,625
6	2500	15,000	36	6,250,000
$\Sigma x = 41$	$\Sigma y = 15{,}468$	$\Sigma xy = 71{,}983$	$\Sigma x^2 = 181$	$\Sigma y^2 = 29{,}089{,}234$

$$r = \frac{n\Sigma xy - (\Sigma x)(\Sigma y)}{\sqrt{n\Sigma x^2 - (\Sigma x)^2}\sqrt{n\Sigma y^2 - (\Sigma y)^2}}$$

$$= \frac{11(71{,}983) - (41)(15{,}468)}{\sqrt{11(181) - (41)^2}\sqrt{11(29{,}089{,}234) - (15{,}468)^2}} = \frac{157{,}625}{\sqrt{310}\sqrt{80{,}722{,}550}}$$

$$\approx 0.996$$

(c) Strong positive linear correlation

18. (a)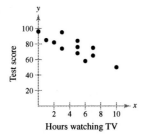

(b)

x	y	xy	x^2	y^2
0	96	0	0	9216
1	85	85	1	7225
2	82	164	4	6724
3	74	222	9	5476
3	95	285	9	9025
5	68	340	25	4624
5	76	380	25	5776
5	84	420	25	7056
6	58	348	36	3364
7	65	455	49	4225
7	75	525	49	5625
10	50	500	100	2500
$\Sigma x = 54$	$\Sigma y = 908$	$\Sigma xy = 3724$	$\Sigma x^2 = 332$	$\Sigma y^2 = 70{,}836$

$$r = \frac{n\Sigma xy - (\Sigma x)(\Sigma y)}{\sqrt{n\Sigma x^2 - (\Sigma x)^2}\sqrt{n\Sigma y^2 - (\Sigma y)^2}}$$

$$= \frac{12(3724) - (54)(908)}{\sqrt{12(332) - (54)^2}\sqrt{12(70836) - (908)^2}} = \frac{-4344}{\sqrt{1068}\sqrt{25{,}568}}$$

$$\approx -0.831$$

(c) Strong negative linear correlation

20. (a)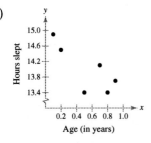

(b)

x	y	xy	x^2	y^2
0.1	14.9	1.49	0.01	222.01
0.2	14.5	2.90	0.04	210.25
0.5	13.4	6.70	0.25	179.56
0.7	14.1	9.87	0.49	198.81
0.8	13.4	10.72	0.64	179.56
0.9	13.7	12.33	0.81	187.69
$\Sigma x = 3.2$	$\Sigma y = 84$	$\Sigma xy = 44.01$	$\Sigma x^2 = 2.24$	$\Sigma y^2 = 1177.88$

$$r = \frac{n\Sigma xy - (\Sigma x)(\Sigma y)}{\sqrt{n\Sigma x^2 - (\Sigma x)^2}\sqrt{n\Sigma y^2 - (\Sigma y)^2}} = \frac{6(44.01) - (3.2)(84)}{\sqrt{6(2.24) - (3.2)^2}\sqrt{6(1177.88) - (84)^2}}$$

$$= \frac{-4.74}{\sqrt{3.20}\sqrt{11.28}} \approx -0.789$$

(c) Negative linear correlation

22. (a)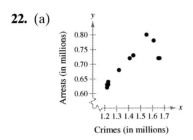

(b)

x	y	xy	x^2	y^2
1.66	0.72	1.195	2.756	0.518
1.65	0.72	1.188	2.723	0.518
1.61	0.78	1.256	2.592	0.608
1.55	0.80	1.240	2.403	0.640
1.44	0.73	1.051	2.074	0.533
1.41	0.72	1.015	1.988	0.518
1.32	0.68	0.898	1.742	0.462
1.23	0.64	0.787	1.513	0.410
1.22	0.63	0.769	1.488	0.397
1.23	0.63	0.775	1.513	0.397
1.22	0.62	0.756	1.488	0.384
$\Sigma x = 15.54$	$\Sigma y = 7.67$	$\Sigma xy = 10.930$	$\Sigma x^2 = 22.279$	$\Sigma y^2 = 5.387$

$$r = \frac{n\Sigma xy - (\Sigma x)(\Sigma y)}{\sqrt{n\Sigma x^2 - (\Sigma x)^2}\sqrt{n\Sigma y^2 - (\Sigma y)^2}}$$

$$= \frac{11(10.930) - (15.54)(7.67)}{\sqrt{11(22.279) - (15.54)^2}\sqrt{11(5.387)(7.67)^2}} = \frac{1.0382}{\sqrt{3.577}\sqrt{0.428}}$$

$$= 0.843$$

(c) Strong positive correlation

24. $r \approx 0.955$

$n = 8$ and $\alpha = 0.05$

$cv = 0.707$

$|r| = 0.955 > 0.707 \Rightarrow$ The correlation is significant.

or

$H_0: p = 0; H_a: p \neq 0$

$\alpha = 0.05$

d.f. $= n - 2 = 6$

$cu = \pm 2.447$; Reject H_0 if $t < -2.447$ or $t > 2.447$.

$$t = \frac{r}{\sqrt{\frac{1-r^2}{n-2}}} = \frac{0.955}{\sqrt{\frac{1-(0.955)^2}{8-2}}} = \frac{0.955}{\sqrt{0.01466}} = 7.887$$

Reject H_0. There is a significant linear correlation between vehicle weight and variability in braking distance.

26. $r \approx 0.831$

$n = 12$ and $\alpha = 0.05$

$cv = 0.576$

$|r| = 0.831 > 0.576 \Rightarrow$ The correlation is significant.

or

$H_0: \rho = 0; H_a: \rho \neq 0$

$\alpha = 0.05$

d.f. $= n - 2 = 10$

$cu = \pm 2.228$; Reject H_0 if $t < -2.228$ or $t > 2.228$.

$r \approx -0.831$

$t = \dfrac{r}{\sqrt{\dfrac{1-r^2}{n-2}}} = \dfrac{-0.831}{\sqrt{\dfrac{1-(-0.831)^2}{12-2}}} \approx -4.724$

Reject H_0. There is a significant linear correlation between the data.

28. $r \approx 0.842$

$n = 11$ and $\alpha = 0.05$

$cv = 0.602$

$|r| = 0.842 > 0.602 \Rightarrow$ The correlation is significant.

or

$H_0: \rho = 0; H_a: \rho \neq 0$

$\alpha = 0.05$

d.f. $= n - 2 = 9$

$cu = \pm 2.262$; Reject H_0 if $t < -2.262$ or $t > 2.262$.

$r \approx 0.842$

$t = \dfrac{r}{\sqrt{\dfrac{1-r^2}{n-2}}} = \dfrac{0.832}{\sqrt{\dfrac{1-(-0.842)^2}{11-2}}} = 4.682$

Reject H_0. There is a significant correlation between the data.

30. The correlation coefficient remains unchanged when the x-values and y-values are switched.

9.2 LINEAR REGRESSION

9.2 EXERCISE SOLUTIONS

2. a **4.** b **6.** e

8. f **10.** b **12.** d

14.

x	y	xy	x^2
3	1100	3,300	9
4	1300	5,200	16
4	1500	6,000	16
5	2100	10,500	25
6	2600	15,600	36
2	460	920	4
3	1200	3,600	9
$\Sigma x = 27$	$\Sigma y = 10{,}260$	$\Sigma xy = 45{,}120$	$\Sigma x^2 = 115$

$$m = \frac{n\Sigma xy - (\Sigma x)(\Sigma y)}{n\Sigma x^2 - (\Sigma x)^2} = \frac{7(45{,}120) - (27)(10{,}260)}{7(115) - (27)^2} = \frac{38{,}820}{76} = 510.789$$

$$b = \bar{y} - m\bar{x} = \left(\frac{10{,}260}{7}\right) - (510.789)\left(\frac{27}{7}\right) = -504.474$$

$\hat{y} = 510.789x - 504.474$

(a) $\hat{y} = 510.789(2) - 504.474 \approx 517.105$ (b) $\hat{y} = 510.789(3) - 504.474 \approx 1027.895$

(c) $\hat{y} = 510.789(6) - 504.474 \approx 2560.263$

(d) It is not meaningful to predict the value of y for $x = 12$ because $x = 12$ is outside the range of the original data.

16.

x	y	xy	x^2
0	96	0	0
1	85	85	1
2	82	164	4
3	74	222	9
3	95	285	9
5	68	340	25
5	76	380	25
5	84	420	25
6	58	348	36
7	65	455	49
7	75	525	49
10	50	500	100
$\Sigma x = 54$	$\Sigma y = 908$	$\Sigma xy = 3724$	$\Sigma x^2 = 332$

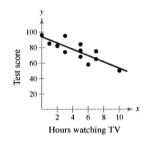

$$m = \frac{n\Sigma xy - (\Sigma x)(\Sigma y)}{n\Sigma x^2 - (\Sigma x)^2} = \frac{12(3724) - (54)(908)}{12(332) - (54)^2} = \frac{-4344}{1068} \approx -4.067$$

$$b = \bar{y} - m\bar{x} = \left(\frac{908}{12}\right) - (-4.067)\left(\frac{54}{12}\right) \approx 93.970$$

$\hat{y} = -4.067x + 93.970$

(a) $\hat{y} = -4.067(4) + 93.970 \approx 77.7$ (b) $\hat{y} = -4.067(8) + 93.970 \approx 61.4$

(c) $\hat{y} = -4.067(9) + 93.970 \approx 57.4$

(d) It is not meaningful to predict the values of y for $x = 15$ because $x = 15$ is outside the range of the original data.

18.

x	y	xy	x^2
70	8.3	581.0	4900
72	10.5	756.0	5184
75	11	825.0	5625
76	11.4	866.4	5776
85	12.9	1096.5	7225
78	14	1092.0	6084
77	16.3	1255.1	5929
80	18	1440.0	6400
$\Sigma x = 613$	$\Sigma y = 102.4$	$\Sigma xy = 7912.0$	$\Sigma x^2 = 47{,}123$

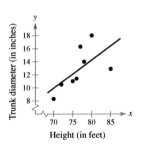

$$m = \frac{n\Sigma xy - (\Sigma x)(\Sigma y)}{n\Sigma x^2 - (\Sigma x)^2} = \frac{8(7912) - (613)(102.4)}{8(47{,}123) - (613)^2} = \frac{524.8}{1215} = 0.432$$

$$b = \bar{y} - m\bar{x} = \left(\frac{102.4}{8}\right) - 0.432\left(\frac{613}{8}\right) = -20.297$$

$\hat{y} = 0.432x - 20.297$

(a) $\hat{y} = 0.432(74) - 20.297 = 11.666$ (b) $\hat{y} = 0.432(81) - 20.297 = 14.690$

(c) It is not meaningful to predict the value of y for $x = 87$ because $x = 87$ is outside the range of the original data.

(d) $\hat{y} = 0.432(79) - 20.297 = 13.826$

20.

x	y	xy	x^2
0.1	14.9	1.49	0.01
0.2	14.5	2.90	0.04
0.4	13.9	xxx	xxx
0.7	14.1	9.87	0.49
0.6	13.9	xxx	xxx
0.9	13.7	12.33	0.81
2.0	14.3	xxx	xxx
0.6	13.9	8.34	0.36
0.5	14.0	7.00	0.25
0.1	14.1	xxx	xxx
$\Sigma x = 4.5$	$\Sigma y = 141.3$	$\Sigma xy = 62.92$	$\Sigma x^2 = 2.61$

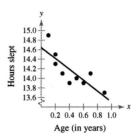

$m = \dfrac{n\Sigma xy - (\Sigma x)(\Sigma y)}{n\Sigma x^2 - (\Sigma x)^2} = \dfrac{10(62.92) - (4.5)(141.3)}{10(2.61) - (4.5)^2} = \dfrac{-6.65}{5.85} \approx -1.137$

$b = \bar{y} - m\bar{x} = \left(\dfrac{141.3}{10}\right) - (-1.137)\left(\dfrac{4.5}{10}\right) \approx 14.642$

$\hat{y} = -1.137 + 14.642$

(a) $\hat{y} = -1.137(0.3) + 14.642 \approx 14.301$

(b) It is not meaningful to predict the value of y for $x = 3.9$ because $x = 3.9$ is outside the range of the original data.

(c) $\hat{y} = -1.137(0.6) + 14.642 \approx 13.959$ (d) $\hat{y} = -1.137(0.8) + 14.642 \approx 13.732$

22.

x	y	xy	x^2
5720	3.78	21,622	32,718,400
4050	2.43	9841.5	16,402,500
6130	4.63	28,381.9	37,576,900
5000	2.88	14,400	25,000,000
5010	3.25	16,282.5	25,100,100
4270	2.76	11,785.2	18,232,900
5500	3.42	18,810	30,250,000
5550	3.51	19,480.5	30,802,500
$\Sigma x = 41,230$	$\Sigma y = 26.66$	$\Sigma xy = 140,603.2$	$\Sigma x^2 = 216,083,300$

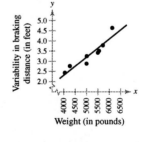

$m = \dfrac{n\Sigma xy - (\Sigma x)(\Sigma y)}{n\Sigma x^2 - (\Sigma x)^2} = \dfrac{8(140,603.2) - (41,230)(26.66)}{8(216,083,300) - (41,230)^2} = \dfrac{25,632.2}{28,753,500} \approx 0.000891502$

$b = \bar{y} - m\bar{x} = \left(\dfrac{26.66}{8}\right) - (-0.000891502)\left(\dfrac{41,230}{8}\right) = -1.26208$

$\hat{y} = 0.000892x - 1.262$

(a) $\hat{y} = 0.000892(4800) - 1.262 = 3.020$ (b) $\hat{y} = 0.000892(5850) - 1.262 = 3.956$

(c) $\hat{y} = 0.000892(4075) - 1.262 = 2.373$

(d) It is not meaningful to predict the value of y for $x = 3000$ because $x = 3000$ is outside the range of the original data.

24. Prediction values are meaningful only for x-values in (or close to) the range of the data.

26. (a)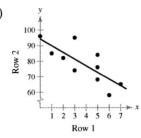
$\hat{y} = -4.297x + 94.200$

(b)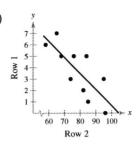
$\hat{y} = -0.1413x + 14.763$

(c) The slope of the line keeps the same sign, but the values of m and b change.

28. $\hat{y} = -0.294 + 70.652$

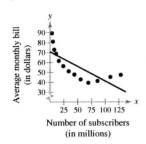

30. $r \approx -0.756$

$H_0: \rho = 0;\ H_a: \rho \neq 0$

d.f. $= n - 2 = 13 - 2 = 11$

Critical values: $t_0 = \pm 3.106$

$$t = \frac{r}{\sqrt{\frac{1-r^2}{n-2}}} = \frac{-0.756}{\sqrt{\frac{1-(-0.756)^2}{13-2}}} = \frac{-0.756}{\sqrt{0.03895}} \approx -3.831$$

$t \approx -3.831 < -3.106$

or

$|r| \approx 0.756 > 0.684$

Reject H_0. There is enough evidence to conclude that a significant linear correlation exists.

32. $\hat{y} = ab^x = 69.481(0.995)^x$

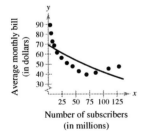

34. $\hat{y} = a + b \ln x = 25.035 + 19.599 \ln x$

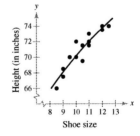

36. The logarithmic equation is a better fitting model to the data.

298 CHAPTER 9 | CORRELATION AND REGRESSION

9.3 MEASURES OF REGRESSION AND PREDICTION INTERVALS

9.3 EXERCISE SOLUTIONS

2. Explained variation = $\Sigma(\hat{y}_i - \bar{y})^2$; the sum of the squares of the differences between the predicted y-values and the mean of the y-values of the ordered pairs.

4. Coefficient of Determination: $r^2 = \dfrac{\Sigma(\hat{y}_i - \bar{y})^2}{\Sigma(y_i - \bar{y})^2}$

   r^2 is the ratio of the explained variation to the total variation and is the percent of variation of y that is explained by the relationship between x and y.

6. $r^2 = (-0.375)^2 \approx 0.141$

   14.1% of the variation is explained. 85.9% of the variation is unexplained.

8. $r^2 = (0.964)^2 \approx 0.929$

   92.9% of the variation is explained. 7.1% of the variation is unexplained.

10. (a) $r^2 = \dfrac{\Sigma(\hat{y}_i - \bar{y})^2}{\Sigma(y_i - \bar{y})^2} \approx 0.995$

    99.5% of the variation in cigarettes exported from the United States can be explained by the variation in the number of cigarettes consumed in the United States and 0.5% of the variation is unexplained.

    (b) $s_e = \sqrt{\dfrac{\Sigma(y_i - \hat{y}_i)^2}{n-2}} = \sqrt{\dfrac{46.996}{5}} \approx 3.066$

    The standard deviation of the number of cigarettes exported for a specific number of cigarettes consumed is about 3,066,000,000.

12. (a) $r^2 = \dfrac{\Sigma(\hat{y}_i - \bar{y})^2}{\Sigma(y_i - \bar{y})^2} \approx 0.578$

    57.8% of the variation in the median number of leisure hours per week can be explained by the variation in the median number of work hours per week and 42.2% of the variation is unexplained.

    (b) $s_e = \sqrt{\dfrac{\Sigma(y_i - \hat{y}_i)^2}{n-2}} = \sqrt{\dfrac{33.218}{8}} \approx 2.013$

    The standard deviation of the median number of leisure hours/week for a specific median number of hours/week is about 2.013 hours.

14. (a) $r^2 = \dfrac{\Sigma(\hat{y}_i - \bar{y})^2}{\Sigma(y_i - \bar{y})^2} \approx 0.888$

    88.8% of the variation in the turnout for federal elections can be explained by the variation in the voting age population and 11.2% of the variation is unexplained.

    (b) $s_e = \sqrt{\dfrac{\Sigma(y_i - \hat{y}_i)^2}{n-2}} = \sqrt{\dfrac{83.585}{6}} \approx 3.732$

    The standard deviation of the turnout in a federal election for a specified voting age population is about 3,732,000.

16. (a) $r^2 = \dfrac{\Sigma(\hat{y}_i - \bar{y})^2}{\Sigma(y_i - \bar{y})^2} \approx 0.877$

    87.7% of the variation in bond and income funds can be explained by the variation in equity funds and 12.3% of the variation is unexplained.

(b) $s_e = \sqrt{\dfrac{\Sigma(y_i - \hat{y}_i)^2}{n-2}} = \sqrt{\dfrac{91{,}350.849}{7}} \approx 114.237$

The standard deviation of the bond and income funds for a specified equity fund is about $114,237,000,000.

18. $n = 7$, d.f. $= 5$, $t_c = 2.571$, $s_e = 3.066$

$\hat{y} = 1.528x + 500.435 = 1.528(490) - 500.435 \approx 248.285$

$E = t_c s_e \sqrt{1 + \dfrac{1}{n} + \dfrac{n(x - \bar{x})^2}{n(\Sigma x^2) - (\Sigma x)^2}}$

$= (2.571)(3.066)\sqrt{1 + \dfrac{1}{7} + \dfrac{7[490 - (3060/7)]^2}{7(1{,}341{,}700) - (3060)^2}}$

$= (2.571)(3.066)\sqrt{1.83392} \approx 10.675$

$\hat{y} \pm E \rightarrow (237.610, 258.960) \rightarrow (237{,}610{,}000{,}000, 258{,}960{,}000{,}000)$

You can be 95% confident that the number of cigarettes exported will be between 237,610,000,000 and 258,960,000,000 when the number of cigarettes consumed is 490 billion.

20. $n = 10$, d.f. $= 8$, $t_c = 1.860$, $s_e \approx 2.013$

$\hat{y} = -0.646x + 50.734 = -0.646(45.1) + 50.734 \approx 21.599$

$E = t_c s_e \sqrt{1 + \dfrac{1}{n} + \dfrac{n(x - \bar{x})^2}{n(\Sigma x^2) - (\Sigma x)^2}}$

$= (1.860)(2.013)\sqrt{1 + \dfrac{1}{10} + \dfrac{10[45.1 - (475.5/10)]^2}{10(22{,}716.29) - (475.5)^2}}$

$= (1.860)(2.013)\sqrt{1.15649} \approx 4.026$

$\hat{y} \pm E \rightarrow (17.573, 25.625)$

You can be 90% confident that the median number of leisure hours/week will be between 17.573 and 25.625 when the median number of work hours/week is 45.1.

22. $n = 8$, d.f. $= 6$, $t_c = 3.707$, $s_e \approx 3.747$

$\hat{y} = 0.373x + 26.373 = 0.373(190) + 26.373 \approx 97.243$

$E = t_c s_e \sqrt{1 + \dfrac{1}{n} + \dfrac{n(x - \bar{x})^2}{n(\Sigma x^2) - (\Sigma x)^2}}$

$= (3.707)(3.732)\sqrt{1 + \dfrac{1}{8} + \dfrac{8[190 - (1321.3/8)]^2}{8(222973.8) - (1321.3)^2}}$

$= (3.707)(3.732)\sqrt{1.25502} \approx 15.498$

$\hat{y} \pm E \rightarrow (81.745, 112.741)$

You can be 99% confident that the voter turnout in federal elections will be between 81.745 million and 112.741 million when the voting age population is 190 million.

300 CHAPTER 9 | CORRELATION AND REGRESSION

24. $n = 9$, d.f. $= 7$, $t_c = 1.895$, $s_e \approx 114.237$

$\hat{y} = 0.689x + 68.861 = 0.689(900) + 68.861 \approx 688.961$

$E = t_c s_e \sqrt{1 + \dfrac{1}{n} + \dfrac{n(x - \bar{x})^2}{n(\Sigma x^2) - (\Sigma x)^2}}$

$= (1.895)(114.237)\sqrt{1 + \dfrac{1}{9} + \dfrac{9[900 - (3130.3/9)]^2}{9(2458011) - (3130.3)^2}}$

$= (1.895)(114.237)\sqrt{1.33380} \approx 250.012$

$\hat{y} \pm E \rightarrow (438.949, 938.973)$

You can be 90% confident that the total assets in bond and income funds will be between $438.949 billion and $938.973 billion when the total assets in equity funds is $900 billion.

26. $\hat{y} = 0.693x + 2.280$

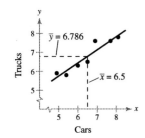

28. (a) Explained variation $= \Sigma(\hat{y}_i - \bar{y})^2 \approx 3.930$

(b) Unexplained variation $= \Sigma(y_i - \hat{y}_i)^2 \approx 0.498$

(c) Total variation $= \Sigma(y_i - \bar{y})^2 \approx 4.429$

30. $s_e \approx 0.316$; 31.6% of the variation in the median age of trucks can be explained by the variation in the median age of cars and 68.4% is unexplained.

32. The slope and correlation coefficient will have the same sign because the denominators of both formulas are always positive while the numerators are always equal.

9.4 MULTIPLE REGRESSION

9.4 EXERCISE SOLUTIONS

2. $\hat{y} = 859 + 5.76x_1 + 3.82x_2$

(a) $\hat{y} = 859 + 5.76(2532) + 3.82(2255) = 24{,}057.42$

(b) $\hat{y} = 859 + 5.76(3581) + 3.82(3021) = 33{,}025.78$

(c) $\hat{y} = 859 + 5.76(3213) + 3.82(3065) = 31{,}074.18$

(d) $\hat{y} = 859 + 5.76(2758) + 3.82(2714) = 27{,}112.56$

4. $\hat{y} = 0.688 + 0.046x_1 - 0.015x_2$

(a) $\hat{y} = 0.668 + 0.046(11.4) - 0.015(8.6) = 1.063$

(b) $\hat{y} = 0.668 + 0.046(8.1) - 0.015(6.2) = 0.948$

(c) $\hat{y} = 0.668 + 0.046(10.7) - 0.015(8.5) = 1.033$

(d) $\hat{y} = 0.668 + 0.046(7.3) - 0.015(5.1) = 0.927$

6. $\hat{y} = -1.661 + 0.156x_1 + 0.031x_2$

(a) $s = 0.5485$ (b) $r^2 = 0.997$

(c) The standard deviation of the predicted equity given a specific net sales and total assets is $0.5485 billion. The multiple regression model explains 99.7% of the variation in y.

8. $n = 6, k = 2, r = 0.998$

$$r^2_{adj} = 1 - \left[\frac{(1-r^2)(n-1)}{n-k-1}\right] = 0.996$$

CHAPTER 9 REVIEW EXERCISE SOLUTIONS

2.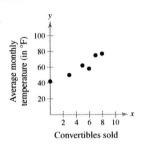

$r \approx 0.944$

Positive linear correlation; The average monthly temperatures increase with convertibles sold.

4.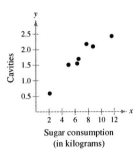

$r \approx 0.953$

Positive linear correlation; The average number of cavities increases as the annual per capita sugar consumption increases.

6. H_0: $\rho = 0$; H_a: $\rho \neq 0$

$\alpha = 0.05$, d.f. $= n - 2 = 20$

$t_0 = \pm 2.086$

$$t = \frac{r}{\sqrt{\frac{1-r^2}{n-2}}} = \frac{-0.55}{\sqrt{\frac{1-(-0.55)^2}{22-2}}} = \frac{-0.55}{\sqrt{0.03488}} \approx -2.945$$

Reject H_0. There is enough evidence to conclude that a significant linear correlation exists.

8. H_0: $\rho = 0$; H_a: $\rho \neq 0$

$\alpha = 0.05$, d.f. $= n - 2 = 4$

$t_0 = \pm 2.776$

$$t = \frac{r}{\sqrt{\frac{1-r^2}{n-2}}} = \frac{0.944}{\sqrt{\frac{1-(0.944)^2}{4}}} = 5.722$$

Reject H_0. There is enough evidence to conclude that a significant linear correlation exists between average monthly temperatures and convertible sales.

10. H_0: $\rho = 0$; H_a: $\rho \neq 0$

$\alpha = 0.01$, d.f. $= n - 2 = 5$

$t_0 = \pm 4.032$

$$t = \frac{r}{\sqrt{\frac{1-r^2}{n-2}}} = \frac{0.953}{\sqrt{\frac{1-(0.953)^2}{7-2}}} = \frac{0.953}{\sqrt{0.01836}} \approx 7.034$$

Reject H_0. There is enough evidence at the 1% level to conclude that there is a linear correlation between sugar consumption and tooth decay.

12. $\hat{y} = 2.883x + 1.163x$

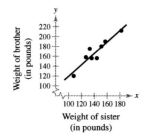

$r \approx 0.932$ (Strong positive linear correlation)

14. $\hat{y} = -0.090x + 44.675$

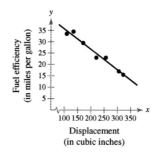

$r \approx -0.984$ (Strong negative linear correlation)

16. (a) $\hat{y} = 2.883 + 1.163(140) = 165.703$

(b) Not meaningful because $x = 100$ lbs. is outside the range of the original data.

(c) $\hat{y} = 2.883 + 1.163(170) = 200.593$

(d) Not meaningful because $x = 200$ lbs. is outside the range of the original data.

18. (a) Not meaningful because $x = 86$ in.^{3} is outside the range of the original data.

(b) $\hat{y} = -0.090(198) + 44.675 = 26.855$

(c) $\hat{y} = -0.090(289) + 44.675 = 18.665$

(d) Not meaningful because $x = 407$ in.^{3} is outside the range of the original data.

20. $r^2 = (-0.962)^2 = 0.925$

92.5% of the variation in y is explained by the model.
7.5% of the variation in y is unexplained by the model.

22. $r^2 = (0.740)^2 = 0.548$

54.8% of the variation in y is explained by the model.
45.2% of the variation in y is unexplained by the model.

24. (a) $r^2 \approx 0.397$

39.7% of the variation in y is explained by the model.
60.3% of the variation in y is unexplained by the model.

(b) $s_e = 101.0$

The standard error of the price for a specific area is $101.

26. $\hat{y} = 2.883 + 1.163(125) = 148.258$

$$E = t_c s_e \sqrt{1 + \frac{1}{n} + \frac{n(x - \bar{x})^2}{n\Sigma x^2 - (\Sigma x)^2}}$$

$$= (1.943)(10.739)\sqrt{1 + \frac{1}{8} + \frac{8[125 - (1138/8)]^2}{8(165,288) - (1138)^2}}$$

$$= (1.943)(10.739)\sqrt{1.2123} = 22.974$$

$$\hat{y} - E < y < \hat{y} + E$$

$$148.258 - 22.974 < y < 148.258 + 22.974$$

$$125.284 < y < 171.232$$

28. $\hat{y} = -0.090(265) + 44.675 = 20.825$

$$E = t_c s_e \sqrt{1 + \frac{1}{n} + \frac{n(x - \bar{x})^2}{n\Sigma x^2 - (\Sigma x)^2}}$$

$$= (2.571)(1.476) \sqrt{1 + \frac{1}{7} + \frac{7(265 - 216.6)^2}{7(369,382) - 1516^2}}$$

$$= (2.571)(1.476) \sqrt{1.20133} \approx 4.159$$

$$\hat{y} - E < y < \hat{y} + E$$
$$20.825 - 4.159 < y < 20.825 + 4.159$$
$$16.668 < y < 24.982$$

30. $\hat{y} = -209.506 + 1.305x = -209.506 + 1.305(400) = 312.494$

$$E = t_c s_e \sqrt{1 + \frac{1}{n} + \frac{n(x - \bar{x})^2}{n\Sigma x^2 - (\Sigma x)^2}}$$

$$= 2.977(101) \sqrt{1 + \frac{1}{16} + \frac{16(400 - 375.75)^2}{16(2,314,270) - 6012^2}}$$

$$= (2.977)(101) \sqrt{1.07314} \approx 311.479$$

$$\hat{y} - E < y < \hat{y} + E$$
$$312.494 - 311.479 < y < 312.494 + 311.479$$
$$1.015 < y < 623.973$$

32. $s_e = 0.9139$
$r^2 = 0.942$

94.2% of the variation in y is explained by the model.

34. (a) 16.116   (b) 8.450   (c) 15.734   (d) 9.847

CHAPTER 9 QUIZ SOLUTIONS

1.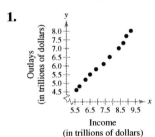

The data appear to have a positive correlation. The outlays increase as the incomes increase.

2. $r \approx 0.996 \rightarrow$ Strong positive linear correlation

3. $H_0: \rho = 0; H_a: \rho \neq 0$

$\alpha = 0.05$, d.f. $= n - 2 = 9$

$t_0 = \pm 2.262$

$$t = \frac{r}{\sqrt{\frac{1 - r^2}{n - 2}}} = \frac{0.996}{\sqrt{\frac{1 - (0.996)^2}{11 - 2}}} = \frac{0.996}{\sqrt{0.00088711}} \approx 33.44$$

Reject H_0. There is enough evidence to conclude that a significant correlation exists.

4. $\Sigma y^2 = 441.140$
   $\Sigma y = 68.600$
   $\Sigma xy = 522.470$
   $b = -0.359$
   $m = 0.891$
   $\hat{y} = 0.891x - 0.359$

5. $\hat{y} = 0.891(5.3) - 0.359 = 4.363$

6. $r^2 = 0.992$

   99.2% of the variation in y is explained by the regression model.
   0.8% of the variation in y is unexplained by the regression model.

7. $s_e = 0.110$

   The standard deviation of personal outlays for a specified personal income is $0.110 trillion.

8. $\hat{y} = 0.891(6.4) - 0.359$
   $= 5.343$

$$E = t_c s_e \sqrt{1 + \frac{1}{n} + \frac{n(x - \bar{x})^2}{n\Sigma x^2 - (\Sigma x)^2}}$$

$$= 2.262(0.110)\sqrt{1 + \frac{1}{11} + \frac{11[6.4 - (81.4/11)]^2}{11(619) - 81.4^2}} \approx (2.262)(0.110)\sqrt{1.15101}$$

$$\approx 0.267$$

$\hat{y} - E < y < \hat{y} + E$
$5.343 - 0.267 < y < 5.343 + 0.267$
$5.076 < y < 5.610 \rightarrow$ ($5.076 trillion, $5.610 trillion)

You can be 95% confident that the personal outlays will be between $5.076 trillion and $5.610 trillion when personal income is $6.4 trillion.

9. (a) 1311.150

   (b) 961.110

   (c) 1120.900

   (d) 1386.740; x_2 has the greatest influence on y.

Chi-Square Tests and the F-Distribution

CHAPTER 10

10.1 GOODNESS OF FIT

2. The observed frequencies must be obtained using a random sample and each expected frequency must be greater than or equal to 5.

4. $E_i = np_i = (500)(0.9) = 450$

6. $E_i = np_i = (415)(0.08) = 33.2$

8. (a) Claimed distribution:

Response	Distribution
Home	70%
Work	17%
Commuting	8%
Other	5%

H_0: Distribution of responses is as shown in table above.

H_a: Distribution of responses differs from the claimed distribution.

(b) $\chi_0^2 = 7.815$; Reject H_0 if $\chi^2 > 7.815$.

(c)

Response	Distribution	Observed	Expected	$\frac{(O-E)^2}{E}$
Home	70%	389	406.70	0.770
Work	17%	110	98.77	1.277
Commuting	8%	55	46.48	1.562
Other	5%	27	29.05	0.145
				3.754

$\chi^2 \approx 3.754$

(d) Fail to reject H_0. There is not enough evidence at the 5% level to conclude that the distribution of the responses differs from the claimed distribution.

10. (a) Claimed distribution:

Response	Distribution
Limited advancement	41%
Lack of recognition	25%
Low salary	15%
Unhappy with mgmt.	10%
Bored/don't know	9%

H_0: Distribution of responses is as shown in table above.

H_a: Distribution of responses differs from the claimed distribution.

(b) $\chi_0^2 = 13.277$; Reject H_0 if $\chi^2 > 13.277$.

(c)

Response	Distribution	Observed	Expected	$\frac{(O-E)^2}{E}$
Limited advancement	41%	78	82.00	0.195
Lack of recognition	25%	52	50.00	0.080
Low salary	15%	30	30.00	0.000
Unhappy with mgmt.	10%	25	20.00	1.250
Bored/don't know	9%	15	18.00	0.500
				2.025

$\chi^2 \approx 2.025$

(d) Fail to reject H_0. There is not enough evidence at the 5% level to conclude that the distribution of the responses differs from the claimed distribution.

12. (a) Claimed distribution:

Month	Distribution
January	8.333%
February	8.333%
March	8.333%
April	8.333%
May	8.333%
June	8.333%
July	8.333%
August	8.333%
September	8.333%
October	8.333%
November	8.333%
December	8.333%

H_0: The distribution of fatal bicycle accidents throughout the year is uniform as shown in table above.

H_a: The distribution of fatal bicycle accidents throughout the year is not uniform.

(b) $\chi_0^2 = 17.275$; Reject H_0 if $\chi^2 > 17.275$.

(c)

Month	Distribution	Observed	Expected	$\frac{(O-E)^2}{E}$
January	8.333%	50	83.00	13.120
February	8.333%	48	83.00	14.759
March	8.333%	81	83.00	0.048
April	8.333%	72	83.00	1.458
May	8.333%	90	83.00	0.590
June	8.333%	101	83.00	3.904
July	8.333%	129	83.00	25.494
August	8.333%	122	83.00	18.325
September	8.333%	89	83.00	0.434
October	8.333%	87	83.00	0.193
November	8.333%	73	83.00	1.205
December	8.333%	54	83.00	10.133
				89.663

$\chi^2 = 89.663$

(d) Reject H_0. There is enough evidence at the 10% level to conclude that the distribution of fatal bicycle accidents throughout the year in not uniform.

14. (a) Claimed distribution:

Time of day	Distribution
Midnight–6 A.M.	34%
6 A.M.–Noon	15%
Noon–6 P.M.	22%
6 P.M.–Midnight	29%

H_0: Distribution of the time of day of roadside hazard crash deaths is as shown in table above.

H_a: Distribution of the time of day of roadside hazard crash deaths differs from the claimed distribution.

(b) $\chi_0^2 = 11.345$; Reject H_0 if $\chi^2 > 11.345$.

(c)

Time of day	Distribution	Observed	Expected	$\frac{(O-E)^2}{E}$
Midnight–6 A.M.	34%	224	213.18	0.549
6 A.M.–Noon	15%	128	94.05	12.255
Noon–6 P.M.	22%	115	137.94	3.815
6 P.M.–Midnight	29%	160	181.83	2.621
				19.240

$\chi^2 = 19.240$

(d) Reject H_0. There is enough evidence at the 1% level to conclude that the distribution of the time of day of roadside hazard crash deaths has changed.

16. (a) Claimed distribution:

Response	Distribution
Married, husband present	25%
Married, husband absent	25%
Widowed/divorced	25%
Never married	25%

H_0: Distribution of the responses is uniform as shown in table above.

H_a: Distribution of the responses is not uniform.

(b) $\chi_0^2 = 9.348$; Reject H_0 if $\chi^2 > 9.348$.

(c)

Response	Distribution	Observed	Expected	$\frac{(O-E)^2}{E}$
Married, husband present	25%	20	25.25	1.092
Married, husband absent	25%	19	25.25	1.547
Widowed/divorce	25%	23	25.25	0.200
Never married	25%	39	25.25	7.488
		101		10.327

$\chi^2 = 10.327$

(d) Reject H_0. There is enough evidence at the 2.5% level to reject the counselor's claim.

18. (a) Claimed distribution:

Response	Distribution
First marriage of bride & groom	50%
First marriage of bride, remarriage of groom	12%
First marriage of groom, remarriage of bride	14%
Remarriage of bride & groom	24%

H_0: Distribution of the responses is as shown in table above.

H_a: Distribution of the responses differs from the claimed distribution.

(b) $\chi_0^2 = 11.345$; Reject H_0 if $\chi^2 > 11.345$.

(c)

Response	Distribution	Observed	Expected	$\frac{(O-E)^2}{E}$
First marriage of bride & groom	50%	55	51.50	0.238
First marriage of bride, remarriage of groom	12%	12	12.36	0.010
First marriage of groom, remarriage of bride	14%	12	14.42	0.406
Remarriage of bride & groom	24%	24	24.72	0.021
		103		0.675

$\chi^2 = 0.675$

(d) Fail to reject H_0. There is not enough evidence at the 1% level to reject the counselor's claim.

20. (a) Frequency distribution: $\mu = 74.775$, $\sigma = 9.822$

Lower Boundary	Upper Boundary	Lower z-score	Upper z-score	Area	Class Boundaries	Distribution	Frequency	Expected	$\frac{(O-E)^2}{E}$
50.5	60.5	−2.47	−1.45	0.0668	50.5–60.5	6.68%	28	27	0.037
60.5	70.5	−1.45	−0.44	0.2564	60.5–70.5	25.64%	106	103	0.087
70.5	80.5	−0.44	0.58	0.3891	70.5–80.5	38.91%	151	156	0.160
80.5	90.5	0.58	1.60	0.2262	80.5–90.5	22.62%	97	90	0.544
90.5	100.5	1.60	2.62	0.0504	90.5–100.5	5.04%	18	20	0.2
							400		1.028

H_0: Variable has a normal distribution.

H_a: Variable does not have a normal distribution.

(b) $\chi_0^2 = 9.488$; Reject H_0 if $\chi^2 > 9.488$.

(c) $\chi^2 \approx 1.028$

(d) Fail to reject H_0. There is not enough evidence at the 5% level to conclude that the distribution of test scores is not normal.

10.2 INDEPENDENCE

10.2 EXERCISE SOLUTIONS

2. A large chi-square test statistic is evidence for rejecting the null hypothesis. So, the chi-square independence test is always a right-tailed test.

4. False. If the two variables of a chi-square test of independence are dependent, then you can expect to find large differences between the observed frequencies and the expected frequencies.

6.

	Rating		
Size	Excellent	Fair	Poor
Seats 100 or less	165.917	235.583	148.500
Seats over 100	196.083	278.417	175.500

8.

	Type of car			
Gender	Compact	Full size	SUV	Truck/van
Male	28.60	39.05	22.55	19.80
Female	23.40	31.95	18.45	16.20

10. (a) H_0: Attitudes about safety are independent of the type of school.

    H_a: Attitudes about safety are dependent on the type of school. (claim)

    (b) d.f. $= (r-1)(c-1) = 1$

    $\chi_0^2 = 6.635$; Reject H_0 if $\chi^2 > 6.635$.

    (c) $\chi^2 \approx 8.691$

    (d) Reject H_0. There is enough evidence at the 1% level to conclude that attitudes about safety are dependent on the type of school.

12. (a) H_0: Adults' ratings are independent of the type of school.

    H_a: Adults' ratings are dependent on the type of school. (claim)

    (b) d.f. $= (r-1)(c-1) = 3$

    $\chi_0^2 = 7.815$; Reject H_0 if $\chi^2 > 7.815$.

    (c) $\chi^2 \approx 148.389$

    (d) Reject H_0. There is enough evidence at the 5% level to conclude that adults' ratings are dependent on the type of school.

14. (a) H_0: Results are independent of the type of treatment.

    H_a: Results are dependent on the type of treatment. (claim)

    (b) d.f. $= (r-1)(c-1) = 1$

    $\chi_0^2 = 2.706$; Reject H_0 if $\chi^2 > 2.706$.

    (c) $\chi^2 \approx 1.032$

    (d) Fail to reject H_0. There is not enough evidence at the 10% level to conclude that results are dependent on the type of treatment. I would not recommend using the drug.

16. (a) H_0: Blood alcohol concentration is independent of age.

    H_a: Blood alcohol concentration is dependent on age. (claim)

    (b) d.f. $= (r-1)(c-1) = 2$

    $\chi_0^2 = 9.210$; Reject H_0 if $\chi^2 > 9.210$.

    (c) $\chi^2 \approx 10.762$

    (d) Reject H_0. There is enough evidence at the 1% level to conclude that blood alcohol concentration is dependent on age.

18. (a) H_0: Age is independent of gender.

    H_a: Age is dependent on gender. (claim)

    (b) d.f. $= (r-1)(c-1) = 5$

    $\chi_0^2 = 11.071$; Reject H_0 if $\chi^2 > 11.071$.

    (c) $\chi^2 \approx 2.834$

    (d) Fail to reject H_0. There is not enough evidence at the 5% level to conclude that age is dependent on gender in such alcohol-related accidents.

20. H_0: The proportions are equal. (claim)

H_a: At least one of the proportions is different from the others.

d.f. $= (r - 1)(c - 1) = 7$

$\chi_0^2 = 14.067$; Reject H_0 if $\chi^2 > 14.067$.

$\chi^2 \approx 3.853$

Fail to reject H_0. There is not enough evidence at the 5% level to reject the claim that the proportions are equal.

10.3 COMPARING TWO VARIANCES

10.3 EXERCISE SOLUTIONS

2. (1) The F-distribution is a family of curves determined by two types of degrees of freedom, d.f.$_N$ and d.f.$_D$.

(2) F-distributions are positively skewed.

(3) The area under the F-distribution curve is equal to 1.

(4) F-values are always greater than or equal to zero.

(5) For all F-distributions, the mean value of F is approximately equal to 1.

4. $F = 4.72$ **6.** $F = 2.63$ **8.** $F = 14.62$

10. H_0: $\sigma_1^2 = \sigma_2^2$ (claim); H_a: $\sigma_1^2 \neq \sigma_2^2$

d.f.$_N = 6$

d.f.$_D = 7$

$F_0 = 5.12$; Reject H_0 if $F > 5.12$.

$F = \dfrac{s_1^2}{s_2^2} = \dfrac{310}{297} \approx 1.044$

Fail to reject H_0. There is not enough evidence to reject the claim.

12. H_0: $\sigma_1^2 = \sigma_2^2$; H_a: $\sigma_1^2 \neq \sigma_2^2$ (claim)

d.f.$_N = 14$

d.f.$_D = 13$

$F_0 = 3.08$; Reject H_0 if $F > 3.08$.

$F = \dfrac{s_1^2}{s_2^2} = \dfrac{141}{117} \approx 1.205$

Fail to reject H_0. There is not enough evidence to support the claim.

14. H_0: $\sigma_1^2 \leq \sigma_2^2$; H_a: $\sigma_1^2 > \sigma_2^2$ (claim)

d.f.$_N = 15$

d.f.$_D = 11$

$F_0 = 2.72$; Reject H_0 if $F > 2.72$.

$F = \dfrac{s_1^2}{s_2^2} = \dfrac{44.6}{39.3} \approx 1.135$

Fail to reject H_0. There is not enough evidence to support the claim.

16. (a) Population 1: Competitor
Population 2: Auto Manufacturer

$$H_0: \sigma_1^2 \leq \sigma_2^2; H_a: \sigma_1^2 > \sigma_2^2 \text{ (claim)}$$

(b) $\text{d.f.}_N = 21$
$\text{d.f.}_D = 18$
$F_0 = 2.18$; Reject H_0 if $F > 2.18$.

(c) $F = \dfrac{s_1^2}{s_2^2} = \dfrac{4.5}{4.2} \approx 1.071$

(d) Fail to reject H_0. There is not enough evidence at the 5% level to conclude that the variance of the fuel consumption for the company's luxury sedan is less than that of the competitor's luxury sedan.

18. (a) Population 1: District 1
Population 2: District 2

$$H_0: \sigma_1^2 = \sigma_2^2 \text{ (claim)}; H_a: \sigma_1^2 \neq \sigma_2^2$$

(b) $\text{d.f.}_N = 9$
$\text{d.f.}_D = 12$
$F_0 = 5.20$; Reject H_0 if $F > 5.20$.

(c) $F = \dfrac{s_1^2}{s_2^2} = \dfrac{(28.8)^2}{(26.8)^2} \approx 1.155$

(d) Fail to reject H_0. There is not enough evidence at the 1% level to reject the claim that the standard deviation of eighth-grade student test scores is the same for Districts 1 and 2.

20. (a) Population 1: 2nd city
Population 2: 1st city

$$H_0: \sigma_1^2 = \sigma_2^2 \text{ (claim)}; H_a: \sigma_1^2 \neq \sigma_2^2$$

(b) $\text{d.f.}_N = 14$
$\text{d.f.}_D = 12$
$F_0 = 4.77$; Reject H_0 if $F > 4.77$.

(c) $F = \dfrac{s_1^2}{s_2^2} = \dfrac{(29.75)^2}{(27.50)^2} \approx 1.170$

(d) Fail to reject H_0. There is not enough evidence at the 1% level to reject the claim that the standard deviations of hotel room rates for the two cities are the same.

22. (a) Population 1: Connecticut
Population 2: Colorado

$$H_0: \sigma_1^2 \leq \sigma_2^2; H_a: \sigma_1^2 > \sigma_2^2 \text{ (claim)}$$

(b) $\text{d.f.}_N = 21$
$\text{d.f.}_D = 23$
$F_0 = 2.04$; Reject H_0 if $F > 2.04$.

(c) $F = \dfrac{s_1^2}{s_2^2} = \dfrac{(9900)^2}{(7800)^2} \approx 1.611$

(d) Fail to reject H_0. There is not enough evidence at the 5% level to support the claim that the standard deviation of the annual salaries for public relations managers is greater in Connecticut than in Colorado.

24. Right-tailed: $F = 2.33$

Left-tailed: (1) d.f.$_N$ = 17 (use 15 since d.f.$_N$ = 17 is not in the table) and d.f.$_D$ = 20

(2) $F = 2.20$

(3) Critical value is $\dfrac{1}{F} = \dfrac{1}{2.20} \approx 0.455$.

26. $\dfrac{s_1^2}{s_2^2} F_L < \dfrac{\sigma_1^2}{\sigma_2^2} < \dfrac{s_1^2}{s_2^2} F_R \rightarrow \dfrac{4.84}{3.24} 0.32 < \dfrac{\sigma_1^2}{\sigma_2^2} < \dfrac{4.84}{3.24} 3.36 \rightarrow 0.478 < \dfrac{\sigma_1^2}{\sigma_2^2} < 5.019$

10.4 ANALYSIS OF VARIANCE

10.4 EXERCISE SOLUTIONS

2. Each sample must be randomly selected from a normal, or approximately normal, population. The samples must be independent of each other. Each population must have the same variance.

4. H_{0A}: There is no difference among the treatment means of Factor A.

H_{aA}: There is at least one difference among the treatment means of Factor A.

H_{0B}: There is no difference among the treatment means of Factor B.

H_{aB}: There is at least one difference among the treatment means of Factor B.

H_{0AB}: There is no interaction between Factor A and Factor B.

H_{aAB}: There is no interaction between Factor A and Factor B.

6. (a) $H_0: \mu_1 = \mu_2 = \mu_3$

H_a: At least one mean is different from the others. (claim)

(b) d.f.$_N$ = $k - 1 = 2$
d.f.$_D$ = $N - k = 13$
$F_0 = 3.81$; Reject H_0 if $F > 3.81$.

(c)

Variation	Sum of Squares	Degrees of Freedom	Mean Squares	F
Between	513	2	257	2.15
Within	1554	13	120	

$F \approx 2.15$

(d) Fail to reject H_0. There is not enough evidence at the 5% level to conclude that at least one of the mean battery prices is different from the others.

8. (a) $H_0: \mu_1 = \mu_2 = \mu_3 = \mu_4$ (claim)

H_a: At least one mean is different from the others.

(b) d.f.$_N$ = $k - 1 = 3$
d.f.$_D$ = $N - k = 23$
$F_0 = 2.34$; Reject H_0 if $F > 2.34$.

(c)

Variation	Sum of Squares	Degrees of Freedom	Mean Squares	F
Between	20,085	3	6995	1.61
Within	95,857	23	4168	

$F \approx 1.61$

(d) Fail to reject H_0. There is not enough evidence at the 10% level to reject the claim that the mean annual amounts are the same in all regions.

10. (a) $H_0: \mu_1 = \mu_2 = \mu_3 = \mu_4$
H_a: At least one mean is different from the others. (claim)

(b) d.f.$_N = k - 1 = 3$
d.f.$_D = N - k = 48$

$F_0 = 2.20$; Reject H_0 if $F > 2.20$.

(c)
Variation	Sum of Squares	Degrees of Freedom	Mean Squares	F
Between	146.3	3	48.8	2.29
Within	1024.5	48	21.3	

$F \approx 2.29$

(d) Reject H_0. There is enough evidence at the 10% level to conclude that the mean square footage for at least one of the four regions is different.

12. (a) $H_0: \mu_1 = \mu_2 = \mu_3 = \mu_4$
H_a: At least one mean is different from the others. (claim)

(b) d.f.$_N = k - 1 = 3$
d.f.$_D = N - k = 16$

$F_0 = 3.24$; Reject H_0 if $F > 3.24$.

(c)
Variation	Sum of Squares	Degrees of Freedom	Mean Squares	F
Between	83,662,620	3	27,874,206.67	1.870
Within	223,558,480	16	14,909,905	

$F \approx 1.870$

(d) Fail to reject H_0. There is not enough evidence at the 5% level to conclude that the mean salary for an individual is different in at least one of the areas.

14. (a) $H_0: \mu_1 = \mu_2 = \mu_3 = \mu_4$
H_a: At least one mean is different from the others. (claim)

(b) d.f.$_N = k - 1 = 3$
d.f.$_D = N - k = 36$

$F_0 = 4.38$; Reject H_0 if $F > 4.38$.

(c)
Variation	Sum of Squares	Degrees of Freedom	Mean Squares	F
Between	61,131	3	20,377	8.46
Within	86,713	36	2409	

$F \approx 8.46$

(d) Reject H_0. There is enough evidence at the 1% level to conclude that the mean energy consumption for at least one region is different.

16. (a) $H_0: \mu_1 = \mu_2 = \mu_3 = \mu_4$ (claim)
H_a: At least one mean is different from the others.

(b) d.f.$_N = k - 1 = 3$
d.f.$_D = N - k = 33$

$F_0 = 4.51$ (using 3 and 30 degrees of freedom); Reject H_0 if $F > 4.51$.

(c)

Variation	Sum of Squares	Degrees of Freedom	Mean Squares	F
Between	1,065,057.38	3	355,019.127	1.408
Within	8,319,244.73	33	252,097.719	

$F \approx 1.408$

(d) Fail to reject H_0. There is not enough evidence at the 1% level to reject the claim that the mean amounts spent are equal for all regions.

18. H_0: Type of vehicle has no effect on the mean number of vehicles sold.
H_a: Type of vehicle has an effect on the mean number of vehicles sold.

H_0: Gender has no effect on the mean number of vehicles sold.
H_a: Gender has an effect on the mean number of vehicles sold.

H_0: There is no interaction effect between type of vehicle and gender on the mean number of vehicles sold.
H_a: There is an interaction effect between type of vehicle and gender on the mean number of vehicles sold.

Source	d.f.	SS	MS	F	P
Type of vehicle	2	84.08	42.04	34.01	0.000
Gender	1	0.38	0.38	0.30	0.589
Interaction	2	12.25	6.12	4.96	0.019
Error	18	22.25	1.24		
Total	23	118.96			

There appears to be an interaction effect between type of vehicle and gender on the mean number of vehicles sold. Also, there appears that type of vehicle has an effect on the mean number of vehicles sold.

20. H_0: Technicians have no effect on mean repair time.
H_a: Technicians have an effect on mean repair time.

H_0: Brand has no effect on mean repair time.
H_a: Brand has an effect on mean repair time.

H_0: There is no interaction effect between technicians and brand on mean repair time.
H_a: There is an interaction effect between technicians and brand on mean repair time.

Source	d.f.	SS	MS	F	P
Technicians	3	714	238	1.74	0.185
Brand	2	382	191	1.40	0.266
Interaction	6	2007	334	2.45	0.054
Error	24	3277	137		
Total	35	6381			

There appears to be an interaction effect between technicians and brand on the mean repair time at the 10% level.

22.

	Mean	Size
Pop 1	139.35	12
Pop 2	93.05	11
Pop 3	107.72	12
Pop 4	130.75	12

$SS_w = 73{,}350$

$\Sigma(n_i - 1) = N - k = 43$

$F_0 = 2.22 \to CV_{\text{Scheffé}} = 2.22(4 - 1) = 6.660$

$$\frac{(\bar{x}_1 - \bar{x}_2)^2}{\dfrac{SS_W}{\Sigma(n_i - 1)}\left[\dfrac{1}{n_1} + \dfrac{1}{n_2}\right]} \approx 7.212 \to \text{Significant difference}$$

$$\frac{(\bar{x}_1 - \bar{x}_3)^2}{\dfrac{SS_W}{\Sigma(n_i - 1)}\left[\dfrac{1}{n_1} + \dfrac{1}{n_3}\right]} \approx 3.519 \to \text{No difference}$$

$$\frac{(\bar{x}_1 - \bar{x}_4)^2}{\dfrac{SS_W}{\Sigma(n_i - 1)}\left[\dfrac{1}{n_1} + \dfrac{1}{n_4}\right]} \approx 0.260 \to \text{No difference}$$

$$\frac{(\bar{x}_2 - \bar{x}_3)^2}{\dfrac{SS_W}{\Sigma(n_i - 1)}\left[\dfrac{1}{n_2} + \dfrac{1}{n_3}\right]} \approx 0.724 \to \text{No difference}$$

$$\frac{(\bar{x}_2 - \bar{x}_4)^2}{\dfrac{SS_W}{\Sigma(n_i - 1)}\left[\dfrac{1}{n_2} + \dfrac{1}{n_4}\right]} \approx 4.782 \to \text{No difference}$$

$$\frac{(\bar{x}_3 - \bar{x}_4)^2}{\dfrac{SS_W}{\Sigma(n_i - 1)}\left[\dfrac{1}{n_3} + \dfrac{1}{n_4}\right]} \approx 1.866 \to \text{No difference}$$

24.

	Mean	Size
Pop 1	40,916	8
Pop 2	39,565	7
Pop 3	31,688	9
Pop 4	46,800	6

$\dfrac{SS_w}{\Sigma(n_i - 1)} \approx 45{,}988{,}138$

$F_0 = 2.31 \to CV_{\text{Scheffé}} = 2.31(4 - 1) = 6.930$

$$\frac{(\bar{x}_1 - \bar{x}_2)^2}{\dfrac{SS_W}{\Sigma(n_i - 1)}\left[\dfrac{1}{n_1} + \dfrac{1}{n_2}\right]} \approx 0.148 \to \text{No difference}$$

$$\frac{(\bar{x}_1 - \bar{x}_3)^2}{\dfrac{SS_W}{\Sigma(n_i - 1)}\left[\dfrac{1}{n_1} + \dfrac{1}{n_3}\right]} \approx 7.842 \to \text{Significant difference}$$

$$\frac{(\bar{x}_1 - \bar{x}_4)^2}{\frac{SS_W}{\Sigma(n_i - 1)}\left[\frac{1}{n_1} + \frac{1}{n_4}\right]} \approx 2.581 \rightarrow \text{No difference}$$

$$\frac{(\bar{x}_2 - \bar{x}_3)^2}{\frac{SS_W}{\Sigma(n_i - 1)}\left[\frac{1}{n_2} + \frac{1}{n_3}\right]} \approx 5.311 \rightarrow \text{No difference}$$

$$\frac{(\bar{x}_2 - \bar{x}_4)^2}{\frac{SS_W}{\Sigma(n_i - 1)}\left[\frac{1}{n_2} + \frac{1}{n_4}\right]} \approx 3.678 \rightarrow \text{No difference}$$

$$\frac{(\bar{x}_3 - \bar{x}_4)^2}{\frac{SS_W}{\Sigma(n_i - 1)}\left[\frac{1}{n_3} + \frac{1}{n_4}\right]} \approx 17.877 \rightarrow \text{Significant difference}$$

CHAPTER 10 REVIEW EXERCISE SOLUTIONS

2. Claimed distribution:

Advertisers	Distribution
Retail goods and services	26%
Financial services	18%
Web media	17%
Other	39%

H_0: Distribution of advertisers is as shown in table above.

H_a: Distribution of advertisers differs from the claimed distribution.

$\chi_0^2 = 6.251$

Advertisers	Distribution	Observed	Expected	$\frac{(O - E)^2}{E}$
Retail goods and services	26%	88	94.90	0.502
Financial services	18%	73	65.70	0.811
Web media	17%	66	62.05	0.251
Other	3%	138	142.35	0.133
		365		1.697

$\chi^2 = 1.697$

Fail to reject H_0. There is not enough evidence at the 10% level to reject the claimed distribution.

4. Claimed distribution:

Age	Distribution
21–29	20.5%
30–39	21.7%
40–49	18.1%
50–59	17.3%
60+	22.4%

H_0: Distribution of ages is as shown in table above.

H_a: Distribution of ages differs from the claimed distribution.

$\chi_0^2 = 13.277$

Age	Distribution	Observed	Expected	$\frac{(O-E)^2}{E}$
21–29	20.5%	45	205	124.878
30–39	21.7%	128	217	36.502
40–49	18.1%	244	181	21.928
50–59	17.3%	224	173	15.035
60+	22.4%	359	224	81.362
		1000		279.705

$\chi^2 \approx 279.705$

Reject H_0. There is enough evidence at the 1% level to support the claim that the distribution of ages of the jury differs from the age distribution of available jurors.

6. (a) Expected frequencies:

	Types of vehicles owned				
Gender	Car	Truck	SUV	Van	Total
Male	94.25	82.65	50.75	4.35	232
Female	100.75	88.35	54.25	4.65	248
Total	195	171	105	9	480

(b) H_0: Type of vehicle is independent of gender.

H_a: Type of vehicle is dependent on gender.

$\chi_0^2 = 7.815$

$\chi^2 = \sum \frac{(O-E)^2}{E} = 8.403$

Reject H_0.

(c) There is enough evidence at the 5% level to conclude that type of vehicle owned is dependent on gender.

8. (a) Expected frequencies:

	Time of day				
Gender	12 A.M.–5:59 A.M.	6 A.M.–11:59 A.M.	12 P.M.–5:59 P.M.	6 P.M.–11:59 p.m.	Total
Male	654.07	490.21	654.07	1470.64	3269
Female	307.93	230.79	307.93	692.36	1539
Total	962	721	962	2163	4808

(b) H_0: Type of vehicle is independent of gender.

H_a: Type of vehicle is dependent on gender.

$\chi_0^2 = 6.251$

$\chi^2 = 0.000631$

Fail to reject H_0.

(c) There is enough evidence at the 10% level to conclude that time and gender are dependent.

10. $F_0 = 4.71$

12. $F_0 = 2.01$

14. $H_0: \sigma_1^2 = \sigma_2^2; H_a: \sigma_1^2 \neq \sigma_2^2$ (claim)

d.f.$_N$ = 5

d.f.$_D$ = 10

$F_0 = 3.33$; Reject H_0 if $F > 3.33$.

$F = \dfrac{s_1^2}{s_2^2} = \dfrac{112{,}676}{49{,}572} \approx 2.273$

Fail to reject H_0. There is not enough evidence at the 10% level to support the claim.

16. Population 1: Nontempered
Population 2: Tempered

$H_0: \sigma_1^2 \leq \sigma_2^2; H_a: \sigma_1^2 > \sigma_2^2$ (claim)

d.f.$_N$ = 8

d.f.$_D$ = 8

$F_0 = 3.44$; Reject H_0 if $F > 3.44$.

$F = \dfrac{s_1^2}{s_2^2} = \dfrac{(25.4)^2}{(13.1)^2} \approx 3.759$

Reject H_0. There is enough evidence at the 5% level to conclude that the yield strength of the nontempered couplings is more variable than that of the tempered couplings.

18. Population 1: Current $\to s_1^2 = 0.00146$
Population 2: New $\to s_2^2 = 0.00050$

$H_0: \sigma_1^2 \leq \sigma_2^2; H_a: \sigma_1^2 > \sigma_2^2$ (claim)

d.f.$_N$ = 11

d.f.$_D$ = 11

$F_0 = 2.82$; Reject H_0 if $F > 2.82$.

$F = \dfrac{s_1^2}{s_2^2} = \dfrac{0.00146}{0.00050} \approx 2.92$

Reject H_0. There is enough evidence at the 5% level to support the claim that the new mold produces inserts that are less variable in diameter than the current mold.

20. $H_0: \mu_1 = \mu_2 = \mu_3 = \mu_4$

H_a: At least one mean is different from the others. (claim)

d.f.$_N$ = $k - 1 = 3$

d.f.$_D$ = $N - k = 18$

$F_0 = 3.16$; Reject H_0 if $F > 3.16$.

Variation	Sum of Squares	Degrees of Freedom	Mean Squares	F
Between	301,144,226	3	100,381,409	0.64
Within	2,821,986,334	18	156,777,019	

$F \approx 0.64$

Fail to reject H_0. There is not enough evidence a the 5% level to conclude that the average annual income is different for the four regions.

CHAPTER 10 QUIZ SOLUTIONS

1. (a) Population 1: San Jose $\rightarrow s_1^2 \approx 429.984$

Population 2: Dallas $\rightarrow s_2^2 \approx 112.779$

$H_0: \sigma_1^2 = \sigma_2^2$; $H_a: \sigma_1^2 \neq \sigma_2^2$ (claim)

(b) $\alpha = 0.01$

(cd) $\text{d.f.}_N = 12$

$\text{d.f.}_D = 17$

$F_0 = 3.97$; Reject H_0 if $F > 3.97$.

(e) $F = \dfrac{s_1^2}{s_2^2} = \dfrac{429.984}{112.779} \approx 3.813$

(f) Fail to reject H_0.

(g) There is not enough evidence at the 1% level to conclude that the variances in annual wages for San Jose, CA and Dallas, TX are different.

2. (a) $H_0: \mu_1 = \mu_2 = \mu_3$ (claim)

H_a: At least one mean is different from the others.

(b) $\alpha = 0.10$

(cd) $\text{d.f.}_N = k - 1 = 2$

$\text{d.f.}_D = N - k = 44$

$F_0 = 2.43$; Reject H_0 if $F > 2.43$.

Variation	Sum of Squares	Degrees of Freedom	Mean Squares	F
Between	2676	2	1338	7.39
Within	7962	44	181	

(e) $F \approx 7.39$

(f) Reject H_0.

(g) There is enough evidence at the 10% level to conclude that the mean annual wages for the three cities are not all equal.

3. (a) Claimed distribution:

Education	25 & Over
Not a HS graduate	16.6%
HS graduate	33.3%
Some college, no degree	17.3%
Associate degree	7.5%
Bachelor's degree	17.0%
Advanced degree	8.2%

H_0: Distribution of educational achievement for people in the United States aged 35–44 is as shown in table above.

H_a: Distribution of educational achievement for people in the United States aged 35–44 differs from the claimed distribution.

(b) $\alpha = 0.01$

(cd) $\chi_0^2 = 15.086$; Reject H_0 if $\chi^2 > 15.086$.

(e)

Education	25 & Over	Observed	Expected	$\frac{(O-E)^2}{E}$
Not a HS graduate	16.6%	37	49.966	3.365
HS graduate	33.3%	102	100.233	0.0312
Some college, no degree	17.3%	59	52.073	0.921
Associate degree	7.5%	27	22.575	0.867
Bachelor's degree	17.0%	51	51.17	0.001
Advanced degree	8.2%	25	24.682	0.004
		301		5.189

$\chi^2 \approx 5.189$

(f) Fail to reject H_0.

(g) There is not enough evidence at the 1% level to conclude that the distribution of educational achievement for people in the United States aged 35–44 differs from the claimed distribution.

4. (a) Claimed distribution:

Education	25 & Over
Not a HS graduate	16.6%
HS graduate	33.3%
Some college, no degree	17.3%
Associate degree	7.5%
Bachelor's degree	17.0%
Advanced degree	8.2%

H_0: Distribution of educational achievement for people in the United States aged 65–74 is as shown in table above.

H_a: Distribution of educational achievement for people in the United States aged 65–74 differs from the claimed distribution.

(b) $\alpha = 0.05$

(c) $\chi_0^2 = 11.071$

(d) Reject H_0 if $\chi^2 > 11.071$.

(e)

Education	25 & Over	Observed	Expected	$\frac{(O-E)^2}{E}$
Not a HS graduate	16.6%	124	67.562	47.146
HS graduate	33.3%	148	135.531	1.147
Some college, no degree	17.3%	61	70.411	1.258
Associate degree	7.5%	15	30.525	7.896
Bachelor's degree	17.0%	37	69.19	14.976
Advanced degree	8.3%	22	33.781	4.109
		407		76.532

$\chi^2 \approx 76.299$

(f) Reject H_0.

(g) There is enough evidence at the 5% level to conclude that the distribution of educational achievement for people in the United States aged 65–74 differs from the claimed distribution.

Nonparametric Tests

CHAPTER 11

11.1 THE SIGN TEST

11.1 EXERCISE SOLUTIONS

2. Identify the claim and state H_0 and H_a. Identify the level of significance and sample size. Find the critical value using Table 8 (if $n \leq 25$) or Table 4 ($n > 25$). Calculate the test statistic. Make a decision and interpret in the context of the problem.

4. (a) H_0: median = 72 (claim); H_a: median $\neq$ 72

   (b) The critical value is 1.

   (c) $x = 6$

   (d) Fail to reject H_0.

   (e) There is not enough evidence at the 1% level to reject the claim that the daily median temperature for the month of July in Pittsburgh is 72°F.

6. (a) H_0: median = 58 (claim); H_a: median $\neq$ 58

   (b) The critical value is 2.

   (c) $x = 6$

   (d) Fail to reject H_0.

   (e) There is not enough evidence at the 1% level to reject the claim that the daily median temperature for the month of January in San Diego is 58°F.

8. (a) H_0: median $\geq$ 38,800; H_a: median < 38,800 (claim)

   (b) The critical value is $z_0 = -1.96$.

   (c) $x = 28$

   $$z = \frac{(x + 0.5) - 0.5(n)}{\frac{\sqrt{n}}{2}} = \frac{(28 + 0.5) - 0.5(70)}{\frac{\sqrt{70}}{2}} = \frac{-6.5}{4.183} \approx -1.554$$

   (d) Fail to reject H_0.

   (e) There is not enough evidence at the 2.5% level to support the claim that the median amount of financial debt for families holding such debts is less than $38,800.

10. (a) H_0: median $\geq$ 30; H_a: median < 30 (claim)

    (b) The critical value is 5.

    (c) $x = 8$

    (d) Fail to reject H_0.

    (e) There is not enough evidence at the 5% level to support the claim that the median age of recipients of physical science doctorates is less than 30 years.

322 CHAPTER 11 | NONPARAMETRIC TESTS

12. (a) H_0: median = 1300 (claim); H_a: median ≠ 1300

    (b) The critical value is 5.

    (c) $x = 9$

    (d) Fail to reject H_0.

    (e) There is not enough evidence at the 10% level to reject the claim that the median square footage of renter-occupied units is 1300 square feet.

14. (a) H_0: median ≤ $9.89 (claim); H_a: median > $9.89

    (b) The critical value is 5.

    (c) $x = 9$

    (d) Fail to reject H_0.

    (e) There is not enough evidence at the 5% level to reject the claim that the median hourly earnings of female workers paid hourly rates is at most $9.89.

16. (a) H_0: The headache hours have not decreased.
    H_a: The headache hours have decreased. (claim)

    (b) The critical value is 1.

    (c) $x = 4$

    (d) Fail to reject H_0.

    (e) There is not enough evidence at the 1% level to support the claim that the daily headache hours have decreased.

18. (a) H_0: The SAT scores have not improved.
    H_a: The SAT scores have improved. (claim)

    (b) The critical value is 1.

    (c) $x = 3$

    (d) Fail to reject H_0.

    (e) There is not enough evidence at the 1% level to support the claim that the verbal SAT scores have improved.

20. (a) H_0: The proportion of credit card holders almost always paying off their balances is equal to the proportion of credit card holders that do not. (claim)
    H_a: The proportion of credit card holders almost always paying off their balances is not equal to the proportion of credit card holders that do not.

    The critical value is 6. ($\alpha = 0.05$).

    $x = 9$

    Fail to reject H_0.

    (b) There is not enough evidence at the 5% level to reject the claim that the proportion of credit card holders almost always paying off their credit card balances is equal to the proportion of credit card holders that do not.

22. (a) H_0: median ≤ $680; H_a: median > $680 (claim)

    (b) The critical value is $z_0 = 2.33$.

(c) $x = 48$

$$z = \frac{(x - 0.5) - 0.5(n)}{\frac{\sqrt{n}}{2}} = \frac{(48 - 0.5) - 0.5(69)}{\frac{\sqrt{69}}{2}} = \frac{13}{4.153} \approx 3.130$$

(d) Reject H_0. There is enough evidence at the 1% level to support the claim that the median weekly earnings of male workers is greater than $680.

24. (a) H_0: median ≤ 26.9 (claim); H_a: median > 26.9

(b) The critical value is $z_0 = 1.645$.

(c) $x = 33$

$$z = \frac{(x - 0.5) - 0.5(n)}{\frac{\sqrt{n}}{2}} = \frac{(33 - 0.5) - 0.5(56)}{\frac{\sqrt{56}}{2}} = \frac{4.5}{3.742} \approx 1.203$$

(d) Fail to reject H_0. There is not enough evidence at the 5% level to reject the claim that the median age of first-time grooms is less than or equal to 26.9.

11.2 THE WILCOXON TESTS

11.2 EXERCISE SOLUTIONS

2. The sample size of both samples must be at least 10.

4. (a) H_0: There is no difference in salaries. (claim)
H_a: There is a difference in salaries.

(b) Wilcoxon rank sum test

(c) The critical value is $z_0 = \pm 1.645$.

(d) $R = 74.5$

$$\mu_R = \frac{n_1(n_1 + n_2 + 1)}{2} = \frac{10(10 + 10 + 1)}{2} = 105$$

$$\sigma_R = \sqrt{\frac{n_1 n_2 (n_1 + n_2 + 1)}{12}} = \sqrt{\frac{(10)(10)(10 + 10 + 1)}{12}} = 13.229$$

$$z = \frac{R - \mu_R}{\sigma_R} = \frac{74.5 - 105}{13.229} \approx -2.306$$

(e) Reject H_0.

(f) There is enough evidence at the 10% level to reject the claim that there is no difference in the salaries.

6. (a) H_0: There is no difference in the number of months mothers breast-feed their babies. (claim)
H_a: There is a difference in the number of months mothers breast-feed their babies.

(b) Wilcoxon rank sum test

(c) The critical value is $z_0 = \pm 2.575$.

(d) $R = 104$

$$\mu_R = \frac{n_1(n_1 + n_2 + 1)}{2} = \frac{11(11 + 12 + 1)}{2} = 132$$

$$\sigma_R = \sqrt{\frac{n_1 n_2 (n_1 + n_2 + 1)}{12}} = \sqrt{\frac{(11)(12)(11 + 12 + 1)}{12}} = 16.248$$

$$z = \frac{R - \mu_R}{\sigma_R} = \frac{104 - 132}{16.248} \approx -1.723$$

(e) Fail to reject H_0.

(f) There is not enough evidence at the 1% level to reject the claim that there is no difference in the number of months mothers breast-feed their babies.

8. (a) H_0: The new drug does not affect the number of headache hours.

   H_a: The new drug does affect the number of headache hours. (claim)

   (b) Wilcoxon signed-rank test

   (c) The critical value is 2.

   (d)

Before	After	Difference	Absolute value	Rank	Signed rank
0.8	1.6	−0.8	0.8	3	−3
2.4	1.3	1.1	1.1	4	4
2.8	1.6	1.2	1.2	6	6
2.6	1.4	1.2	1.2	6	6
2.7	1.5	1.2	1.2	6	6
0.9	1.6	−0.7	0.7	2	−2
1.2	1.7	−0.5	0.5	1	−1

   The sum of the negative ranks is $-3 + (-2) + (-1) = -6$.

   The sum of the positive ranks is $4 + 6 + 6 + 6 = 22$.

   Because $|-6| < |22|$, the test statistic is $w_s = 6$.

   (e) Fail to reject H_0.

   (f) There is not enough evidence at the 5% level to conclude that the drug affects the number of headache hours.

10. H_0: The fuel additive does not improve gas mileage.

    H_a: The fuel additive does improve gas mileage. (claim)

    The critical value is $z_0 = -1.645$.

    $$w_s = 0$$

    $$z = \frac{w_s - \frac{n(n + 1)}{4}}{\sqrt{\frac{n(n + 1)(2n + 1)}{24}}} = \frac{0 - 264}{\sqrt{2860}} = -4.937$$

    Reject H_0. There is enough evidence at the 5% level to conclude that the fuel additive improves gas mileage.

11.3 THE KRUSKAL-WALLIS TEST

11.3 EXERCISE SOLUTIONS

2. The Kruskal-Wallis test is always a right-tailed test because the null hypothesis is only rejected when H is significantly large.

4. (a) H_0: There is no difference in the premiums.
      H_a: There is a difference in the premiums. (claim)

   (b) The critical value is 5.991.

   (c) $H \approx 12.55$

   (d) Reject H_0.

   (e) There is enough evidence at the 5% level to support the claim that the distributions of the annual premiums in the three states are different.

6. (a) H_0: There is no difference in the salaries.
      H_a: There is a difference in the salaries. (claim)

   (b) The critical value is 6.251.

   (c) $H \approx 10.57$

   (d) Reject H_0.

   (e) There is enough evidence at the 10% level to support the claim that the distributions of the annual salaries in the four states are different.

8. (a) H_0: There is no difference in the mean energy consumptions.
      H_a: There is a difference in the mean energy consumptions. (claim)
      The critical value is 11.345.
      $H \approx 16.26$
      Reject H_0. There is enough evidence to support the claim.

   (b)

Variation	Sum of Squares	Degrees of Freedom	Mean Squares	F
Between	32,116	3	10,705	8.18
Within	45,794	35	1308	

   For $\alpha = 0.01$, the critical value is about 4.41. Because $F = 8.18$ is greater than the critical value, the decision is to reject H_0. There is enough evidence to support the claim.

   (c) Both tests come to the same decision, which is, that the mean energy consumptions are different.

11.4 RANK CORRELATION

11.4 EXERCISE SOLUTIONS

2. The ranks of corresponding data pairs are exactly identical.
   The ranks of corresponding data pairs are in reverse order.
   The ranks of corresponding data pairs have ho relationship.

4. (a) H_0: $\rho_s = 0$; H_a: $\rho_s \neq 0$ (claim)

   (b) The critical value is 0.818.

   (c) $\Sigma d^2 = 127$
   $$r_s = 1 - \frac{6\Sigma d^2}{n(n^2 - 1)} \approx 0.423$$

   (d) Fail to reject H_0.

   (e) There is not enough evidence at the 1% level to support the claim that there is a correlation between the overall score and price.

6. (a) H_0: $\rho_s = 0$; H_a: $\rho_s \neq 0$ (claim)

   (b) Critical value is 0.497.

   (c) $\Sigma d^2 = 123.5$
   $$r_s = 1 - \frac{6\Sigma d^2}{n(n^2 - 1)} \approx 0.568$$

   (d) Reject H_0.

   (e) There is enough evidence at the 10% level to support the claim that there is a correlation between overall score and price.

8. H_0: $\rho_s = 0$; H_a: $\rho_s \neq 0$ (claim)

   The critical value is 0.700.

   $\Sigma d^2 = 118$
   $$r_s = 1 - \frac{6\Sigma d^2}{n(n^2 - 1)} \approx 0.017$$

   Fail to reject H_0. There is not enough evidence at the 5% level to conclude that there is a correlation between mathematics achievement scores and GNP.

10. There is not enough evidence at the 5% level to conclude that there is a significant correlation between eighth grade science achievement scores and the GNP of a country, between eighth grade math achievement scores and the GNP of a country, or between eighth grade science and math achievement scores. But this also means that there is not enough evidence to conclude that there's not a significant correlation between the variables. More evidence is needed to state whether or not a significant correlation exists. (Answers will vary.)

12. H_0: $\rho_s = 0$; H_a: $\rho_s \neq 0$ (claim)

    Critical value $= \frac{\pm z}{\sqrt{n-1}} = \frac{\pm 1.96}{\sqrt{34-1}} \approx \pm 0.341$.

    $\Sigma d^2 = 7310.5$
    $$r_s = 1 - \frac{6\Sigma d^2}{n(n^2 - 1)} \approx -0.017$$

    Fail to reject H_0. There is not enough evidence to support the claim.

11.5 THE RUNS TEST

11.5 EXERCISE SOLUTIONS

2. Number of runs = 9
Run lengths = 2, 2, 1, 1, 2, 2, 1, 1, 2

4. Number of runs = 10
Run lengths = 3, 3, 1, 2, 6, 1, 2, 1, 1, 2

6. n_1 = number of Us = 8
n_2 = number of Ds = 6

8. n_1 = number of As = 13
n_2 = number of Bs = 9

10. n_1 = number of Ms = 9
n_2 = number of Fs = 3
cv = 2 and 8

12. n_1 = number of Xs = 7
n_2 = number of Ys = 14
cv = 5 and 15

14. (a) H_0: The selection of members was random.
H_a: The selection of members was not random. (claim)

(b) n_1 = number of Rs = 18
n_2 = number of Ds = 16
cv = 11 and 25

(c) G = 19 runs

(d) Fail to reject H_0.

(e) At the 5% level, there is not enough evidence to support the claim that the selection of members was not random.

16. (a) H_0: The births are random by gender. (claim)
H_a: The births are not random by gender.

(b) n_1 = number of Ms = 9
n_2 = number of Fs = 20
cv = 8 and 18

(c) G = 12 runs

(d) Fail to reject H_0.

(e) At the 5% level, there is not enough evidence to reject the claim that births are random by gender.

18. (a) H_0: The sequence of arrivals is random.
H_a: The sequence of arrivals is not random. (claim)

(b) n_1 = number of Ts = 42
n_2 = number of Ls = 16
cv = ±1.96

(c) G = 32 runs

$$\mu_G = \frac{2n_1n_2}{n_1+n_2} + 1 = \frac{2(42)(16)}{42+16} + 1 = 24.2$$

$$\sigma_G = \sqrt{\frac{2n_1n_2(2n_1n_2 - n_1 - n_2)}{(n_1+n_2)^2(n_1+n_2-1)}} = \sqrt{\frac{2(42)(16)(2(42)(16) - 42 - 16)}{(42+16)^2(42+16-1)}} = 3.0$$

$$z = \frac{G - \mu_G}{\sigma_G} = \frac{32 - 24.2}{3.0} = 2.6$$

(d) Reject H_0.

(e) At the 5% level, there is enough evidence to support the claim that the sequence of arrivals is not random.

20. H_0: The selection of students' GPAs was random.
H_a: The selection of students' GPAs was not random.

median = 1.8

n_1 = number above median = 12
n_2 = number below median = 3

cv = 2 and 8
G = 3 runs

Fail to reject H_0.

At the 5% level, there is not enough evidence to support the claim that the selection of students' GPAs was not random.

CHAPTER 11 REVIEW EXERCISE SOLUTIONS

2. (a) H_0: median ≤ $1200; H_a: median > $1200 (claim)

(b) The critical value is 1.

(c) $x = 6$

(d) Fail to reject the claim.

(e) There is not enough evidence at the 1% level to support the claim that the median credit card debt is more than $1200.

4. (a) H_0: There is no reduction in diastolic blood pressure. (claim)
H_a: There is a reduction in diastolic blood pressure.

(b) The critical value is 1.

(c) $x = 4$

(d) Fail to reject H_0.

(e) There is not enough evidence at the 5% level to reject the claim that there is no reduction in diastolic blood pressure.

6. (a) H_0: median = $27,900 (claim); H_a: median ≠ $27,900

(b) The critical values are $z_0 = \pm 1.96$.

(c) $x = 21$

$$z = \frac{(x + 0.5) - 0.5(n)}{\frac{\sqrt{n}}{2}} = \frac{(21 + 0.5) - 0.5(54)}{\frac{\sqrt{54}}{2}} = \frac{-5.5}{3.674} \approx -1.497$$

(d) Fail to reject H_0.

(e) There is not enough evidence at the 5% level to reject the claim that the median starting salary is $27,900.

8. (a) Dependent; Wilcoxon Signed Rank Test

(b) H_0: The new drug does not affect the number of headache hours experienced.
H_a: The new drug does affect the number of headache hours experienced. (claim)

(c) The critical value is 4.

(d) $w_s = 1$

(e) Reject H_0.

(f) There is enough evidence at the 5% level to support the claim that the new drug does affect the number of headache hours experienced.

10. (a) H_0: There is no difference in the amount of time to earn a doctorate between the fields of study.
H_a: There is a difference in the amount of time to earn a doctorate between the fields of study. (claim)

(b) The critical value is 5.991.

(c) $H \approx 22.28$

(d) Reject H_0.

(e) There is enough evidence at the 5% level to reject the claim that there is a difference in the amount of time to earn a doctorate between the fields of study.

12. (a) H_0: $\rho_s = 0$; H_a: $\rho_s \neq 0$ (claim)

(b) The critical value is 0.700.

(c) $\Sigma d^2 = 91.5$
$$r_s = 1 - \frac{6\Sigma d^2}{n(n^2 - 1)} \approx 0.238$$

(d) Fail to reject H_0.

(e) There is not enough evidence at the 5% level to support the claim that there is a correlation between overall score and price.

14. (a) H_0: The departure status of buses is random.
H_a: The departure status of buses is not random. (claim)

(b) n_1 = number of Ts = 11
n_2 = number of Ls = 7
cv = 5 and 14

(c) $G = 5$ runs

(d) Reject H_0.

(e) There is enough evidence at the 5% level to support the claim that the departure status of buses is not random.

CHAPTER 11 QUIZ SOLUTIONS

1. (a) H_0: There is no difference in the salaries between genders.
H_a: There is a difference in the salaries between genders. (claim)

(b) Wilcoxon Rank Sum Test

(c) Critical values are $z_0 = \pm 1.645$.

(d) $R = 66$

$$\mu_R = \frac{n_1(n_1 + n_2 + 1)}{2} = \frac{9(9 + 9 + 1)}{2} = 85.500$$

$$\sigma_R = \sqrt{\frac{n_1 n_2(n_1 + n_2 + 1)}{12}} = \sqrt{\frac{(9)(9)(9 + 9 + 1)}{12}} = 11.325$$

$$z = \frac{R - \mu_R}{\sigma_R} = \frac{66 - 85.5}{11.325} \approx -1.722$$

(e) Reject H_0.

(f) There is enough evidence at the 10% level to support the claim that there is a difference in the salaries between genders.

2. (a) H_0: median $= 28$ (claim); H_a: median $\neq 28$

(b) Sign Test

(c) Critical value is 6.

(d) $x = 10$

(e) Fail to reject H_0.

(f) There is not enough evidence at the 5% level to reject the claim that the median age in Puerto Rico is 28 years.

3. (a) H_0: There is no difference in the annual premiums between the states.
H_a: There is a difference in the annual premiums between the states. (claim)

(b) Kruskal-Wallis Test

(c) Critical value is 5.991.

(d) $H \approx 1.43$

(e) Fail to reject H_0.

(f) There is not enough evidence at the 5% level to conclude that there is a difference in the annual premiums between the states.

4. (a) H_0: The days with rain are random.
H_a: The days with rain are not random. (claim)

(b) The Runs test

(c) $n_1 = $ number of Ns $= 15$
$n_2 = $ number of Rs $= 15$

$cv = 10$ and 22

(d) $G = 16$ runs

(e) Fail to reject H_0.

(f) There is not enough evidence at the 5% level to conclude that days with rain are not random.

CHAPTER 1

CASE STUDY: RATING TELEVISION SHOWS IN THE UNITED STATES

1. Yes. A rating of 8.4 is equivalent to 9,105,600 households which is twice the number of households at a rating of 4.2.

2. $\dfrac{5000}{108,400,000} \approx 0.00004613 \Rightarrow 0.005\%$

3. Program Name and Network

4. Rank and Rank Last Week
   Data can be arranged in increasing or decreasing order.

5. Day, Time
   Data can be chronologically ordered.
   Hours or minutes

6. Rating, Share, and Audience

7. Shows are ranked by rating. Share is not arranged in decreasing order.

8. A decision of whether a program should be cancelled or not can be made based on the Nielsen ratings.

CHAPTER 2

CASE STUDY: SUNGLASS SALES IN THE UNITED STATES

1. (a) Using the grouped data sample mean formula (f = number of sunglasses sold and x = midpoint of class prices):
   $$\bar{x} = \frac{\Sigma xf}{n} = \$76.03$$

   (b) $\bar{x} = \$67.42$

   (c) $\bar{x} = \$28.36$

2. Total revenue $= \Sigma xf$

   The outlet with the greatest total revenue is the Optical Store with $735,030.00.

3. $\dfrac{\text{Revenue}}{\text{location}} = \dfrac{\text{Total revenue}}{\text{\# of locations}}$

   The outlet with the greatest revenue per location is the Sunglass Specialty store with $243,509.70.

4. (a) Using the grouped data sample standard deviation formula: $s = \sqrt{\dfrac{\Sigma(x-\bar{x})^2 f}{n-1}} = \$39.43$

   (b) $s = \$43.39$

   (c) $s = \$20.51$

5. The outlet with the greatest standard deviation is the Sunglass Specialty store with $43.39.

6. By creating a histogram for each outlet, it can be shown that the Sunglass Specialty store has an approximate bell-shaped distribution.

CHAPTER 3

CASE STUDY: PROBABILITY AND PARKING LOT STRATEGIES

1. No, each parking space is not equally likely to be empty. Spaces near the entrances are more likely to be filled.

2. Pick-a-Lane strategy appears to take less time (in both time to find a parking space and total time in parking lot).

3. Cycling strategy appears to give a shorter average walking distance to the store.

4. In order to assume that each space is equally likely to be empty, drivers must be able to see which spaces are available as soon as they enter a row.

5. $P(\text{1st row}) = \frac{1}{7}$ since there are seven rows.

6. $P(\text{in one of 20 closest spaces}) = \frac{20}{72}$

7. Answers will vary. Using the Pick-a-Lane strategy, the probability that you can find a parking space in the most desirable category is low. Using the Cycling strategy, the probability that you can find a parking space in the most desirable category depends on how long you cycle. The longer you cycle, the probability that you can find a parking space in the most desirable category will increase.

CHAPTER 4

CASE STUDY: BINOMIAL DISTRIBUTION OF AIRPLANE ACCIDENTS

1. $P(\text{crash in 1999}) = \frac{6}{9,000,000} = 0.00000067$

2. (a) $P(4) = 0.191$

   (b) $P(10) = 0.003$

   (c) $P(1 \leq x \leq 5) = P(1) + P(2) + P(3) + P(4) + P(5)$
   $$= 0.098 + 0.177 + 0.213 + 0.191 + 0.138 = 0.817$$

3. $n = 9,000,000$, $p = 0.0000008$

   (a) $P(4) \approx 0.083598$

   (b) $P(10) \approx 0.077027$

   (c) $P(1 \leq x \leq 5) \approx 0.275151$

x	P(x)
0	0.000747
1	0.005375
2	0.019351
3	0.046444
4	0.083598
5	0.120382
6	0.144458
7	0.148586
8	0.133727
9	0.106982
10	0.077027

4. No, because each flight is not an independent trial. There are many flights that use the same plane.

5. Assume $p = 0.0000004$. If a person flies around the clock for 438 years, we might estimate the number of flights to be 438 years · 365 days per year · 10 flights per day, or 1,598,700 flights. The average number of fatal accidents for that number of flights would be 0.64, or approximately one.

CHAPTER 5

CASE STUDY: BIRTH WEIGHTS IN AMERICA

1. (a) 37 to 39 weeks ($\mu = 7.33$)

   (b) Over 42 weeks ($\mu = 7.65$)

   (c) 28 to 31 weeks ($\mu = 4.07$)

2. (a) $P(x < 5.5) = P(z < 3.04) = 0.9988 \to 99.88\%$

   (b) $P(x < 5.5) = P(z < -0.16) = 0.4364 \to 43.64\%$

   (c) $P(x < 5.5) = P(z < -1.68) = 0.0465 \to 4.65\%$

   (d) $P(x < 5.5) = P(z < -1.92) = 0.0274 \to 2.74\%$

3. (a) top 10% $\to$ 90th percentile $\to z \approx 1.285$
   $x = \mu + z\sigma = 7.33 + (1.285)(1.09) \approx 8.731$

   Birth weights must be at least 8.731 pounds to be in the top 10%.

   (b) top 10% $\to$ 90th percentile $\to z \approx 1.285$
   $x = \mu + z\sigma = 7.65 + (1.285)(1.12) \approx 9.089$

   Birth weights must be at least 9.089 pounds to be in the top 10%.

4. (a) $P(6 < x < 9) = P(0.18 < z < 2.21) = 0.9864 - 0.5714 = 0.415$

   (b) $P(6 < x < 9) = P(-1.22 < z < 1.53) = 0.9370 - 0.1112 = 0.8258$

   (c) $P(6 < x < 9) = P(-1.47 < z < 1.21) = 0.8869 - 0.0708 = 0.8161$

5. (a) $P(x < 3.3) = P(z < 1.19) = 0.8830$

   (b) $P(x < 3.3) = P(z < -1.64) = 0.0505$

   (c) $P(x < 3.3) = P(z < -3.70) = 0.000425$ (using technology)

CHAPTER 6

CASE STUDY: SHOULDER HEIGHTS OF APPALACHIAN BLACK BEARS

1. (a) $\bar{x} = 79.7$

   (b) $\bar{x} = 75.7$

2. (a) $s = 12.1$

   (b) $s = 7.6$

3. (a) $\bar{x} \pm z_c \dfrac{s}{\sqrt{n}} \Rightarrow 79.7 \pm 1.96 \dfrac{12.1}{\sqrt{40}} = (76.0, 83.5)$

   (b) $\bar{x} \pm t_c \dfrac{s}{\sqrt{n}} \Rightarrow 75.7 \pm 2.048 \dfrac{7.6}{\sqrt{28}} = (72.7, 78.6)$

4. $\bar{x} \pm z_c \dfrac{s}{\sqrt{n}} \Rightarrow 78.1 \pm 1.96 \dfrac{10.6}{\sqrt{68}} = (75.5, 80.6)$

CHAPTER 7

CASE STUDY: HUMAN BODY TEMPERATURE: WHAT'S NORMAL?

1. (a) (see part d)

   (b) $z_0 = \pm 1.96$ (see part d)

   (c) Rejection regions: $z < -1.96$ and $z > 1.96$ (see part d)

   (d) $\bar{x} \approx 98.25$, $s \approx 0.73$

   $$z = \dfrac{\bar{x} - \mu}{\dfrac{s}{\sqrt{n}}} = \dfrac{98.25 - 98.6}{\dfrac{0.73}{\sqrt{130}}} = \dfrac{-0.35}{0.0640} \approx -5.469$$

   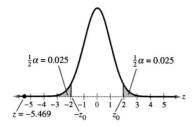

   (e) Reject H_0.

   (f) There is sufficient evidence at the 5% level to reject the claim that the mean body temperature of adult humans is 98.6° F.

2. No, $\alpha = 0.01 \rightarrow z_0 = \pm 2.575 \rightarrow$ Reject H_0

3. H_0: $\mu_m = 98.6$ and H_a: $\mu_m \neq 98.6$

   $\alpha = 0.01 \rightarrow z_0 = \pm 2.575$

   $\bar{x} \approx 98.105$, $s \approx 0.699$

   $$z = \dfrac{\bar{x} - \mu}{\dfrac{s}{\sqrt{n}}} = \dfrac{98.105 - 98.6}{\dfrac{0.699}{\sqrt{65}}} = \dfrac{-0.495}{0.0867} \approx -5.709$$

   Reject H_0. There is sufficient evidence to reject the claim.

4. H_0: $\mu_w = 98.6$ and H_a: $\mu_w \neq 98.6$

   $\alpha = 0.01 \rightarrow z_0 = \pm 2.575$

   $\bar{x} \approx 98.394$, $s \approx 0.743$

   $z = \dfrac{\bar{x} - \mu}{\frac{s}{\sqrt{n}}} = \dfrac{98.394 - 98.6}{\frac{0.743}{\sqrt{65}}} = \dfrac{-0.206}{0.0922} \approx -2.234$

   Fail to reject H_0. There is insufficient evidence to reject the claim.

5. $\bar{x} \approx 98.25$, $s \approx 0.73$

   $\bar{x} \pm z_c \dfrac{s}{\sqrt{n}} = 98.25 \pm 2.575 \dfrac{0.73}{\sqrt{130}} \; 98.25 \pm 0.165 \approx (98.05, 98.415)$

6. Wunderlich may have sampled more women than men, thus causing an overestimated body temperature.

CHAPTER 8

CASE STUDY: DASH DIET AND BLOOD PRESSURE

1. H_0: $\mu_1 = \mu_2$; H_a: $\mu_1 \neq \mu_2$ (claim)

   $z_0 = \pm 1.96$

   $z = \dfrac{(\bar{x}_1 - \bar{x}_2) - (\mu_1 - \mu_2)}{\sqrt{\dfrac{s_1^2}{n_1} + \dfrac{s_2^2}{n_2}}} = \dfrac{(123.9 - 124.5) - (0)}{\sqrt{\dfrac{(9.9)^2}{269} + \dfrac{(10.1)^2}{268}}} = \dfrac{-0.6}{\sqrt{0.745}} \approx -0.695$

   Fail to reject H_0. There is enough evidence to support the claim.

2. H_0: $\mu_1 = \mu_2$; H_a: $\mu_1 \neq \mu_2$ (claim)

   $z_0 = \pm 1.96$

   $z = \dfrac{(\bar{x}_1 - \bar{x}_2) - (\mu_1 - \mu_2)}{\sqrt{\dfrac{s_1^2}{n_1} + \dfrac{s_2^2}{n_2}}} = \dfrac{(123.9 - 128.4) - (0)}{\sqrt{\dfrac{(9.9)^2}{269} + \dfrac{(9.2)^2}{273}}} = \dfrac{-4.5}{\sqrt{0.674}} \approx -5.480$

   Reject H_0. There is enough evidence to support the claim.

3. H_0: $\mu_1 = \mu_2$; H_a: $\mu_1 \neq \mu_2$ (claim)

   $z_0 = \pm 1.96$

   $z = \dfrac{(\bar{x}_1 - \bar{x}_2) - (\mu_1 - \mu_2)}{\sqrt{\dfrac{s_1^2}{n_1} + \dfrac{s_2^2}{n_2}}} = \dfrac{(78.6 - 79.5) - (0)}{\sqrt{\dfrac{(6.8)^2}{269} + \dfrac{(6.7)^2}{268}}} = \dfrac{-0.9}{\sqrt{0.339}} \approx -1.545$

   Fail to reject H_0. There is not enough evidence to support the claim.

4. H_0: $\mu_1 = \mu_2$; H_a: $\mu_1 \neq \mu_2$ (claim)

   $z_0 = \pm 1.96$

   $z = \dfrac{(\bar{x}_1 - \bar{x}_2) - (\mu_1 - \mu_2)}{\sqrt{\dfrac{s_1^2}{n_1} + \dfrac{s_2^2}{n_2}}} = \dfrac{(76.8 - 81.2) - (0)}{\sqrt{\dfrac{(6.8)^2}{269} + \dfrac{(6.3)^2}{273}}} = \dfrac{-2.6}{\sqrt{0.317}} \approx -4.616$

   Reject H_0. There is enough evidence to support the claim.

5. (a) H_0: $\mu_1 = \mu_2$; H_1: $\mu_1 \neq \mu_2$ (claim)

$z_0 = \pm 2.575$

$$z = \frac{(\bar{x}_1 - \bar{x}_2) - (\mu_1 - \mu_2)}{\sqrt{\frac{s_1^2}{n_1} + \frac{s_2^2}{n_2}}} = \frac{(123.9 - 122) - (0)}{\sqrt{\frac{(9.9)^2}{269} + \frac{(9)^2}{100}}} = \frac{1.9}{\sqrt{1.174}} \approx 1.753$$

Fail to reject H_0. There is not enough evidence to support the claim.

(b) H_0: $\mu_1 = \mu_2$; H_1: $\mu_1 \neq \mu_2$ (claim)

$z_0 = \pm 2.575$

$$z = \frac{(\bar{x}_1 - \bar{x}_2) - (\mu_1 - \mu_2)}{\sqrt{\frac{s_1^2}{n_1} + \frac{s_2^2}{n_2}}} = \frac{(78.6 - 83) - (0)}{\sqrt{\frac{(6.8)^2}{269} + \frac{(6)^2}{100}}} = \frac{-4.4}{\sqrt{0.532}} \approx -6.033$$

Reject H_0. There is enough evidence to support the claim.

6. There was not a significant difference in the systolic and diastolic blood pressure between those in the established plus DASH group and those in the established group. There was a significant difference in systolic and diastolic blood pressure between those in the established plus DASH group and those in the advice only group. There was not a significant difference in systolic blood pressure while there was a significant difference in diastolic blood pressure when the established plus DASH group was compared to the group given blood pressure reducing medicine.

CHAPTER 9

CASE STUDY: CORRELATION OF BODY MEASUREMENTS

1. Answers will vary.

2.
	r
a	0.698
b	0.746
c	0.351
d	0.953
e	0.205
f	0.580
g	0.844
h	0.798
i	0.116
j	0.954
k	0.710
l	0.710

3. d, g, and j have strong correlations ($r > 0.8$).

   (d) $\hat{y} = 1.129x - 8.817$

   (g) $\hat{y} = 1.717x + 23.144$

   (j) $\hat{y} = 1.279x + 10.451$

4. (a) $\hat{y} = 0.086(180) + 21.923 = 37.403$

   (b) $\hat{y} = 1.026(100) - 15.323 = 87.277$

5. (weight, chest) → $r = 0.888$
   (weight, hip) $r = 0.930$
   (neck, wrist) $r = 0.849$
   (chest, hip) $r = 0.953$
   (chest, thigh) $r = 0.936$
   (chest, knee) $r = 0.855$
   (chest, ankle) $r = 0.908$
   (abdom, thigh) $r = 0.863$
   (abdom, knee) $r = 0.903$
   (hip, thigh) $r = 0.894$
   (hip, ankle) $r = 0.908$
   (thigh, knee) $r = 0.954$
   (thigh, ankle) $r = 0.874$
   (ankle, knee) $r = 0.857$
   (forearm, wrist) $r = 0.854$

CHAPTER 10

CASE STUDY: *TRAFFIC SAFETY FACTS*

1. $41{,}611(0.24) \approx 9986.64$

2. $\dfrac{1555}{6117} \approx 0.254 > 0.24 \rightarrow$ Central United States and

   $\dfrac{2373}{9533} \approx 0.249 > 0.24 \rightarrow$ Western United States

3. The number of fatalities in the Western United States for the 25–34 age group exceeded the expected number of fatalities.

4. H_0: Region is independent of age.

   H_a: Region is dependent on age.

   d.f. $= (r - 1)(c - 1) = 14$

   $\chi_0^2 = 23.685 \rightarrow$ Reject H_0 if $\chi^2 > 23.685$.

   $\chi^2 \approx 54.948$

   Reject H_0. There is enough evidence at the 5% level to conclude that region is dependent on age.

5. H_0: Eastern United States has same distribution as the entire United States.
   H_a: Eastern United States has a different distribution than the entire United States.

   $\chi_0^2 = 14.067 \rightarrow$ Reject H_0 if $\chi^2 > 14.067$.

   $\chi^2 \approx 8.473$

   Fail to reject H_0. There is not enough evidence at the 5% level to conclude that the eastern United States has a different distribution than the entire United States.

6. H_0: Central United States has same distribution as the entire United States.
   H_a: Central United States has a different distribution than the entire United States.

   $\chi_0^2 = 14.067 \rightarrow$ Reject H_0 if $\chi^2 > 14.067$.

   $\chi^2 \approx 10.289$

   Fail to reject H_0. There is not enough evidence at the 5% level to conclude that the central United States has a different distribution than the entire United States.

7. H_0: Western United States has same distribution as the entire United States.

H_a: Western United States has a different distribution than the entire United States.

$\chi_0^2 = 14.067 \rightarrow$ Reject H_0 if $\chi^2 > 14.067$.

$\chi^2 \approx 51.070$

Reject H_0. There is enough evidence at the 5% level to conclude that the western United States has a different distribution than the entire United States.

8. Alcohol, vehicle speed, driving conditions, etc.

CHAPTER 11

CASE STUDY: EARNINGS BY COLLEGE DEGREE

1. (a), (b), (c), (d)

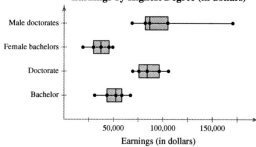

The doctorate groups appear to make more money than the bachelor groups.

2. H_0: median ≤ \$85,000 (claim); H_a: median > \$85,000

The critical value is 1.

$x = 5$

Fail to reject H_0. There is not enough evidence to reject the claim.

3. H_0: median = \$95,500 (claim); H_a: median ≠ \$95,500

The critical value is 1.

$x = 3$

Fail to reject H_0. There is not enough evidence to reject the claim.

4. H_0: median ≥ \$55,500 (claim); H_a: median < \$55,500

The critical value is 1.

$x = 4$

Fail to reject H_0. There is not enough evidence to reject the claim.

5. H_0: median = \$40,000 (claim); H_a: median ≠ \$40,000

The critical value is 1.

$x = 4$

Fail to reject H_0. There is not enough evidence to reject the claim.

6. H_0: There is no difference in calcium intake. (claim)

H_a: There is a difference in calcium intake.

The critical values are $z_0 = \pm 2.575$.

$R = 98$

$$\mu_R = \frac{n_1(n_1 + n_2 + 1)}{2} = \frac{10(10 + 10 + 1)}{2} = 105$$

$$\sigma_R = \sqrt{\frac{n_1 n_2(n_1 + n_2 + 1)}{12}} = \sqrt{\frac{(10)(10)(10 + 10 + 1)}{12}} \approx 13.2$$

$$z = \frac{R - \mu_R}{\sigma_R} = \frac{98 - 105}{13.2} \approx -0.530$$

Fail to reject H_0. There is not enough evidence to reject the claim.

7. H_0: There is no difference in calcium intake.

H_a: There is a difference in calcium intake. (claim)

The critical values are $z_0 = \pm 2.575$.

$R = 137.0$

$$\mu_R = \frac{n_1(n_1 + n_2 + 1)}{2} = \frac{10(10 + 10 + 1)}{2} = 105$$

$$\sigma_R = \sqrt{\frac{n_1 n_2(n_1 + n_2 + 1)}{12}} = \sqrt{\frac{(10)(10)(10 + 10 + 1)}{12}} \approx 13.2$$

$$z = \frac{R - \mu_R}{\sigma_R} = \frac{137.0 - 105}{13.2} \approx 2.424$$

Fail to reject H_0. There is not enough evidence to support the claim.

CASE STUDY

CHAPTER 1

USES AND ABUSES

1. Answers will vary.
2. Answers will vary.

CHAPTER 2

USES AND ABUSES

1. Answers will vary.
2. Answers will vary.

CHAPTER 3

USES AND ABUSES

1. (a) $P(\text{winning Tues and Wed}) = P(\text{winning Tues}) \cdot P(\text{winning Wed})$
$$= \left(\frac{1}{1000}\right) \cdot \left(\frac{1}{1000}\right)$$
$$= 0.000001$$

   (b) $P(\text{winning Wed given won Tues}) = P(\text{winning Wed})$
$$= \frac{1}{1000}$$
$$= 0.001$$

   (c) $P(\text{winning Wed given didn't win Tues}) = P(\text{winning Wed})$
$$= \frac{1}{1000}$$
$$= 0.001$$

2. Answers will vary.

   $P(\text{pick-up or SUV}) \leq 0.55$ since $P(\text{pick-up}) = 0.25$ and $P(\text{SUV}) = 0.30$, but a person may own both a pick-up and an SUV (ie not mutually exclusive events). Thus $P(\text{pick-up and SUV}) \geq 0$ and this probability would have to be subtracted from 0.55.

   If the events were mutually exclusive, the value of the probability would be 0.55. Otherwise, the probability will be less than 0.55 (not 0.60).

CHAPTER 4

USES AND ABUSES

1. 40

   $P(40) = 0.0812$

2. $P(35 \leq x \leq 45) = P(35) + P(36) + \cdots + P(45)$

   $= 0.7386$

3. The probability of finding 36 adults out of 100 who prefer Brand A is 0.059. So the manufacturer's claim is hard to believe.

4. The probability of finding 25 adults out of 100 who prefer Brand A is 0.000627. So the manufacturer's claim would not be believable.

CHAPTER 5

USES AND ABUSES

1. You are not certain to get exactly 95 people due to the fact this is a sample of 100 and sampling variability will sometimes produce a sample with more or fewer than 95 people lying within 2 standard deviations of the mean. (Answers will vary.)

2. It is more likely that all 10 people lie within 2 standard deviations of the mean. This can be shown by using the Emporical Rule and the Multiplication Rule.

   (a) By the Emporical Rule, the probability of lying within 2 standard deviations of the mean is 0.95.

   Let x = number of people selected who lie within 2 standard deviations of the mean.

   $P(x = 10) = (0.95)^{10} \approx 0.599$

   (b) P(at least one person does not lie within 2 standard deviations of the mean)

   $= 1 - P(x = 10)$

   $\approx 1 - 0.599$

   $= 0.401$

CHAPTER 6

USES AND ABUSES

1. Answers will vary.

2. Answers will vary.

CHAPTER 7

USES AND ABUSES

1. Randomly sample car dealerships and gather data necessary to answer question. (Answers will vary.)

2. H_0: $p = 0.47$
   We cannot prove that $p = 0.47$. We can only show that there is insufficient evidence in our sample to reject H_0: $p = 0.47$. (Answers will vary.)

3. If we gather sufficient evidence to reject the null hypothesis when it is really true, then a Type I error would occur. (Answers will vary.)

4. If we fail to gather sufficient evidence to reject the null hypothesis when it is really false, then a Type II error would occur. (Answers will vary.)

CHAPTER 8

USES AND ABUSES

1. Age and health. (Answers will vary.)

2. Blind: The patients do not know which group (medicine or placebo) they belong to.

   Double Blind: Both the researcher and patient do not know which group (medicine or placebo) that the patient belongs to.

   (Answers will vary.)

CHAPTER 9

USES AND ABUSES

1. Answers will vary.

2. Answers will vary. One example would be temperature and output of sulfuric acid. When sulfuric acid is manufactured, the amount of acid that is produced depends on the temperature at which the industrial process is run. As the temperature increases, so does the output of acid, up to a point. Once the temperature passes that point, the output begins to decrease.

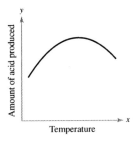

344 | USES AND ABUSES

CHAPTER 10

USES AND ABUSES

1. Answers will vary. An example would be a biologist who is analyzing effectiveness of three types of pesticides. The biologists sprays one pesticide on five different acres during week one. At the end of week one, he calculates the mean number of insects in the acres. He sprays the second pesticide on the same five acres during week two. At the end of week two, he calculates the mean number of insects in the acres. He follows a similar procedure with the third pesticide during week three. He wants to determine whether there is a difference in the mean number of insects per acre. Since the same five acres are treated each time, it is unclear as to the effectiveness of each of the pesticides.

2. Answers will vary. An example would be another biologist who is analyzing the effectiveness of three types of pesticides. He has fifteen acres to test. He randomly assigns each pesticide to five acres and treats all of the acres for three weeks. At the end of three weeks, the biologist calculates the mean number of insects for each of the pesticides. He wants to determine whether there is a difference in the mean number of insects per acre.

    Rejecting the null hypothesis would mean that at least one of the means differs from the others.

    In this example, rejecting the null hypothesis means that there is at least one pesticide whose mean number of insects per acre differs from the other pesticides.

CHAPTER 11

USES AND ABUSES

1. (Answers will vary.)

    H_0: median $\geq$ 10
    H_0: median $<$ 10

    Using $n = 5$ and $\alpha \approx 0.05$, the $cu = 0$ ($\alpha = 0.031$). Thus, every item in the sample would have to be less than 10 in order to reject the H_0. (i.e. 100% of the sample would need to be less than 10.)

    However, using $n = 20$ and $\alpha \approx 0.05$, the $cu = 6$ ($\alpha = 0.057$). Now only 14 items (or 70%) of the sample would need to be less than 10 in order to reject H_0.

2.

Nonparametric Test	Parametric Test
(a) Sign Test	z-test
(b) Paired Sample Sign Test	t-test
(c) Wilcoxon Signed-Rank Test	t-test
(d) Wilcoxon Rank Sum Test	z-test or t-test
(e) Kruskal-Wallis Test	One-way Anova
(f) Spearmans Rank Correlation Coefficient	Pearson Correlation Coefficient

CHAPTER 1

REAL STATISTICS–REAL DECISIONS

1. (a) If the survey identifies the type of reader that is responding, then stratified sampling ensures that representatives from each group of readers (engineers, manufacturers, researchers, and developers) are included in the sample.

   (b) Yes

   (c) After dividing the population of readers into their subgroups (strata), a simple random sample of readers of the same size is taken from each subgroup.

   (d) You may take too large a percentage of your sample from a subgroup of the population that is relatively small.

2. (a) "Fuel" is a variable consisting of labels-Qualitative.

   Once the results have been compiled you have quantitative data, because "Percent Responding" is a variable consisting of numerical measurements.

   "Time" is a variable consisting of labels-Qualitative.

   Once the results have been compiled you have quantitative data, because "Percent Responding" is a variable consisting of numerical measurements.

   (b) "Fuel" is a nominal level of measurement since the data uses only labels.

   "Percent Responding" is a ratio level of measurement since the data can be ordered and you can calculate meaningful differences between data entries.

   "Time" is an ordinal level of measurement since the data can be arranged in order, but differences between data entries are not meaningful.

   "Percent Responding" is a ratio level of measurement since the data can be ordered and you can calculate meaningful differences between data entries.

   (c) Sample

   (d) Statistics

3. (a) Using a mailed survey introduces bias into the sampling process. The data was not collected randomly and may not adequately represent the population.

   (b) Stratified sampling would have been more appropriate assuming the population of readers consists of various subgroups that should be represented in the sample.

CHAPTER 2

REAL STATISTICS–REAL DECISIONS

1. (a) Examine data from the four cities and make comparisons using various statistical methodologies. For example, compare the average price of gas for each city.

   (b) Calculate: Mean, range, and population standard deviation for each city.

2. (a) Construct a Pareto chart because the data in use are quantitative and a Pareto chart positions data in order of decreasing height, with the tallest bar positioned at the left.

(b)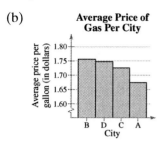

(c) Yes

3. (a) Find the mean, range, and population standard deviation for each city.

(b) City A: $\bar{x} = 1.675$ $\sigma \approx 0.03$ range = 0.09

City B: $\bar{x} = 1.756$ $\sigma \approx 0.02$ range = 0.07

City C: $\bar{x} = 1.726$ $\sigma \approx 0.02$ range = 0.08

City D: $\bar{x} = 1.748$ $\sigma \approx 0.03$ range = 0.08

(c) The mean from Cities B, C, and D appear to be similar. However, City A appears to have a lower price for a gallon of gas. These measurements support the conclusion in Exercise 2.

4. (a) You would tell your readers that on average, the cost for gas is higher in this city than in other cities.

(b) Location of city, national inflation of gas prices, store's economic condition.

CHAPTER 3

REAL STATISTICS–REAL DECISIONS

1. (a) Answers will vary. Investigate the probability of not matching any of the 5 white balls selected.

(b) You could use the Multiplication Rule, the Fundamental Counting Principle, and Combinations.

2. If you played only the red ball, the probability of it is $\frac{1}{42}$. However, because you must pick 5 white balls, you must get the white balls wrong. So, using the Multiplication Rule, we get

$P\begin{pmatrix}\text{matching only the red ball and not} \\ \text{matching any of the 5 white balls}\end{pmatrix} = P(\text{matching red ball}) \cdot P(\text{not matching any white balls})$

$$= \frac{1}{42} \cdot \frac{44}{49} \cdot \frac{43}{48} \cdot \frac{42}{47} \cdot \frac{41}{46} \cdot \frac{40}{45}$$

$$\approx 0.01356$$

$$\approx \frac{1}{74}.$$

CHAPTER 4

REAL STATISTICS–REAL DECISIONS

1. (a) Answers will vary. For example, calculate the probability of obtaining zero clinical pregnancies out of 10 randomly selected ART cycles.

(b) Binomial
Discrete, we are counting the number of successes (clinical pregnancies).

2. Using a binomial distribution where $n = 10$, $p = 0.305$, we find $P(0) = 0.0263$.

It is not impossible to obtain zero clinical pregnancies out of 10 ART cycles. However, it is rather unlikely to occur.

3. (a) Using $n = 10$, $p = 0.226$, $P(8) = 0.000183$

Suspicious, because the probability is very small.

(b) Using $n = 10$, $p = 0.182$, $P(0) = 0.1341$

Not suspicious, because the probability is not that small.

CHAPTER 5

REAL STATISTICS–REAL DECISIONS

1. (a) $z = \dfrac{x - \mu}{\sigma} = \dfrac{25.6 - 25.4}{0.2} = 1$

$P(\text{not detecting shift}) = P(x < 25.6) = P(z < 1) = 0.8413$

(b) $n = 12 \quad p = 0.8413 \quad q = 0.1587$

$np = (12)(0.8413) \approx 10.0956 > 5$ and $nq = (12)(0.1587) \approx 1.90 < 5$

$\sigma = \sqrt{npq} = \sqrt{(12)(0.8413)(0.1587)} = \sqrt{1.602} \approx 1.2658 \quad \mu = 10.0956$

$x = 0.5 \leftarrow$ correction for continuity

$z = \dfrac{0.5 - 10.0956}{1.2658} \approx \dfrac{-9.5956}{1.2658} \approx -7.58$

$P(x \geq 0.5) = P(z \geq -7.58) \approx 1 - 0 = 1$

2. (a) $z = \dfrac{x - \mu}{\dfrac{\sigma}{\sqrt{n}}} = \dfrac{25.6 - 25.4}{\dfrac{0.2}{\sqrt{\mu}}} = \dfrac{0.2}{0.1} = 2$

$P(\text{not detecting shift}) = P(\bar{x} < 25.6) = P(z < 2) = 0.9772$

(b) $n = 3 \quad p = 0.9772 \quad q = 0.0228$

$np = (3)(0.9772) \approx 2.9316 < 5$ and $nq = (3)(0.0228) \approx 0.0684 < 5$

$\mu = 2.9316 \quad \sigma = \sqrt{(3)(0.9772)(.0228)} \approx \sqrt{0.0668} \approx 0.2585$

$x = 5 \leftarrow$ correction for continuity

$z = \dfrac{0.5 - 2.9316}{0.2585} \approx -9.41$

$P(x \geq 0.5) = P(z \geq -9.41) \approx 1 - 0 \approx 1$

(c) The mean is more sensitive to change.

3. Answers will vary.

Using 3 samples of size 4 will be more likely to detect the shift due to the nature of the distribution of the sample means as demonstrated in problems 1 and 2.

CHAPTER 6

REAL STATISTICS–REAL DECISIONS

1. (a) No, there has not been a change in the mean concentration levels because the confidence interval for year 1 overlaps with the confidence interval for year 2.

   (b) No, there has not been a change in the mean concentration levels because the confidence interval for year 2 overlaps with the confidence interval for year 3.

   (c) Yes, there has been a change in the mean concentration levels because the confidence interval for year 1 does not overlap with the confidence interval for year 3.

2. Due to the fact the CI from year 1 does not overlap the CI from year 3, it is very likely that the efforts to reduce the concentration of acetylene in the air are significant over the 3-year period.

3. (a) Used the sampling distribution of the sample means due to the fact that the "mean concentration" was used. The point estimate is the most unbiased estimate of the population mean.

   (b) No, it is more likely the sample standard deviation of the current year's sample was used. Typically σ is unknown.

CHAPTER 7

REAL STATISTICS–REAL DECISIONS

1. (a) Take a random sample to get a diverse group of people.
      Stratified random sampling

   (b) Random sample

   (c) Answers will vary.

2. H_0: $p \geq 0.40$; H_a: $p < 0.40$ (claim)

$z_0 = -1.282$

$\hat{p} = \dfrac{x}{n} = \dfrac{15}{32} \approx 0.469$

$z = \dfrac{\hat{p} - p}{\sqrt{\dfrac{pz}{n}}} = \dfrac{0.469 - 0.40}{\sqrt{\dfrac{(0.40)(0.60)}{32}}} = \dfrac{0.069}{0.087} = 0.794$

Fail to reject H_0. There is not enough evidence to support the claim.

3. H_0: $\mu \geq 60$ (claim); H_a: $\mu < 60$

$\bar{x} = 53.067$ $s = 19.830$ $n = 15$

$t_0 = -1.345$

$t = \dfrac{\bar{x} - \mu}{\dfrac{s}{\sqrt{n}}} = \dfrac{53.067 - 60}{\dfrac{19.83}{\sqrt{15}}} = \dfrac{-6.933}{5.120} = -1.354$

Reject H_0. There is enough evidence to reject the claim.

4. There is not enough evidence to conclude the proportions of Americans who think Social Security will have money to provide their benefits is less than 40%. There is enough evidence to conclude the mean age of those individuals responding "yes" is less than 60. (Answers will vary.)

CHAPTER 8

REAL STATISTICS–REAL DECISIONS

1. (a) Take a simple random sample of records today and 10 years ago from a random sample of hospitals.

Cluster sampling

(b) Answers will vary.

(c) Answers will vary.

2. t-test; Independent

3. H_0: $\mu_1 = \mu_2$; H_a: $\mu_1 \neq \mu_2$ (claim)

d.f. $= \min\{n_1 - 1, n_2 - 1\} = 19$

$t_0 = \pm 2.093$

$t = \dfrac{(\bar{x}_1 - \bar{x}_2) - (\mu_1 - \mu_2)}{\sqrt{\dfrac{s_1^2}{n_1} + \dfrac{s_2^2}{n_2}}} = \dfrac{(4.3 - 3.4) - (0)}{\sqrt{\dfrac{(1.2)^2}{20} + \dfrac{(0.7)^2}{25}}} = \dfrac{0.90}{\sqrt{0.0916}} \approx 2.974$

Reject H_0. There is enough evidence to support the claim.

REAL STATISTICS–REAL DECISIONS

1. (a) 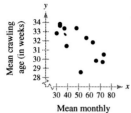 Negative linear correlation

(b)

Mean monthly temperature, x	Mean crawling age, y	xy	x^2	y^2
33	33.64	1110.12	1089.00	1131.65
30	32.82	984.60	900.00	1077.15
33	33.83	1116.39	1089.00	1144.47
37	33.35	1233.95	1369.00	1112.22
48	33.38	1602.24	2304.00	1114.22
57	32.32	1842.24	3249.00	1044.58
66	29.84	1969.44	4356.00	890.43
73	30.52	2227.96	5329.00	931.47
72	29.70	2138.40	5184.00	882.09
63	31.84	2005.92	3969.00	1013.79
52	28.58	1486.16	2704.00	816.82
39	31.44	1226.16	1521.00	988.47
$\Sigma x = 603$	$\Sigma y = 381.26$	$\Sigma xy = 18{,}943.58$	$\Sigma x^2 = 33{,}063$	$\Sigma y^2 = 12{,}147.36$

$$r = \frac{n(\Sigma xy) - (\Sigma x)(\Sigma y)}{\sqrt{n\Sigma x^2 - (\Sigma x)^2}\sqrt{n\Sigma y^2 - (\Sigma y)^2}} = \frac{12(18{,}943.58) - (603)(381.26)}{\sqrt{12(33{,}063) - (603)^2}\sqrt{12(12{,}147.36) - (381.26)^2}} = -0.700$$

(c) $H_0: \rho = 0$; $H_a: \rho \neq 0$

$n = 12 \Rightarrow cv = 0.576$

$|r| = 0.700 > 0.576 \Rightarrow$ Reject H_0. There is a significant linear correlation.

(d) $m = \dfrac{n(\Sigma xy) - (\Sigma x)(\Sigma y)}{n(\Sigma x^2) - (\Sigma x)^2} = \dfrac{12(18{,}943.58) - (603)(381.26)}{12(33{,}063) - (603)^2} = -0.078$

$b = \bar{y} - m\bar{x} = \left(\dfrac{381.26}{12}\right) - (-0.078)\left(\dfrac{603}{12}\right) = 35.678$

$\hat{y} = -0.078x + 35.678$

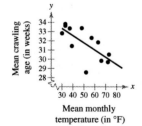

The regression line appears to be a good fit.

(e) Since the regression line is a good fit, it is reasonable to use temperature to predict mean crawling age.

(f) $r^2 = 0.490$

$s_e = 1.319$

49% of the variation in y is explained by the regression line. The standard deviation of mean crawling age for a specific mean monthly temperature is 1.319 weeks.

2. Agree. Due to high correlation, parents can "consider" the temperature when determining when their baby will crawl. However, the parents should not "conclude" that temperature is the other variable related to mean crawling age.

CHAPTER 10

REAL STATISTICS–REAL DECISIONS

1.

Ages	Distribution	Observed	Expected	$\frac{(O-E)^2}{E}$
Under 20	1%	30	10	40.000
20–29	13%	200	130	37.692
30–39	16%	300	160	122.500
40–49	19%	270	190	33.684
50–59	16%	150	160	0.625
60–69	13%	40	130	62.308
70+	22%	10	220	200.455
				$\chi^2 = 497.26$

H_0: Distribution of ages is as shown in the table above. (claim)

H_a: Distribution of ages differs from the claimed distribution.

d.f. $= n - 1 = 6$

$\chi_0^2 = 16.812$

Reject H_0. There is enough evidence at the 1% level to conclude the distribution of ages of telemarketing fraud victims differs from the survey.

2. (a)

Type of Fraud	Age								Total
	Under 20	20–29	30–39	40–49	50–59	60–69	70–79	80+	
Sweepstakes	10 (15)	60 (120)	70 (165)	130 (185)	90 (135)	160 (115)	280 (155)	200 (110)	1000
Credit cards	20 (15)	180 (120)	260 (165)	240 (185)	180 (135)	70 (115)	30 (155)	20 (110)	1000
Total	30	240	330	370	270	230	310	220	2000

(b) H_0: The type of fraud is independent of the victim's age.

H_a: The type of fraud is dependent on the victim's age.

d.f. $= (r-1)(c-1) = (1)(7) = 7$

$\chi_0^2 = 18.475$

$\chi^2 = \Sigma \frac{(O-E)^2}{E} \approx 619.533$

Reject H_0. There is enough evidence at the 1% level to conclude that the victim's ages and type of fraud are dependent.

CHAPTER 11

REAL STATISTICS–REAL DECISIONS

1. (a) random sample

   (b) Answers will vary.

   (c) Answers will vary.

2. (a) Answers will vary. For example, ask "Is the data a random sample?" and "Is the data normally distributed?"

   (b) Sign test; You need to use the nonparametric test since nothing is known about the shape of the population.

   (c) H_0: median ≥ 3.5; H_0: median < 3.5 (claim)

   (d) $\alpha = 0.05$
   $n = 19$
   $cv = 5$
   $x = 7$

   Fail to reject H_0. There is not enough evidence at the 5% level to support the claim that the median tenure is less than 3.5 years.

3. (a) Since the data from each sample appears non-normal, the Wilcoxon Rank Sum test should be used.

   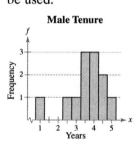
   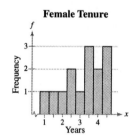

   (b) H_0: The median tenure for male workers is less than or equal to the median tenure for female workers.

   H_a: The median tenure for male workers is greater than the median tenure for female workers. (claim)

   (c) $R = 177$

   $$\mu_R = \frac{n_1(n_1 + n_2 + 1)}{2} = \frac{12(12 + 14 + 1)}{2} = 162$$

   $$\sigma_R = \sqrt{\frac{n_1 n_2(n_1 + n_2 + 1)}{12}} = \sqrt{\frac{(12)(14)(12 + 14 + 1)}{12}} \approx 19.442$$

   $$z = \frac{R - \mu_R}{\sigma_R} = \frac{177 - 162}{19.442} \approx 0.772$$

   Using $\alpha = 0.05$, $z_0 = \pm 1.645$

   Fail to reject H_0. There is not enough evidence at the 5% level to conclude that the median tenure for male workers is greater than the median tenure for female workers.

CHAPTER 1

TECHNOLOGY: USING TECHNOLOGY IN STATISTICS

1. From a list of numbers ranging from 1 to 74, randomly select eight of them. Answers will vary.

2. From a list of numbers ranging from 1 to 200, randomly select 20 of them. Answers will vary.

3. From a list of numbers ranging from 0 to 9, randomly select five of them. Repeat this three times. Average these three numbers and compare with the population average of 4.5. Answers will vary.

4. The average of the numbers 0 to 40 is 20. From a list of numbers ranging from 0 to 40, randomly select seven of them. Repeat this three times. Average these three numbers and compare with the population average of 20. Answers will vary.

5. Answers will vary.

6. No, we would anticipate 10-1's, 10-2's, 10-3's, 10-4's, 10-5's, and 10-6's. An inference that we might draw from the results is that the die is not a fair die.

7. Answers will vary.

8. No, we would anticipate 50 heads and 50 tails. An inference that we might draw from the results is that the coin is not fair.

CHAPTER 2

TECHNOLOGY: MONTHLY MILK PRODUCTION

1. $\bar{x} = 2270.5$

2. $s = 653.2$

3. | Lower limit | Upper limit | Frequency |
    |---|---|---|
    | 1147 | 1646 | 7 |
    | 1647 | 2146 | 15 |
    | 2147 | 2646 | 13 |
    | 2647 | 3146 | 11 |
    | 3147 | 3646 | 3 |
    | 3647 | 4146 | 0 |
    | 4147 | 4646 | 1 |

4. Monthly Milk Production

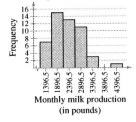

The distribution does not appear to be bell shaped.

5. 68%, 95%
   74% of the entries were within one standard deviation of the mean (1617.3, 2923.7).
   98% of the entries were within two standard deviations of the mean (964.1, 3576.9).

6. $\bar{x} = 2316.5$

7. $s \approx 641.75$

8. Using grouped data formulas tend to give good approximations of the actual mean and standard deviation.

CHAPTER 3

TECHNOLOGY: SIMULATION: COMPOSING MOZART VARIATIONS WITH DICE

1. $2 + 11 = 13$ phrases were written.

2. $(11)^7 \cdot 2 \cdot (11)^7 \cdot 2 = 1.518999334333 \times 10^{15}$

3. (a) $\dfrac{1}{11} \approx 0.091$

   (b) Results will vary.

4. (a) $P(\text{option 6, 7, or 8 for 1st bar}) = \dfrac{3}{11} = 0.273$

   $P(\text{option 6, 7, or 8 for all 14 bars}) = \left(\dfrac{3}{11}\right)^{14} \approx 0.000000012595$

   (b) Results will vary.

5. (a)
   $P(1) = \dfrac{1}{36}$ $P(7) = \dfrac{5}{36}$

   $P(2) = \dfrac{2}{36}$ $P(8) = \dfrac{4}{36}$

   $P(3) = \dfrac{3}{36}$ $P(9) = \dfrac{3}{36}$

   $P(4) = \dfrac{4}{36}$ $P(10) = \dfrac{2}{36}$

   $P(5) = \dfrac{5}{36}$ $P(11) = \dfrac{1}{36}$

   $P(6) = \dfrac{6}{36}$

   (b) Results will vary.

6. (a) $P(\text{option 6, 7, or 8 for 1st bar}) = \dfrac{15}{36} \approx 0.417$

   $P(\text{option 6, 7, or 8 for all 14 bars}) = \left(\dfrac{15}{36}\right)^{14} \approx 0.00000475$

   (b) Results will vary.

CHAPTER 4

TECHNOLOGY: USING POISSON DISTRIBUTIONS AS QUEUING MODELS

1.

x	P(x)
0	0.0183
1	0.0733
2	0.1465
3	0.1954
4	0.1954
5	0.1563
6	0.1042
7	0.0595
8	0.0298
9	0.0132
10	0.0053
11	0.0019
12	0.0006
13	0.0002
14	0.0001
15	0.0000
16	0.0000
17	0.0000
18	0.0000
19	0.0000
20	0.0000

2. (a) (See part b) 1, 2, 4, 7

(b)

Minute	Customers Entering Store	Total Customers in Store	Customers Serviced	Customers Remaining
1	3	3	3	0
2	3	3	3	0
3	3	3	3	0
4	3	3	3	0
5	5	5	4	1
6	5	6	4	2
7	6	8	4	4
8	7	11	4	7
9	3	10	4	6
10	6	12	4	8
11	3	11	4	7
12	5	12	4	8
13	6	14	4	10
14	3	13	4	9
15	4	13	4	9
16	6	15	4	11
17	2	13	4	9
18	2	11	4	7
19	4	11	4	7
20	1	8	4	4

3. Answers will vary.

4. 20 customers (an additional 1 customer would be forced to wait in line/minute)

5. Answers will vary.

6. $P(10) = 0.0181$

7. (a) It makes no difference if the customers were arriving during the 1st minute or the 3rd minute. The mean number of arrivals will still be 4 customers/minute.

$P(3 \leq x \leq 5) = P(3) + P(4) + P(5) \approx 0.1954 + 0.1954 + 0.1563 = 0.5471$

(b) $P(x > 4) = 1 - P(x \leq 4) = 1 - [P(0) + P(1) + \cdots + P(4)] = 1 - [0.6289] = 0.3711$

(c) $P(x > 4$ during *each* of the first four minutes$) = (0.3711)^4 \approx 0.019$

CHAPTER 5

TECHNOLOGY: AGE DISTRIBUTION IN THE UNITED STATES

1. $\mu \approx 36.565$

2. $\bar{x}$ of the 36 sample means ≈ 36.209
This agrees with the Central Limit Theorem.

3. No, the distribution of ages appears to be positively skewed.

4.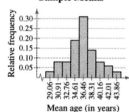

The distribution is approximately bell-shaped and symmetrical. This agrees with the Central Limit Theorem.

5. $\sigma \approx 22.421$

6. σ of the 36 sample means $= 3.502$.

$$\frac{\sigma}{\sqrt{n}} = \frac{22.381}{\sqrt{40}} \approx 3.539$$

This agrees with the Central Limit Theorem.

CHAPTER 6

TECHNOLOGY: MOST ADMIRED POLLS

1. $\hat{p} = \frac{x}{n} = \frac{40}{1004} = 0.040$

$\hat{p} \pm z_c \sqrt{\frac{\hat{p}\hat{q}}{n}} = 0.040 \pm 1.96 \sqrt{\frac{0.040 \cdot 0.960}{1004}} = 0.040 \pm 0.012 \approx (0.028, 0.052)$

2. No, the proportion is 4% plus or minus 1.2%.

3. $\hat{p} = 0.07$

$$\hat{p} \pm z_c \sqrt{\frac{\hat{p}\hat{q}}{n}} = 0.07 \pm 1.96 \sqrt{\frac{0.07 \cdot 0.93}{1004}} = 0.07 \pm 0.016 \approx (0.054, 0.086)$$

4. Answers will vary.

5. No, because the 95% CI does not contain 10%.

CHAPTER 7

TECHNOLOGY: THE CASE OF THE VANISHING WOMEN

1. Reject H_0 with a P-value less than 0.0005.

2. Type I error

3. Sampling process was non-random.

4. (a) H_0: $p = 0.2914$ (claim); H_a: $p \neq 0.2914$

   (b) Test and Confidence Interval for One Proportion

   Test of $p = 0.2914$ vs p not $= 0.2914$

Exact Sample	X	N	Sample p	99.0% CI	Z-Value	P-Value
1	9	100	0.090000	(0.016284, 0.163716)	−4.43	0.000

   (c) Reject H_0.

   (d) It was highly unlikely that random selection produced a sample of size 100 that contained only 9 women.

CHAPTER 8

TECHNOLOGY: TAILS OVER HEADS

1. Test and Confidence Interval for One Proportion

   Test of $p = 0.5$ vs. p not $= 0.5$

Exact Sample	X	N	Sample p	95.0% CI	P-Value
1	5772	11902	0.484961	(0.47598, 0.49394)	0.001

   Reject H_0: $p = 0.5$

2. Yes, obtaining 5772 heads is a very uncommon occurrence (see sampling distribution). The coins might not be fair.

3. $\frac{1}{500} \cdot 100 = 0.2\%$

4. $z = 8.802 \rightarrow$ Reject H_0: $\mu_1 = \mu_2$ There is sufficient evidence at the 5% level to reject the claim that there is no difference in the mint dates in Philadelphia and Denver. (Answers will vary.)

5. $z = 1.010 \rightarrow$ Fail to reject H_0: $\mu_1 = \mu_2$ There is insufficient evidence at the 5% level to reject the claim that there is no difference in the value of the coins in Philadelphia and Denver. (Answers will vary.)

CHAPTER 9

TECHNOLOGY: NUTRIENTS IN BREAKFAST CEREALS

1.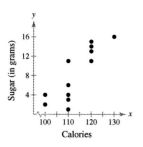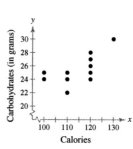

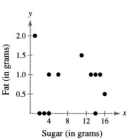 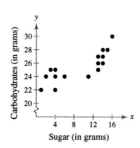 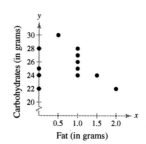

2. (calories, sugar), (calories, carbohydrates), (sugar, carbohydrates)

3. (calories, sugar): $r \approx 0.873$
   (calories, fat): $r \approx 0.327$
   (calories, carbohydrates): $r \approx 0.689$
   (sugar, fat): $r \approx 0.285$
   (sugar, carbohydrates): $r \approx 0.755$
   (fat, carbohydrates): $r \approx -0.156$

   Largest r: (calories, sugar)

4. (a) $\hat{y} = -60.305 + 0.607x$

   (b) $\hat{y} = 4.627 + 0.181x$

5. (a) $\hat{y} = -60.305 + 0.607(120) = 12.508$

   (b) $\hat{y} = 4.627 + 0.181(120) = 26.288$

6. (a) $C = 82.970 + 0.945S + 2.247F + 0.842R$

   (b) $C = 97.070 + 1.180S + 0.268R$

7. $C = 97.070 + 1.180(10) + 0.268(25) = 115.753$

CHAPTER 10

TECHNOLOGY: TEACHER SALARIES

1. Yes, since the salaries are from different states, it is safe to assume that the samples are independent.

2. Using a technology tool, it appears that all three samples were taken from approximately normal populations.

3. H_0: $\sigma_1^2 = \sigma_2^2$; H_a: $\sigma_1^2 \neq \sigma_2^2$ ($F_0 = 2.86$)

   (CA, OH) → $F \approx 1.033$ → Fail to reject H_0.

   (CA, WY) → $F \approx 1.055$ → Fail to reject H_0.

   (OH, WY) → $F \approx 1.022$ → Fail to reject H_0.

4. The three conditions for a one-way ANOVA test are satisfied.

   H_0: $\mu_1 = \mu_2 = \mu_3$
   H_a: At least one mean is different from the others.

Variation	Sum of Squares	Degrees of Freedom	Mean Square	F
Between	2.213×10^9	2	1.106×10^9	18.19
Within	2.736×10^9	45	60,805,943	

   $F \approx 18.19$

   Reject H_0. There is not enough evidence at the 5% level to conclude that one mean salary is different from the others.

5. The samples are independent.

   H_0: $\sigma_1^2 = \sigma_2^2$; H_a: $\sigma_1^2 \neq \sigma_2^2$ ($F_0 = 2.86$)

   (AK, NV) → $F \approx 1.223$ → Fail to reject H_0.

   (AK, NY) → $F \approx 3.891$ → Reject H_0.

   (NV, NY) → $F \approx 4.760$ → Reject H_0.

   The sample from the New York shows some signs of being drawn from a non-normal population.

   The three conditions for one-way ANOVA are not satisfied.

CHAPTER 11

TECHNOLOGY: U.S. INCOME AND ECONOMIC RESEARCH

1.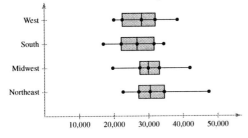

   Annual Income of People (in dollars)

   The median annual incomes do not appear to differ between regions.

2. H_0: median $\geq$ \$25,000 (claim)

   H_a: median $<$ \$25,000

   The critical value is 2.

   $x = 11$

   Fail to reject H_0. There is not enough evidence to reject the claim.

3. H_0: There is no difference in selling prices in the Northeast and South. (claim)

   H_a: There is a difference in selling prices in the Northeast and South.

   The critical values are $z_0 = \pm 1.96$.

   $R = 176.0$

   $\mu_R = \dfrac{n_1(n_1 + n_2 + 1)}{2} = \dfrac{12(12 + 12 + 1)}{2} = 150$

   $\sigma_R = \sqrt{\dfrac{n_1 n_2 (n_1 + n_2 + 1)}{12}} = \sqrt{\dfrac{(12)(12)(12 + 12 + 1)}{12}} \approx 17.3$

   $z = \dfrac{R - \mu_R}{\sigma_R} = \dfrac{176.0 - 150}{17.3} \approx 1.50$

   Fail to reject H_0. There is not enough evidence to reject the claim.

4. H_0: There is no difference in the selling prices for all four regions. (claim)

   H_a: There is a difference in the selling prices for all four regions.

   $H = 3.27 \rightarrow P\text{-value} = 0.351$

   Fail to reject H_0. There is not enough evidence to reject the claim.

5. H_0: There is no difference in the selling prices for all four regions. (claim)

   H_a: There is a difference in the selling prices for all four regions.

   Analysis of Variance

Source	Sum of Squares	Degrees of Freedom	Mean Square	F	P
Factor	259,314,166	3	86,438,055	2.04	0.122
Error	1.862×10^9	44	42,326,290		
Total	2.122×10^9	47			

   $F \approx 2.04 \rightarrow P\text{-value} = 0.122$

   Fail to reject H_0. There is not enough evidence to reject the claim.

6. *(Box-and-whisker plot)*

   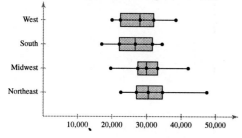

   The median selling prices appear to be the same.

(Wilcoxon rank sum test)

H_0: There is no difference in selling prices in the Northeast and South. (claim)
H_a: There is a difference in selling prices in the Northeast and South.
The critical values are $z_0 = \pm 1.96$.
$R = 275.0$

$$\mu_R = \frac{n_1(n_1 + n_2 + 1)}{2} = \frac{15(15 + 15 + 1)}{2} = 232.5$$

$$\sigma_R = \sqrt{\frac{n_1 n_2(n_1 + n_2 + 1)}{12}} = \sqrt{\frac{(15)(15)(15 + 15 + 1)}{12}} \approx 24.1$$

$$z = \frac{R - \mu_R}{\sigma_R} = \frac{275.0 - 232.5}{24.1} \approx 1.76$$

Fail to reject H_0. There is not enough evidence to reject the claim.

(Kruskal-Wallis test)

H_0: There is no difference in selling prices for all four regions. (claim)
H_a: There is a difference in selling prices for all four regions.
$H = 4.83 \rightarrow p\text{-value} \approx 0.185$
Fail to reject H_0. There is not enough evidence to reject the claim.

(ANOVA test)

H_0: There is no difference in selling prices for all four regions. (claim)
H_a: There is a difference in selling prices for all four regions.

Analysis of Variance

Source	Sum of Squares	Degrees of Freedom	Mean Square	F	P
Factor	970,155,681	3	323,385,227	1.40	0.253
Error	1.295×10^{10}	56	2,313,012,222		
Total	1.392×10^{10}	59			

$F \approx 1.40 \rightarrow p\text{-value} = 0.253$

Fail to reject H_0. There is not enough evidence to reject the claim.